Methods in Cyanobacterial Research

This laboratory reference compiles the essential protocols used in cyanobacterial research. Cyanobacteria are model system for biofuel production and sequestration of carbon dioxide to tackle global climate change due to increasing levels of greenhouse gases. Topics range from identification and growth parameters to biomolecule estimation and isolation. It also covers genome analysis and bioengineering of cyanobacteria for novel products.

Key Features:

- Provides step-by-step procedures and troubleshooting tips.
- Covers identification, growth parameters and pigment estimation methodologies.
- Describes isolation, characterization and purification of microcystin, phycobiliproteins and scytonemin from cyanobacteria.
- Discusses genomics, proteomics and bioengineering of cyanobacteria from novel products.
- Explains sample preparation and visualization protocols for electron microscopy–based analysis of cyanobacteria.

This collection is useful to students and researchers in life sciences. It is also meant for industry experts who are involved in the production of biofuels, biofertilizers, other value-added products and carbon sequestration using cyanobacteria.

Methods in Cyanobacterial Research

Edited by
Shailendra P. Singh, Rajeshwar P. Sinha,
and Donat-P. Häder

CRC Press
Taylor & Francis Group
Boca Raton London New York

CRC Press is an imprint of the
Taylor & Francis Group, an **informa** business

Design cover image: Shutterstock - 10145497

First edition published 2024
by CRC Press
2385 NW Executive Center Drive, Suite 320, Boca Raton FL 33431

and by CRC Press
4 Park Square, Milton Park, Abingdon, Oxon, OX14 4RN

CRC Press is an imprint of Taylor & Francis Group, LLC

ISBN: 978-1-032-49834-8 (hbk)
ISBN: 978-1-032-50418-6 (pbk)
ISBN: 978-1-003-39838-7 (ebk)

DOI: 10.1201/9781003398387

Typeset in Times
by KnowledgeWorks Global Ltd.

Contents

Preface

Cyanobacteria are Gram-negative photosynthetic bacteria that have the unique ability to produce oxygen during photosynthesis similar to eukaryotic algae and higher plants. These prokaryotes are found in a wide range of ecological niches and play a significant role in global nitrogen and carbon dioxide fixation. Cyanobacteria have acted as a model system to study different biological processes such as photosynthesis, photoprotection, nitrogen fixation, circadian rhythm and iron and redox homeostasis. The complete genome sequence of several fresh and marine water cyanobacteria is available in the public database, and tools and techniques are well established for their genetic modification and/or installation of novel metabolic pathways for the production of valuable chemicals. Recently, cyanobacteria have emerged as a model system for biofuel production and sequestration of carbon dioxide to tackle the problem associated with global climate change due to increasing levels of greenhouse gas CO_2. Thus, cyanobacteria are very important components of our ecosystem and can be used for human welfare as well as to achieve the United Nations' Sustainable Development Goals. Several experimental methods are used in cyanobacterial research to study these biological processes or to assess the performance of native or genetically engineered strains of cyanobacteria. However, these research techniques and their detailed experimental procedures and methods are scattered over different publications and sometimes mentioned in brief due to limited space or cross-referencing. Therefore, this edited book aims to collect protocols with step-by-step detailed procedures that are commonly used in cyanobacterial research. However, the protocols included in this book can be applicable to a wide range of organisms. This collection of protocols will be useful to undergraduate, master, PhD, and postdoctoral students and/or scholars of life sciences in addition to scholars working with cyanobacteria. This book will also be very helpful for scientists working in industries involved in the production of biofuels, biofertilizers, other value-added products and carbon sequestration using cyanobacteria.

Editors
Dr. Shailendra P. Singh, Prof. Rajeshwar P. Sinha, and Prof. Donat-P. Häder

About the Editors

Dr. Shailendra P. Singh is a senior assistant professor at the Centre of Advanced Study in Botany, Institute of Science, Banaras Hindu University (BHU), Varanasi, India. He earned his PhD from the Department of Biology, Friedrich-Alexander University, Erlangen-Nuremberg, Germany. He conducted postdoctoral work at Plant Research Laboratory, Plant Biology, Michigan State University, USA. Before joining BHU, Dr. Singh served as a senior scientist at Algae to Oil (A2O), Reliance Industries, Mumbai, India. He is an awardee of the Early Career Research Award, DST-SERB and UGC-BSR startup grants. He is an editorial board member of several international journals. Dr. Singh's research interests include photoprotection, genetics of sunscreens, DNA damage, oxidative stress, photosynthesis, carbon-concentrating mechanisms, developmental biology and complementary chromatic acclimation in cyanobacteria. The outcome of his research has been published in more than 100 research papers, reviews, book chapters and conference proceedings.

Prof. Dr. Rajeshwar P. Sinha, DAAD Fellowship Awardee and Fellow, Society for Applied Biotechnology, India, is a senior professor of molecular biology, Centre of Advanced Study in Botany, Institute of Science, Banaras Hindu University (BHU), Varanasi, India. He earned his PhD in biotechnology from BHU, Varanasi, and visited countries such as Argentina, Austria, Belgium, Canada, China, Germany, Greece, France, Italy, Japan, Luxembourg, Norway, Poland, Republic of Korea, Spain, Switzerland, The Netherlands, UK and USA in the field of academics/research. He is working on the physiological, biochemical, molecular, nanobiotechnological and computational biology aspects of cyanobacteria. His primary research focus is on UV-B radiation impacts on DNA damage and repair, phycobiliproteins, mycosporine-like amino acids, and scytonemin. He has published over 490 research papers, reviews, book chapters and conference proceedings and edited or authored 14 books. He is a lifetime member of several national and international scientific societies and an editorial board member of various national and international journals. He has over 15,000 citations with an h-index of 56 and i10- index of 148. He is among the top 2% most cited scientists in the world.

Prior to his retirement, **Prof. Dr. Donat-P. Häder** was the director of the Botanical Institute, held the chair of Ecophysiology of Plants at the Friedrich-Alexander University, Erlangen-Nürnberg, Germany and was the director of the Botanical Garden. He is the author of over 800 scientific publications, produced 32 books and holds seven national and international patents. He is or has been the editor of several international journals. His research interests are cyanobacteria, gravitational biology of microorganisms and ecophysiology and ecotoxicology of aquatic ecosystems, and he is an appointed member of the Environmental Effects Panel of the United Nations on the Effects of Ozone Depletion and Climate Change.

Contributors

Preeti Agarwal
Department of Botany
Gargi College, University of Delhi
New Delhi, India

Lira A. Gaysina
Department of Bioecology and Biological Education
M. Akmullah Bashkir State Pedagogical University
Ufa, Russia
Banaras Hindu University
Varanasi, India

Amit Gupta
Laboratory of Photobiology and Molecular Microbiology
and
Centre of Advanced Study in Botany
Banaras Hindu University
Varanasi, India

Anjali Gupta
Centre of Advanced Study in Botany
Department of Botany, Institute of Science

Rinkesh Gupta
Department of Botany
Centre of Advanced Study in Botany, Institute of Science
Banaras Hindu University
Varanasi, India

Donat-P. Häder
Department of Biology
Emeritus from Friedrich-Alexander University
Möhrendorf, Germany

Jyoti Jaiswal
Laboratory of Photobiology and Molecular Microbiology
Centre of Advanced Study in Botany, Institute of Science
Banaras Hindu University
Varanasi, India

Pannaga P. Jutur
Omics of Algae Group, Industrial Biotechnology
International Centre for Genetic Engineering and Biotechnology
New Delhi, India

Hakuto Kageyama
Graduate School of Environmental and Human Sciences
and
Department of Chemistry
Faculty of Science and Technology
Meijo University
Nagoya, Japan

Pritam Kaur
Department of Botany
Gargi College, University of Delhi
New Delhi, India

Ashish Kumar
Omics of Algae Group, Industrial Biotechnology
International Centre for Genetic Engineering and Biotechnology
New Delhi, India

Deepak Kumar
Laboratory of Photobiology and Molecular Microbiology
Centre of Advanced Study in Botany, Institute of Science
Banaras Hindu University
Varanasi, India

Naresh Kumar
Department of Botany
Banaras Hindu University
Varanasi, India

Subodh Kumar
Department of Pharmacology
PGIMER
Chandigarh, India

Vinod Kumar
Department of Botany,
Pandit Prithi Nath PG College
Kanpur, India

Neha Kumari
Laboratory of Photobiology and Molecular Microbiology
Centre of Advanced Study in Botany, Institute of Science
Banaras Hindu University
Varanasi, India

Akanksha Madan
Department of Botany
Gargi College, University of Delhi
New Delhi, India

Abhisek Majumdar
Advanced Nanomaterial Lab
Department of Physics
IIT Guwahati
Guwahati Assam, India

Pankaj K. Maurya
Centre of Advanced Study in Botany
Department of Botany, Institute of Science
Banaras Hindu University
Varanasi, India

Bikash Medhi
Department of Pharmacology
PGIMER
Chandigarh, India

Sonal Mishra
Laboratory of Photobiology and Molecular Microbiology
Centre of Advanced Study in Botany, Institute of Science
Banaras Hindu University
Varanasi, India
and
Department of Botany
Acharya Narendra Dev College
Kanpur, India

Soumila Mondal
Centre of Advanced Study in Botany
Department of Botany, Institute of Science
Banaras Hindu University
Varanasi, India
and
Interfaculty Institute of Microbiology and Infection Medicine
Department of Organismic Interactions
Cluster of Excellence 'Controlling Microbes to Fight Infections', Tübingen University
Tübingen, Germany

Asha A. Nesamma
Omics of Algae Group, Industrial Biotechnology
International Centre for Genetic Engineering and Biotechnology
New Delhi, India

Piyush Pachauri
Systems Biology for Biofuel Group, International Centre for Genetic Engineering and Biotechnology (ICGEB)
New Delhi, India

Sagarika Pal
Department of Botany
Banaras Hindu University
Varanasi, India

Harsh Pant
Department of Botany
Banaras Hindu University
Varanasi, India

Priyul Pandey
Department of Botany
Centre of Advanced Study in Botany, Institute of Science
Banaras Hindu University
Varanasi, India

Jainendra Pathak
Department of Botany
Pt. Jawaharlal Nehru College (Affiliated to Bundelkhand University, Jhansi)
Banda, India

Preeti Rai
Systems Biology for Biofuel Group,
International Centre for Genetic
Engineering and Biotechnology (ICGEB)
New Delhi, India

Rajneesh
Laboratory of Photobiology and Molecular
Microbiology
Centre of Advanced Study in Botany,
Institute of Science
Banaras Hindu University
Varanasi, India

Mohammed Rehmanji
Omics of Algae Group, Industrial
Biotechnology
International Centre for
Genetic Engineering and
Biotechnology
New Delhi, India

Niharika Sahu
Laboratory of Photobiology and Molecular
Microbiology
Centre of Advanced Study in Botany
Banaras Hindu University
Varanasi, India

Phulen Sarma
Department of Pharmacology
PGIMER
Chandigarh, India

Bhaskar Sharma
Electron Microscope Facility (SAIF)
Department of Anatomy
All India Institute of Medical Sciences
New Delhi, India

Ashish P. Singh
Laboratory of Photobiology and Molecular
Microbiology
Centre of Advanced Study in Botany
Banaras Hindu University
Varanasi, India

Garvita Singh
Department of Botany, Gargi College
University of Delhi
New Delhi, India

Prabhakar Singh
Electron Microscope Facility (SAIF)
Department of Anatomy
All India Institute of Medical Sciences
New Delhi, India

Prashant Singh
Department of Botany
Banaras Hindu University
Varanasi, India

Prashant R. Singh
Laboratory of Photobiology and Molecular
Microbiology
Centre of Advanced Study in Botany
Banaras Hindu University
Varanasi, India

Rabinder Singh
Omics of Algae Group, Industrial Biotechnology
International Centre for Genetic Engineering
and Biotechnology
New Delhi, India

Shailendra P. Singh
Centre of Advanced Study in Botany
Department of Botany, Institute of Science
Banaras Hindu University
Varanasi, India

Varsha K. Singh
Laboratory of Photobiology and Molecular
Microbiology
Centre of Advanced Study in Botany,
Institute of Science
Banaras Hindu University
Varanasi, India

Rajeshwar P. Sinha
Laboratory of Photobiology and Molecular
Microbiology
Centre of Advanced Study in Botany,
Institute of Science
Banaras Hindu University
Varanasi, India

Shireesh Srivastava
Systems Biology for Biofuel Group,
International Centre for Genetic
Engineering and Biotechnology (ICGEB)
New Delhi, India

Sapna Tiwari
Centre of Advanced Study in Botany
Department of Botany, Institute of Science
Banaras Hindu University
Varanasi, India

Rungaroon Waditee-Sirisattha
Department of Microbiology
Faculty of Science, Chulalongkorn University
Patumwan, Bangkok, Thailand

1 Evaluating the Polyphasic Approach in Cyanobacterial Taxonomy

Sagarika Pal, Naresh Kumar, Harsh Pant, Lira A. Gaysina, and Prashant Singh

1.1 INTRODUCTION

Heterocytous cyanobacteria are a special group of cyanobacteria with the ability to fix atmospheric nitrogen with the help of a specialized cell known as a heterocyte (heterocyst in the old nomenclature). The heterocyte is characterized by the presence of a thick cell wall, lack of pigmentation and presence of polar nodules. Earlier classification systems placed the heterocytous cyanobacteria into two orders, Nostocales and Stigonematales (Fritsch 1945; Desikachary 1959). However, in the recent classification proposed by Komárek et al. (2014), the order Nostocales includes all the heterocytous cyanobacteria. The taxonomical proposals used earlier for cyanobacteria only applied morphological and, rarely, ecological parameters to establish cyanobacterial orders (Thuret 1875; Bornet and Flahault 1888; Gomont 1892). Geitler's collective work was also based on the phenotypic characters of cyanobacteria, where he relied on the morphology of sampled specimens from the field (Geitler 1932, 1942).

This view prompted several researchers to study cyanobacterial taxonomy and led to the formation of the basis of numerous revised classification attempts, such as Elenkin (1936, 1949, 1938); Fritsch (1942, 1945, 1944); Drouet and Daily (1956); Desikachary (1959); and Bourrelly (1970). All these classifications were created under one common criterion, i.e., the morphological characteristics, and emanated under the traditional botanical system of nomenclature. In the 1970s another criterion was implemented in cyanobacterial taxonomy. The criterion was the differences in fatty acid compositions (monounsaturated, polyunsaturated, either alpha or gamma-linolenic acid), which led to the development of "chemotaxonomy" (Kenyon et al. 1972; Murata et al. 1992). A major change occurred in the treatment of cyanobacteria when Waterbury and Stanier (1977), Krumbein (1979) and Rippka et al. (1979) proposed the classification of cyanobacteria under the bacteriological code.

It was established that the earlier designated "cyanophyta" are prokaryotic and the term "cyanobacteria" was introduced. In the bacteriological system, the lowest taxonomic unit is the "strain", implying an axenic culture. The bacteriological system is different from the botanical system, and because the previous works of cyanobacterial taxonomy were based on the botanical code, a comprehensive system was elucidated to incorporate both systems. In a comprehensive system, the naming and description were based on the botanical code, but information on bacteriological and molecular parameters was also employed (Castenholz et al. 2001). A major issue in the classic, phenetic species concept is the lack of resolution on the species level, as the morphological and ecological datasets are found to be insufficient to ascertain the original biodiversity of cyanobacteria (Johansen and Casamatta 2005).

The modern age of cyanobacterial taxonomy began with the comprehensive studies performed by Anagnostidis and Komárek (1985, 1990, 1988) and Komárek and Anagnostidis (1986, 1989). In the 1990s, progress in molecular biology and use of the 16S ribosomal RNA (rRNA) gene drastically revolutionized the field of taxonomy, as the analysis of the 16S rRNA gene supported

DOI: 10.1201/9781003398387-1

the phylogenetic clustering among the morphologically similar taxonomic units in cyanobacteria (Anagnostidis and Komárek 1985, 1990, 1988). Because of the conserved nature of the 16S rRNA gene and its presence in all prokaryotic organisms, it has emerged as an important marker to correctly identify cyanobacteria. Most recently, taxonomic descriptions and revisions have been made by using a combination of tools, including morphological, ecological, genetic, molecular and phylogenetic observations. The combined approach, or the total evidence approach, is referred to as the "polyphasic approach" (Hofmann et al. 2005; Komárek 2003, 2010; Komárek et al. 2014). The introduction of the polyphasic approach has brought a tremendous boost to the cyanobacterial taxonomy, especially in the order Nostocales, and many family-level revisions have also been performed.

This led to the establishment of families like Fortieaceae, Gloeotrichiaceae (Komárek et al. 2014), Godleyaceae, Tolypothrichaceae (Hauer et al. 2014), Dapisostemonaceae (Hentschke et al. 2016), Cyanomargaritaceae (Shalygin et al. 2017), Geitleriaceae (Kilgore et al. 2018), Heteroscytonemataceae (Sendall and McGregor 2018) and Calotrichaceae (Saraf et al. 2019) in the order Nostocales.

In this chapter, we will discuss the different methodologies used in the polyphasic approach, which are used for the correct and accurate identification of heterocytous cyanobacteria. It must be noted that in general, except for the media composition, the identification methods and work plan are the same for all cyanobacteria.

1.1.1 Tools Used in the Polyphasic Approach: Amalgamation of Traditional and Modern Taxonomy

Traditional taxonomic tools include morphological evaluation along with sometimes ecological characters. Since the majority of the classification systems were based on morphological characteristics and formed the foundation for studies of cyanobacterial systematics and taxonomy, it was important to amalgamate the traditional system with modern molecular techniques to form a combined approach (Figure 1.1). The morphological plasticity of cyanobacteria under environmental and laboratory conditions was a major disadvantage that led to erroneous identifications in the past; however, morphological assessments are still an essential criterion and one of the many parameters used in the identification of cyanobacteria. The advent of molecular methods and the use of molecular markers along with morphological and ecological characteristics has now become an essential tool in cyanobacterial systematics and taxonomy.

1.1.2 Morphological Characters: A Foremost Tool in Cyanobacterial Taxonomy

The morphological features of cyanobacteria are more diverse than in any other prokaryotic organism, such as bacteria and Archaea (Wilmotte and Herdman 2001). Thus, morphological characters

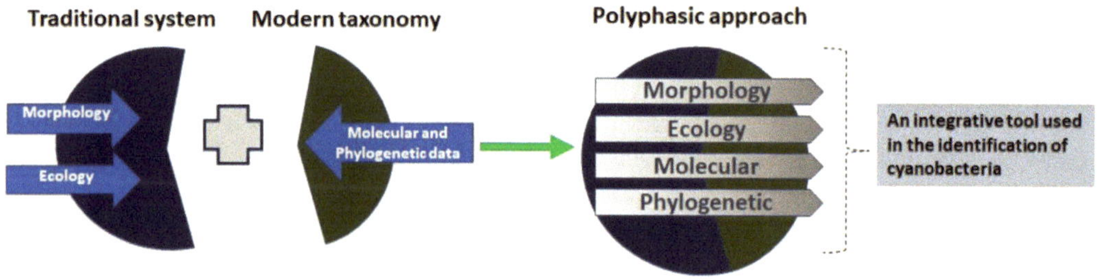

FIGURE 1.1 A schematic diagram showing the amalgamation of both the traditional and modern taxonomy to give rise to an integrative "polyphasic approach" (color figure in e-book).

played a significant role in cyanobacterial taxonomy in the earlier days. The study of morphology in heterocytous cyanobacteria helps us to understand the basic phenotypic features, such as colony morphology, trichome morphology, cell division pattern and the presence of intracellular materials (Figure 1.2). In all heterocytous cyanobacteria, continuous cell connections form filamentous structures. The cystous cyanobacteria can be unbranched or branched. The unbranched cyanobacteria are formed when cell division happens only in the transverse plane. The branched cyanobacteria

FIGURE 1.2 (a) *Nostoc*-like taxon, isopolar filament. (b) *Nostoc*-like taxon, isopolar filament with a conical apical cell at both ends of the filament. (c) and (d) *Dulcicalothrix* sp., heteropolar filaments with tapering apical ends; (e) Tolypothrichaceae cyanobacterium showing heteropolar filaments with a non-tapering apical end. (f) *Neowestiellopsis* sp., showing T-type of true branching. (g) *Brasilonema* sp., showing false branching (color figure in e-book).

are of two types, true-branched and false-branched. The true-branched cyanobacteria are formed when cell division in the transverse plane is followed by lateral division, whereas false branches are formed when transverse plane division is followed by fragmentation, where the lateral protrusion of the main filament is present within the same sheath (Golubić et al. 1996).

Based on morphological features as a primary outline, the last edition of *Bergey's Manual of Systematic Bacteriology* (Castenholz et al. 2001) divided cyanobacteria into five subsections. Subsections I and II included unicellular cyanobacteria, having binary cell division and multiple cell division, respectively. Subsections III, IV and V consisted of the filamentous type, where subsection III comprised a non-heterocytous group of cyanobacteria, subsection IV consisted of heterocytous forms, which can divide in a single plane, and subsection V included cyanobacteria, which were heterocytous in nature and cell division took place in multiple planes. Along with other characteristics, the branching pattern has been an important criterion in many classification systems (Geitler 1932, 1942; Fritsch 1945). In the recent classification system, the unbranched heterocytous cyanobacteria are grouped under families such as Nostocaceae, Aphanizomenonaceae, Gloeotrichiaceae and Rivulariaceae (Komárek et al. 2014). The true-branched cyanobacteria are grouped in families like Hapalosiphonaceae, Stigonemataceae, Symphyonemataceae, Chlorogloeopsidaceae and Capsosiraceae, whereas the false-branched cyanobacteria are grouped in families like Scytonemataceae, Tolypothrichaceae and Godleyaceae (Komárek et al. 2014).

In taxonomy, some specific and unique morphological features sometimes act as an important character to demarcate one taxon from another, such as the presence of some specialized cells or unique stages of the life cycle. The presence of an akinete and the size and shape or position of akinete formation can help to distinguish one species from another (Figure 1.2). In the case of *Cylindrospermum*, three species have been described with an exceptionally high percentage of similarity of the 16S rRNA gene (<99%) but having very distinct morphology. In all three species, the size and shape of the akinete were different from each other and found to be a consistent character (Johansen et al. 2014). Another example, *Westiellopsis akinetica*, a true branching cyanobacterium, was morphologically distinguished by the presence of large akinetes in rows from other members of the genus *Westiellopsis* (Mishra et al. 2020). The formation of necridia can also act as a special character for a species. For example, in the case of *Fortiea necridiiformans*, the presence of necridia and akinete formation in rows distinguished it from other members of the genus *Fortiea* (Saraf et al. 2020). Thus, morphological characters play an important role in cyanobacterial taxonomy.

1.1.3 Ecological Characterization: Understanding the Habitat and Correlation with Modern Taxonomy

The study of "ecology", or the interaction of an organism with its physical surroundings, is an important parameter to uncover the huge amount of heterogeneity and diversity of cyanobacteria (Figure 1.3). As cyanobacteria are a very diverse group of organisms that can survive in extremely adverse conditions, it is important to study the eco-physiological characteristics of their natural habitats. From extreme cold places to hot springs, the ecology is unique and plays an important role in taxonomical studies. The analysis of the geographical distribution of ecotypes is considered a significant marker in taxonomy (Komárek 2018). The ecological criteria have been helpful in many instances for distinguishing cyanobacterial isolates, which aids in the correct identification and classification of cyanobacteria. Some of the case studies are as follows:

1. The case study of *Microchaete*, where the reference strain *M. grisea* was marine in traditional classification systems, has been distinguished from the freshwater type *M. tenera*. (Hauer et al. 2014).
2. The establishment of the genus *Dulcicalothrix*, which is ecologically distinguished from the *Calothrix* clade. The members of the genus *Dulcicalothrix* are from freshwater/terrestrial

FIGURE 1.3 (a) *Nostoc* sp., dark-colored aggregated balls. (b) Dark thick mucilage on the surface of the soil. (c) Greenish cyanobacterial film on the wet and water-lodged surface of the soil. (d) Greenish and brownish oil paint-like spill of cyanobacterial film on the surface of a concrete wall. (e) Dry and bright green cyanobacterial layer on the surface of the concrete fencing wall. (f) Thick, brown, dry, mossy-like mat on the surface of the concrete wall. (g) Thick mat growing on a rock surface under a waterfall. (h) Thick mat floating on the surface of a water trough. (i) Bluish-green cyanobacterial growth on stones near the waterfall. (j) Brownish-colored, spherical balls on the soil surface. (k) Dark, epipelic cyanobacterial film in the eutrophicated wetland. (l) Thick dry mat in a paddy field (color figure in e-book).

ecosystems, while the members of the genus *Calothrix* are from marine/freshwater ecosystems (Saraf et al. 2019).
3. The complexities of *Mastigocladus* and *Fischerella* are well-known, where the ecological marker plays an essential role. The members of the true-branching genus *Mastigocladus* are hot spring dwellers and are distinguished from *Fischerella*. Suggestions were made to change the type species of *Fischerella, F. thermalis* (isolated from hot spring) to *F. muscicola*, which is a soil-dwelling taxon and distinct from a thermal isolate (Kaštovský and Johansen 2008).
4. *Cylindrospermum sensu stricto* and the recently described genera *Johanseniella* are morphologically similar but ecologically distant. While the former genera members are from temperate climatic conditions, the latter includes taxa which are from tropical and subtropical climatic conditions (Johansen et al. 2014; Pal et al. 2022).

Furthermore, there are many more explicit examples where ecology plays a major role in bringing out species-level distinctions. The unique ecological condition can provide an important means to study cyanobacterial species. For example, species like *Nodularia quadrata* are found throughout the coasts of Antarctica. Some species are described based on their ecologically unique habitat names.

Dactylothamnos antarcticus and *Hassallia antarctica* are both found in Antarctica, which is one of the most important polar biogeographical regions (Komárek et al. 2015). *Ewamiania thermalis* gen. et sp. nov. is another example where ecology, along with molecular markers, has played an important role as *Ewamiania* was morphologically similar to Scytonemataceae, but *Ewamiania* has been isolated from the thermal spring complex of the northeastern region of Australia, and hence, the name *thermalis* was given. The species *Roholtiella volcanica*, one of the members of the genus *Roholtiella*, has been described recently. It was isolated from the volcanic soil of Kamchatka. Being loaded with toxic compounds, the volcanic soil thus provides an important study site to understand the succession and colonization by algal communities (Gaysina et al. 2022).

1.1.4 Molecular Markers: The Dawn of New Age Systematics

With the advent of biomolecular and biotechnological techniques, classical systematics based on natural systems have been extensively re-investigated as well as explored at intra-genomic levels. One such boon to taxonomy is the development of molecular markers. To be defined comprehensively, a marker simply means a locus or a point which demarcates a specific entity. If viewed from a morphological point, a marker can be an external feature that helps to distinguish one character from another. At a biochemical level, markers can be viewed as specific enzymes which are localized in specific compartments and thus help to identify the organelles with which they are associated. For example, succinate dehydrogenase is a specific mitochondrial marker.

Molecular markers can be defined as specific nucleotide sequences present in nucleic acids, which are usually associated with a specific gene at a well-defined locus. The characterization of various structural and functional genes has been made possible in closely related prokaryotic and eukaryotic organisms due to the development of these markers. A lot of these markers have been employed in phylogenetic studies to establish a cladistic approach to classification. Taxonomy as of today is never complete without phylogeny. To unravel the complex evolutionary relationships of various similar and dissimilar taxa, one needs to view and analyze the available data through a polyphasic approach, thereby including evidence from diverse realms such as morphology, cytology, ecology and, most importantly, phylogeny, which helps in understanding the evolutionary links. A diverse range of conserved sequences, as well as hypervariable regions, occur in all types of organisms that can serve as potential markers to differentiate between taxa based on the sequence similarity of their nucleotides. DNA barcoding and molecular systematics of higher plants, such as angiosperms, greatly depend on certain sequences present in the chloroplast genome owing to its slow rate of evolution (Dong et al. 2012).

Cyanobacterial systematics, on the other hand, is way more complex and requires a highly stable system of sequences which remain conserved over a long period. These markers can be either structural or functional when studied at their respective loci (Song et al. 2015). Several regions or markers have been employed to gather sequencing data from various cyanobacterial strains, which have greatly bolstered our understanding of the phylogeny of these ancient organisms.

1.1.5 Molecular Markers Used in Cyanobacterial Systematics

Since the inception of molecular biology and bioinformatics techniques, several genetic loci or markers have been identified, validated and characterized for their potential application in phylogenetic studies. A few of the most significant ones are discussed next:

i. *16S rRNA gene*: An integral component of the prokaryotic ribosomal small subunit, the 16S rRNA plays a role in the identification of the Shine-Dalgarno sequence in messenger RNA (mRNA), which is the site for the initiation of translation. The gene encoding for 16S rRNA is viewed as a potential marker and phylogenetic tool for the classification of bacteria and cyanobacteria (Schmidt et al. 1991). About 1.5 kb in length, this gene is one of the most conserved genes in prokaryotic systems (Figure 1.4). Given its universality in prokaryotes and eukaryotes, stability, informative variable areas, extensive representation in sequence databases and low prevalence of horizontal gene transfer, it is regarded by many as the most apt gene locus for researching evolutionary history (Boyer et al. 2001).

   The conserved portions of the 16S rRNA gene can be targeted by universal polymerase chain reaction (PCR) primers, allowing amplification of the gene from a variety of microorganisms. The 16S rRNA gene has both conserved and variable sections which are taxonomically very important. While sequencing the variable sections enables differentiation between specific distinct microbes, the conserved area enables universal amplification (Kolbert and Persing 1999). Large portions of the 16S rRNA gene are highly conserved, whereas occasional changeable sections are also present. However, it has been demonstrated that the 16S rRNA gene lacks resolving power at the species level in prokaryotes (Konstantinidis and Tiedje 2005). It has been determined that 98.65% of 16S rRNA gene sequences match the established average nucleotide identity (ANI) criteria for bacterial species identification (Kim et al. 2014).

   There have been occasions where groups of cyanobacteria with distinct phenotypes had identical 16S rRNA gene sequences while having quite diverse nitrogen metabolism loci (Miller et al. 2006). The *rpo*B, *nif*D, *psb*A and *rbc*L genes are among other potential genetic markers that have been proposed for use in prokaryotic taxonomy (Case et al. 2007). To solve the taxonomic shortcomings of the 16S rRNA gene, multilocus sequence analysis (MLSA), which concatenates the 16S rRNA gene data with additional molecular markers, has become popular in molecular investigations (Wilmotte et al. 2017).

ii. *rbc*L *gene*: A constitutive gene coding for ribulose bisphosphate carboxylase, i.e., *rbc*L, has been extensively used as a PCR-based marker for sequence analysis and further phylogenetic studies in cyanobacterial taxonomy. *rbc*L is an important plastid-encoded gene (~1430 bp) which has shown potential in the phylogenetic analysis due to its slow rate of

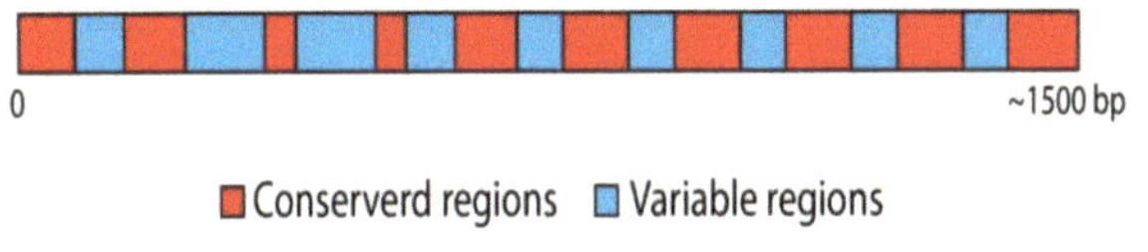

FIGURE 1.4 The linear structure of the 16S rRNA gene showing conserved and variable regions (color figure in e-book).

evolution because the gene is devoid of far-length mutations (Singh et al. 2014). Specific primers for the *rbc*L gene can be retrieved from the literature or can be synthesized using software such as PRIMER 3.0. Using these cyanobacterial-specific primers, the gene can be amplified involving optimized PCR protocols (Daugbjerg and Andersen 1997).

iii. *nif genes*: Nitrogen fixation is a well-established phenomenon in heterocytous cyanobacteria, and the genes encoding for the nitrogenase complex, the primary enzyme involved in nitrogen fixation, are usually members of the *nif* family or homologs (Thiel 2019). According to a study by Esteves-Ferreira et al. (2017), it was revealed that in accordance with their morphology, classification and phylogeny, the sequences of nitrogen-fixing genes (*nif*D) from several morphologically diverse cyanobacterial strains are grouped, and the *nif*D genes from heterocytous strains have a common ancestor. In another study on 41 different cyanobacterial strains from different parts of the Indian subcontinent, *nif* gene analysis was used to understand the relatedness of various taxa based on the aforementioned functional gene locus (Singh et al. 2013).

iv. *Internal transcribed spacer* (*ITS*): The 16S-23S rRNA operon is present in all bacterial genomes (Schleifer and Kandler 1989; Iteman et al. 2000). This operon is often found in numerous copies in heterotrophic bacteria and cyanobacteria, though their exact number varies (Tourova 2003). The ITS region is a hypervariable, regulatory region consisting of various domains such as D1-D1′, D2, D3, $tRNA^{Ile}$, V2, $tRNA^{Ala}$, BOXB, BOXA, D4, V3, etc. ITS regions can be classified as those having transfer RNA (tRNA) regions of both Ala and Ile or either of the two (Villanueva 2018). The most important work on the ITS region was done by Iteman et al. (2000) wherein they compared conserved structural and regulatory domains within divergent 16S rRNA-23S rRNA spacer sequences of cyanobacteria. According to Iteman et al. (2000), the cyanobacterium *Nostoc* sp. PCC 7120 produced three products from PCR amplification of the ITS between the 16S rRNA and 23S rRNA genes.

The first two products were variously sized genuine ITS sections, while the third was a heteroduplex. The $tRNA^{Ile}$ and $tRNA^{Ala}$ genes were carried by the larger spacer (ITS-L), which was made up of 512 nucleotides. A big stem-loop structure (V2) made up of brief, tandemly repeated sequences separated the two genes. The shorter spacer (ITS-S), which had a length of 283 nucleotides and was otherwise nearly identical to ITS-L, was missing both tRNA genes as well as the 5′ half of the intervening stem. Further, the study involved folding of secondary structures of ITS using RNAdraw software. For the deduced transcripts of the *rrn* operons of different cyanobacteria, putative secondary structures were then identified and compared to those of *Escherichia coli*. The locations of sections similar to bacterial antiterminators (box B-box A) and highly conserved motifs vital for the folding and maturation of rRNA transcripts were discovered. Thus, for the detection and identification of cyanobacteria at various taxonomic levels, the conserved and variable portions of the cyanobacterial 16S-23S ITS region are promising targets for PCR primers and oligonucleotide probes (Iteman et al. 2000).

Differences in ITS secondary structures have now evolved into a popular taxonomic approach to describe new species, as morphological features alone cannot serve as apomorphies and other typical molecular markers lack species-level resolution (Iteman et al. 2000). To identify and differentiate between closely related taxa, the usage of ITS and restriction fragment–length polymorphisms of ITS amplicons are suggested (Iteman et al. 2000). Later investigations showed that the ITS secondary structures, which are created by folding conserved hairpin loop domains within the ITS, may be used to distinguish cryptic species, making them an effective tool for phylogenetic studies (Johansen et al. 2013). The conservation of domains across all orders of cyanobacteria suggests that there is an active selection even if the ITS region is regulatory. Additionally, patterns of tRNA inclusion or exclusion in ITS regions suggest that various operons may be subjected to different selective pressures. Within the same genome, several of the helices from ITS operons with various patterns of tRNA inclusion differ significantly from one another. In addition, different cyanobacterial taxa have shown patterns of domain deletions and duplications within the ITS (Villanueva 2018).

1.1.6 Phylogenetics: An Integral Component of the Polyphasic Approach

For a long time, organisms have been categorized into hierarchical systems of classification for ease of understanding and documentation purposes. The sole basis of these classifications was derived from the variations in the morphological characteristics that set one group of organisms apart from another. In particular, the natural system of classification has been in place ever since the concept of taxonomy originated.

Gradually, other characteristics such as anatomical, biochemical, cytological, etc., started coming into the picture, making the classification system more and more exhaustive. However, none of these characters reflected the evolutionary history and relationships or phylogeny among the organisms in question. A key difference between taxonomy and systematics is that the latter is based on a more exhaustive and informative approach called cladistics. Cladistics takes into consideration evolutionary characters which could be categorized as plesiomorphic (primitive) and apomorphic (derived). Together, these two reflect the development of characteristics along a lineage and how they diverge or converge under different and confluent climatic conditions. One of the key features of a cladistic approach is the consideration of DNA sequences as a classic characteristic for the development of what is called a cladogram, or simply a phylogenetic tree.

The convergent phase of such morphological, ecological and molecular data forms the basis of the present-day polyphasic approach to cyanobacterial systematics (Komárek et al. 2014). Phylogeny reflects the true evolutionary history and relationships among the cyanobacterial taxa and forms the basis of modern-era cyanobacterial systematics along with ecological and morphological characteristics. Classification solely based on classical morphological features has resulted in the creation of large polyphyletic taxa. According to Komárek et al. (2014), it is advisable to have a large number of groups at each taxonomic level which are truly monophyletic instead of small groups with a polyphyletic origin.

An overall series of steps that could aid in developing classification using phylogenetic principles are as follows:

i. Standardization of parameters which are more relevant to cladistic and character weighting.
ii. Scoring of sequence data.
iii. Generation of phylogenetic trees using various algorithms based on different principles like maximum parsimony, maximum likelihood, Bayesian inference, etc.

Usually, more than two different algorithms are used to compare their respective trees to ensure that the same result or consistency is maintained throughout the process, irrespective of the principle of the algorithm involved. Finally, it must be realized that the polyphasic approach encompasses a different set of methodologies that reflect various characteristics of cyanobacteria. This also means that there is always a certain degree of freedom associated with the methodologies that can be used for identifying a cyanobacterium. Hence, the components of the polyphasic approach will certainly differ as per the taxa under study.

1.2 EQUIPMENT

1. *Customized culture room having racks with light*: For purification, culturing and subculturing of cyanobacteria.
2. *Microscope*: For studying the morphological characteristics of cyanobacteria.
3. *Autoclave*: For sterilization of glassware, media components and small equipment.
4. *Laminar flow*: For working with the cyanobacteria in an aseptic environment.
5. *Water bath*: Needed in the process of DNA extraction.
6. *Vortex*: Needed in the process of DNA extraction.
7. *Centrifuge*: Needed in the process of DNA extraction.
8. *Gel electrophoretic unit*: Separation of charged DNA molecules.

9. *Thermal cycler*: For performing various DNA amplification processes.
10. *Ultraviolet (UV) transilluminator*: For visualizing DNA.
11. *Small tabletop Biological Oxygen Demand (BOD) incubator*: For incubating plates and tubes while doing cloning and plasmid isolations.
12. *Heating block*: Needed during the cloning procedure.

1.3 RECIPE

BG-11$_0$ medium: All growth medium components can be stored at room temperature except citric acid, which needs to be stored in a refrigerator (Table 1.1).

1.4 MATERIALS

1. FastDNA Spin Kit for Soil (MP Biomedicals, USA) for DNA extraction. All the components can be stored at room temperature.
2. dNTPs, Taq polymerase, Taq buffer, primers, double distilled water: For setting up PCR reactions. All the components of the PCR reaction must be stored at –20°C.
3. *Agarose*: For gel electrophoresis.
4. StrataClone PCR Cloning Kit (Agilent Technologies, Germany): Must be stored at –80°C.
5. QIAprep® Spin Miniprep Kit for plasmid isolation: Stored at room temperature.
6. *EcoRI restriction enzyme*: Stored at –20°C.
7. *DNA ladder*: Stored at –20°C.
8. *Ethidium bromide*: Stored at room temperature.

1.5 METHODS

Figure 1.5 depicts all the steps of a polyphasic approach for cyanobacterial systematics. In the following section, we present all the steps of the polyphasic approach in detail.

1.5.1 Step 1: Sample Collection, Isolation, Cultivation and Maintenance of Purified Strains

Cyanobacterial samples must be collected carefully using sterilized field equipment and should be stored immediately in appropriately labeled Falcon tubes. If possible, samples should be collected along with the natural water in which it is present. In the case of dried samples, sterilized BG-11$_0$ media (Rippka et al. 1979) can be used. In the case of soil crusts, samples are taken from the typical biotopes of the studied territory and the sites with heterogenic ecological conditions (for example,

TABLE 1.1
Composition of BG-11$_0$ Medium Used for Culturing of Heterocytous Cyanobacteria

Macronutrients	g/100 mL	Micronutrients	mg/100 mL
$K_2HPO_4.3H_2O$	0.40	H_3BO_4	0.286
$MgSO_4.7H_2O$	0.75	$MnCl_2.4H_2O$	0.1810
$CaCl_2.2H_2O$	0.36	$ZnSO_4.7H_2O$	0.0222
$NaHCO_3$	0.06	$CuSO_4.2H_2O$	0.0079
Citric acid	0.20	$CO(NO_3)_2.2H_2O$	0.00492
Ferric ammonium citrate	0.06	Sodium molybdate	0.0039
EDTA	0.01		

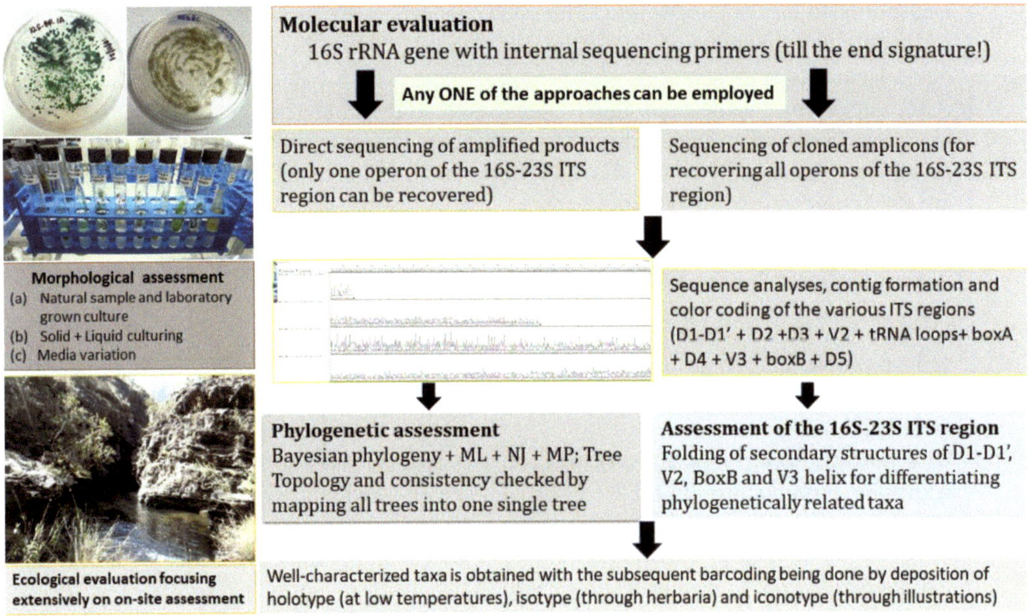

FIGURE 1.5 Basic steps involved in a polyphasic approach: (a) Morphological assessment; (b) ecological evaluation; (c) molecular evaluation; (d) sequence analyses, contig formation and color coding of the ITS region; (e) phylogenetic assessment; (f) assessment of the 16S-23S ITS region; and (g) preparation of holotype, isotype and iconotype (color figure in e-book).

ravines, riversides and paths). Usually, each sample consists of several subsamples collected from a site of around 2 m^2 to a depth of 5 cm (Gaysina et al. 2018).

1. Important ecological and physiochemical characteristics of the habitat must be noted at the time of sampling.
2. Field photographs form an essential component of cyanobacterial taxonomic studies and must be taken at different angles and views so that a good measure of ecological and geographical features of the surrounding area can be obtained.
3. On arrival in the laboratory, the natural sample must be observed as soon as possible under the microscope to assess the overall cyanobacterial diversity. The pictures of the natural samples must be appropriately labeled, and care must be taken in order not to mix the labels.
4. Samples can then be grown in BG-11_0 media, where appropriate cycles of subculturing must be done in both solid and liquid media to obtain unialgal colonies. The number of cycles will vary from case to case.
5. After purification, unialgal strains can then be maintained in a culture room illuminated with 5 µmol photons m^{-2} s^{-1} with a photoperiod of 14:10 hours light/dark cycle and temperature of 26 ± 2°C.

1.5.2 Step 2: Morphological Evaluation

1. An extensive morphological evaluation must be attempted, where all the phenotypic characteristics such as filament polarity, sheath, positioning of heterocyte, cell shape and size and akinete as well as hormogonium formation must be recorded.
2. It is advisable to take the microscopic images in different magnifications such as 200×, 400× and 1000× to properly record the interesting traits of the taxa under investigation.
3. Along with the micrographs, measurements of vegetative cells, specialized cells and branches, if present, must also be recorded.

4. Depending on the taxa under study, life cycle changes must also be observed. For example, while studying members of the *Nostoc* sensu stricto, these life cycle events gain importance. Hence, time must be devoted to regularly observing the strain under the microscope.
5. Also, depending on the taxa under study, interesting colony morphologies can also be observed many times under the microscope. For example, in the case of genera like *Rivularia*, *Calothrix*, *Gloeotrichia*, *Spirirestis* and many others, the colony morphologies are very interesting characteristics.

1.5.3 Step 3: DNA Extraction

1. Extraction of the total genomic DNA must be done from a 10- to 12-day-old culture (basically in the log phase of growth). Older cultures tend to have too many extracellular materials associated with them, and downstream processing could become a bit complicated in those cases.
2. DNA extraction can be done using the FastDNA Spin Kit for Soil (MP biomedicals, USA) with some modifications in the homogenization step (the homogenization should be done for a longer time, as many cultures do not tend to get homogenized very easily).
3. DNA should be then visualized on 0.8% agarose gel and further quantified on a Thermo Fischer NanoDrop™ One/One C (AZY1601774).
4. Amplification of the 16S rRNA gene and the 16S-23S ITS region can then be carried out using the Bio-Rad T1000™ Thermal Cycler (or any other suitable thermal cycler, depending on the availability) in a 25-µL volume mixture containing 250 µM dNTPs, 1.5 mM $MgCl_2$, 0.5 µM of forward and reverse primers, 1U/µL Taq DNA polymerase and 2 µL (25–50 ng) of DNA template.
5. Primers used for the amplification are forward primer 1 (5′-CTCTGTGTGCCT AGGTATCC-3′) and reverse primer 2 (5′-GGGGAATTTTCCGCAATGGG-3′) (Wilmotte et al. 1993; Nübel et al. 1997; Boyer et al. 2002). Reaction conditions must be adjusted to 35 cycles of 94°C for 45 seconds, 57°C for 45 seconds, 72°C for 135 seconds followed by a final extension of 3 minutes at 72°C and finally, the reaction can be held indefinitely at 4°C.
6. After analyzing the amplified PCR product in 1% agarose gel, cloning should be done using the StrataClone PCR Cloning Kit (Agilent Technologies, Germany). It is always better to start the cloning immediately after checking the quality and accuracy of the PCR product.
7. Plasmids should be isolated using the QIAprep® Spin Miniprep plasmid isolation kit (QIAGEN India Pvt. Ltd.), and EcoR1 restriction digestion must be done to ensure that the cloning was successful.
8. Restriction digestion products should be visualized in 1% agarose gel to select the right clones.
9. For complete recovery of the different variants of the ITS operons, it is always advisable to sequence more than eight plasmids. This approach is not cost-effective, but it will ensure coverage of all operons.
10. Sequence analysis must be done carefully, keeping a special eye on PCR errors and then curating the alignments manually to maintain the secondary structure of the 16S rRNA gene.

1.5.4 Step 4: Phylogenetic Analysis

1. A National Center for Biotechnology Information (NCBI) blast search (www.ncbi.nlm.nih.gov/blast) must be performed to identify all the related sequences, and the reference sequences should be downloaded from the NCBI database. The number of related or reference sequences do vary according to the taxon under study, but exhaustive taxon sampling must be performed to get a better idea of the taxonomic position of the taxa of interest.

2. The alignment can be done using MAFFT (Katoh et al. 2019) and should essentially be manually curated to maintain the secondary structure of the 16S rRNA gene.
3. Phylogenetic trees must be constructed using multiple methods to ensure a consistent topology and bootstrap value. This is an essential part of taxonomic investigations and, hence, must be done with utmost care.
4. Phylogenetic trees can be made using Bayesian inference (BI), neighbor-joining (NJ), maximum parsimony (MP) and maximum likelihood (ML) methods.
5. The BI tree can be inferred using Mr Bayes 3.2.2 (Ronquist et al. 2012), and the analysis must be run until the standard deviation of the split frequency goes down below 0.01. This would ensure proper convergence of the tree. Two runs of eight Markov chains can be set with a burn-in value set to 25%, while the print frequency, sampling frequency and diagnostic frequency should be set to 1,000.
6. The ML tree can be reconstructed in IQ-TREE, v6.10 (Trifinopoulos et al. 2016) using a 1000 ultra-fast bootstrap replication method. ML trees can also be run using the MEGAX program (Kumar et al. 2018) using a suitable model and the standard bootstrap method with 1,000 replications.
7. For BI and ML trees, the best model can be determined using jModelTest (Darriba et al. 2012).
8. The MP and the NJ trees can be inferred in MEGAX (Kumar et al. 2018) using a suitable model and the standard bootstrap method with 1000 replications.

1.5.5 Step 5: 16S-23S ITS Secondary Structure Analysis

1. The various regions of 16S-23S ITS must be first color-coded for ease of analysis.
   Attention must be paid to demarcate operons with and without tRNA.
2. Color-coded regions of the D1–D1′, V2, Box-B and V3 helices should be folded on the Mfold web server (Zuker 2003), and the obtained secondary structures must be compared carefully with the secondary structures of the corresponding phylogenetically related taxa.
3. The folded secondary structures corresponding to the D1–D1′ region, V2, Box-B and V3 should then be redrawn in Adobe Illustrator (ver. 24.1.1).
4. To further investigate and perform the lengthwise comparison of the ITS region, a comparative length-wise table of the leader, D1–D1′, spacer + D2, spacer + D3 + spacer, $tRNA^{isoleucine}$, spacer + V2 + spacer, $tRNA^{alanine}$, spacer + boxB + spacer, boxA, D4 + spacer and the V3 + end of ITS can be performed with all the phylogenetically related strains. This ensures clarity and ease of comparison.

1.5.6 Step 6: Estimation of p-distance for the 16S rRNA Gene and 16S-23S ITS Region

1. The p-distance for the 16S rRNA gene and 16S-23S ITS region can be calculated using MEGA X (Kumar et al. 2018).
2. Further, the percent similarity of the 16S rRNA gene for both intra- and inter-clade analyses (depending on the taxa under evaluation) can be determined using the formula (1 – p-distance) * 100.
3. The percent dissimilarity of 16S-23S ITS can be determined using the formula (p-distance * 100).

1.5.7 Step 7: Preparation of Holotype, Isotype and Line Diagram

1. The holotype must be deposited in a recognized culture collection by preserving the metabolically inactivated form to obtain an accession number. It is always advised to submit the holotype before communicating/publishing new taxa, as this is essential for taxonomic

descriptions. There is a difference in the botanical and the bacterial codes regarding the deposition requirements, but still, it is an essential part, irrespective of the code of nomenclature being followed.

2. For isotype, herbaria preparation should be done in duplicate, and for this a Buckner funnel can be used along with quantitative filter paper (wet-strengthened grade-91) to sieve out the excess media, and then the mat remaining on the filter paper can be air-dried to form the herbaria. Further, the herbaria can be kept in 60-m Petri plates and packed in acid-free envelopes along with information such as strain ID, strain name, habitat, locality, collection date, collector name and culture collection accession number.
3. A line diagram, detailing all the morphological details of the taxon under study, can be drawn on transparent sheets using the stippling method.

1.6 DATA ANALYSIS

Cyanobacterial taxonomy has come a long way from traditional morphology-based classifications to modern molecular and bioinformatics-aided taxonomy. Many complexities in cyanobacterial taxonomy arose due to the erroneous identification of cyanobacteria based on morphological criteria. However, the development of an integrative approach has helped to develop a better and more robust system to identify and classify cyanobacteria. It is notable that even when molecular method–based studies seem to provide more resolution, the traditional approaches continue to play a significant role in taxonomy. The use of morphological, ecological, molecular and phylogenetic tools together brings a greater depth and better resolution to untangle the complications in cyanobacterial taxonomy. The classification system proposed by Komárek et al. (2014) has provided a foundation to study cyanobacteria by using a polyphasic approach. A combined approach and the importance of each component of the polyphasic approach are very crucial and cannot be replaced or removed. Starting from the in-depth study of the sampling site and recording the important ecological characteristics to studying the morphological details, followed by molecular analysis and using phylogenetic algorithms are all important to understand the taxonomic position of cyanobacteria. Each component is an integral part of the approach, which is combined to form an impeccable tool used in cyanobacterial taxonomy. Taxonomic interpretations are usually subject to future revisions, as the science of taxonomy is constantly growing with the addition of new data. However, there are some vital points to consider in a taxonomic dataset before drawing any taxonomic inferences. These important points are as follows:

1. Life cycle patterns in some cyanobacteria are very important but difficult to follow. Hence, attention must be paid while dealing with the taxonomy of members of the genus *Nostoc*.
2. With the advances in sequencing technology and an enhanced focus on molecular and phylogenetic inferences, many times a species that may appear to be new and occupying a strongly supported phylogenetic position may simply be a previously not sequenced species already mentioned in various monographs. Hence, it is always advised to carefully compare the morphology of a newly sequenced species with previously reported old species that did not have a sequence.
3. Always pay attention to the positioning of the type species while constructing a 16S rRNA gene phylogeny. In principle, every taxonomic unit should be monophyletic, although a huge amount of discrepancies does occur leading to confusing situations. However, establishing the monophyly of a genus of cyanobacteria is something that must be taken care of while deriving phylogenetic inferences.
4. Be extremely careful while adopting the currently accepted genetic standards for prokaryotes based on 16S rRNA gene similarity (with $<94.5\%$ similarity signifying different genera). Cyanobacteria are different kinds of prokaryotes which have distinct morphology with a huge amount of variations. Hence, most of the time these similarity percentages do not hold much importance.

5. The cutoff values for the delineation of distinct species within a genus are many times difficult to follow, especially in the Nostocalean groups.
6. In the case of 16S-23S ITS, if the foldable regions like the D1–D1′, V2, Box-B and V3 regions are more or less similar to the phylogenetically related strains, the percentage dissimilarity (>7% dissimilarity signifying different species) should be considered for better resolution.

1.7 NOTES/PRECAUTIONS

1. Handling of cultures should be done under axenic conditions.
2. Cyanobacterial purification is a complicated and time-consuming process which needs continuous assessment of emerging colonies. It is very common to have different morphologies of the natural and laboratory-grown cultures. Hence, proper assessment of morphologies is critical and necessary.
3. The presence of extracellular polysaccharides results in poor quality and quantity of gDNA, and therefore, a method for genomic DNA extraction from cyanobacterial samples should be carefully selected (see Chapter 18).
4. Plasmid isolation must be done within 12–16 hours of growth. Longer incubation will result in false colonies without an insert.

ACKNOWLEDGMENTS

We thank the head of the Department of Botany, Banaras Hindu University (BHU), for providing the necessary facilities and encouragement. SP thanks the Department of Science and Technology for the DST-INSPIRE Senior Research Fellowship (SRF). The work was funded by the Department of Science and Technology (DST-SERB), India, through the Core Research Grant Project CRG/2018/004111. Banaras Hindu University is acknowledged for providing a seed grant under the Institution of Eminence (IoE) Scheme.

REFERENCES

Anagnostidis A and Komárek J (1990) Modern approaches to the classification of cyanobacteria, Stigonematales. *Archiv fur Hydrobiologie* 14: 224–286.

Anagnostidis K and Komárek J (1985) Modern approach to the classification system of cyanophytes 1—Introduction. *Archiv fur Hydrobiologie Supplement* 71: 291–302.

Anagnostidis K and Komárek J (1988) Modern approach to the classification system of cyanophytes 3—Oscillatoriales. *Algological Studies* 50–53: 327–472.

Bornet É and Flahault C (1888) Note sur deux nouveaux genres d'algues perforantes. *J Mersch* 2: 161–165.

Bourrelly P (1970) Les algues d'eau douce III. N Boubée & Cie, Paris.

Boyer SL, Flechtner VR and Johansen JR (2001) Is the 16S-23S rRNA internal transcribed spacer region a good tool for use in molecular systematics and population genetics? A case study in cyanobacteria. *Molecular Biology and Evolution* 18(6): 1057–1069.

Boyer SL, Johansen JR, Flechtner VR and Howard GL (2002) Phylogeny and genetic variance in terrestrial *Microcoleus* (Cyanophyceae) species based on sequence analysis of the 16S rRNA gene and associated 16S-23S ITS region. *Journal of Phycology* 38: 1222–1235.

Case RJ, Boucher Y, Dahllöf I, Holmström C, Doolittle WF and Kjelleberg S (2007) Use of 16S *rRNA* and *rpoB* genes as molecular markers for microbial ecology studies. *Applied and Environmental Microbiology* 73: 278–288.

Castenholz RW, Wilmotte A, Herdman M, Rippka R, Waterbury JB, Iteman I and Hoffmann L (2001) Phylum BX. Cyanobacteria. In: Bergey's Manual® of Systematic Bacteriology (Eds. D.R. Boone, R.W. Castenholz and G.M. Garrity). Springer, New York. pp. 473–599.

Darriba D, Taboada GL, Doallo R and Posada D (2012) JModelTest 2: More models, new heuristics and parallel computing. *Nature Methods* 9: 772.

Daugbjerg N and Andersen RA (1997) Phylogenetic analyses of the rbcL sequences from haptophytes and heterokont algae suggest their chloroplasts are unrelated. *Molecular Biology and Evolution* 14(12): 1242–1251.

Desikachary TV (1959) Cyanophyta. In: ICAR Monograph of Algae. ICAR, New Delhi.

Dong W, Liu J, Yu J, Wang L and Zhou S (2012) Highly variable chloroplast markers for evaluating plant phylogeny at low taxonomic levels and for DNA barcoding. *PLoS ONE 7*(4): e35071.

Drouet F and Daily WA (1956) Revision of the coccoid Myxophyceae. *Butler University Botanical Studies 12*(2): 1–218.

Elenkin AA (1936) Monographie algarum cyanophycearum aquidulcium et terrestrium infinibus URSS inventatrum. Vol 1. Acad Nauk URSS Moscow, Leningrad. pp. 675.

Elenkin AA (1938) Monographia algarum cyanophycearum aquidulcium et terrestrium in finibus URSS inventarum. *Sinezelenje vodorosli SSSR* 1: 1–908.

Elenkin AA (1949) Monographia algarum cyanophycearum aquidulcium et terrestrium in finibus URSS inventarum. *Sinezelenye vodorosli SSSR*. Pars Specialis (Systematica) 2: 1908.

Esteves-Ferreira A, Cavalcanti J, Vaz M, Alvarenga L, Nunes-Nesi A and Araújo W (2017) Cyanobacterial nitrogenases: Phylogenetic diversity, regulation and functional predictions. *Genetics and Molecular Biology* 40: 261–275.

Fritsch FE (1942) The interrelations and classification of the Myxophyceae (Cyanophyceae). *The New Phytologist* 41(2): 134–148.

Fritsch FE (1944) Present-day classification of algae. *The Botanical Review* 10: 233–277.

Fritsch FE (1945) The structure and reproduction of the algae. In: Foreword, Phaeophyceae, Rhodophyceae Myxophyceae. Vol II. Cambridge University Press, Cambridge. pp. 1–939.

Gaysina LA, Bohunická M and Hazuková Vand Johansen JR (2018) Biodiversity of terrestrial cyanobacteria of the South Ural region. *Cryptogamie, Algologie* 39(2): 168.

Gaysina LA, Johansen JR, Saraf A, Allaguvatova RZ, Pal S and Singh P (2022) *Roholtiella volcanica* sp. nov., a new species of cyanobacteria from Kamchatkan volcanic soils. *Diversity* 14(8): 620.

Geitler L (1932) Cyanophyceae. In: Rabenhorst's Kryptogamen Flora von Deutschland. Vol 14. Akad Verlagsges, Leipzig. pp. 1–1196

Geitler L (1942) Schizophyta (Klasse Schizophyceae). In: Natürliche Pflanzenfamilien 1b (Eds. A. Engler and K. Prantl). Duncker & Humblot, Berlin. pp. 1–232.

Golubić S, Hernández-Mariné M and Hoffmann L (1996) Developmental aspects of branching in filamentous cyanophyta/cyanobacteria. *Algological Studies/Archiv fur Hydrobiologie Supplement Volumes* 83: 303–329.

Gomont M (1892) Monographie des *Oscillariées. Ann Sci Nat Ser Bot* 15: 263–368.

Hauer T, Bohunická M, Johansen J, Mareš J and Berrendero-Gomez E (2014) Reassessment of the cyanobacterial family Microchaetaceae and establishment of new families Tolypothrichaceae and Godleyaceae. *Journal of Phycology* 50(6): 1089–1100.

Hentschke G, Johansen J, Pietrasiak N, Fiore M, Rigonato J, Sant'anna C and Komárek J (2016) Phylogenetic placement of *Dapisostemon* gen. nov. and *Streptostemon*, two tropical heterocytous genera (Cyanobacteria). *Phytotaxa* 245(2): 129–143.

Hofmann L, Komárek J and Kaštovský J (2005) System of cyanoprokaryotes (cyanobacteria) – State in 2004. *Algological Studies* 117: 95–115.

Iteman I, Rippka R, de Marsac NT and Herdman M (2000) Comparison of conserved structural and regulatory domains within divergent 16S rRNA-23S rRNA spacer sequences of cyanobacteria. *Microbiology* 146: 1275–1286.

Johansen J, Bohunická M, Lukešová A, Hrčková K, Vaccarino M and Chesarino N (2014) Morphological and molecular characterization within 26 strains of the genus *Cylindrospermum* (Nostocaceae, Cyanobacteria), with descriptions of three new species. *Journal of Phycology* 50(1): 187–202.

Johansen J and Casamatta D (2005) Recognizing cyanobacterial diversity through adoption of a new species paradigm. *Algological Studies/Archiv fur Hydrobiologie* 117: 71–93.

Johansen JR, Pietrasiak N, Osorio-Santos K, Bohunická M, Miscoe L, Kováčik L and Martin MP (2013) Has the time come to name cryptic species of cyanobacteria? A case study in *Oculatella*. In: Book of abstracts, *19th Symposium of the International Society for Cyanophyte Research*, Cleveland, Ohio, pp. 15.

Kaštovský J and Johansen JR (2008) *Mastigocladus laminosus* (Stigonematales, Cyanobacteria): Phylogenetic relationship of strains from thermal springs to soil-inhabiting genera of the order and taxonomic implications for the genus. *Phycologia* 47(3): 307–320.

Katoh K, Rozewicki J and Yamada KD (2019) MAFFT online service: Multiple sequence alignment, interactive sequence choice and visualization. *Briefings in Bioinformatics* 20(4): 1160–1166.

Kenyon CN, Rippka R and Stanier RY (1972) Fatty acid composition and physiological properties of some filamentous blue-green algae. *Archiv fur Mikrobiologie* 83(3): 216–236.

Kilgore C, Johansen JR, Mai T, Hauer T, Casamatta DA and Sheil CA (2018) Molecular characterization of *Geitleria appalachiana* sp. nov. (Nostocales, Cyanobacteria) and formation of Geitleriaceae fam. nov. *Fottea* 18(2): 150–163.

Kim M, Oh HS, Park SC and Chun J (2014) Towards a taxonomic coherence between average nucleotide identity and 16S *rRNA* gene sequence similarity for species demarcation of prokaryotes. *International Journal of Systematic and Evolutionary Microbiology* 64: 346–351.

Kolbert CP and Persing DH (1999) Ribosomal DNA sequencing as a tool for identification of bacterial pathogens. *Current Opinion in Microbiology* 2(3): 299–305.

Komárek J (2003) Problem of the taxonomic category. *Algological Studies/Archiv fur Hydrobiologie Supplement* 1: 281–97.

Komárek J (2010) Recent changes (2008) in cyanobacteria taxonomy based on a combination of molecular background with phenotype and ecological consequences (genus and species concept). *Hydrobiologia* 639(1): 245–259.

Komárek J (2018) Several problems of the polyphasic approach in the modern cyanobacterial system. *Hydrobiologia* 811(1): 7–17.

Komárek J and Anagnostidis K (1986) Modern approach to the classification system of cyanophytes 2 chroococcales. *Archiv fur Hydrobiologie* 43: 157–226.

Komárek J and Anagnostidis K (1989) Modern approach to the classification system of cyanophytes 4 nostocales. *Archiv fur Hydrobiologie* 82: 247–345.

Komárek J, Genuário DB, Fiore MF and Elster J (2015) Heterocytous cyanobacteria of the Ulu Peninsula, James Ross Island, Antarctica. *Polar Biology* 38: 475–492.

Komárek J, Kaštovský J, Mareš J and Johansen JR (2014) Taxonomic classification of cyanoprokaryota (cyanobacterial genera) using a polyphasic approach. *Preslia* 86: 295–335.

Konstantinidis KT and Tiedje JM (2005) Towards a genome-based taxonomy for prokaryotes. *Journal of Bacteriology* 187: 6258–6264.

Krumbein WE (1979) Photolithotropic and chemoorganotrophic activity of bacteria and algae as related to beachrock formation and degradation (Gulf of Aqaba, Sinai). *Geomicrobiology Journal* 1(2): 139–203.

Kumar S, Stecher G, Li M, Knyaz C and Tamura K (2018) MEGA X: Molecular evolutionary genetics analysis across computing platforms. *Molecular Biology and Evolution* 35(6): 1547–1549.

Miller SR, Purugganan MD and Curtis SE (2006) Molecular population genetics and phenotypic diversification of two populations of the thermophilic cyanobacterium *Mastigocladus laminosus. Applied and Environmental Microbiology* 72: 2793–2800.

Mishra D, Suradkar A, Saraf A and Singh P (2020) Phylogenetic evaluation of the true-branched heterocytous cyanobacteria and description of soil dwelling *Westiellopsis akinetica* sp. nov. *FEMS Microbiology Letters* 367(5): fnaa046.

Murata N, Wada H and Gombos Z (1992) Modes of fatty acid desaturation in cyanobacteria. *Plant and Cell Physiology* 33: 933–941.

Nübel U, Garcia-Pichel F and Muyzer G (1997) PCR primers to amplify 16S *rRNA* genes from cyanobacteria. *Applied and Environmental Microbiology* 63: 3327–3332.

Pal S, Saraf A, Kumar N, Singh A, Talukdar U, Kohar N and Singh P (2022) Digging deeper into the taxonomy *of Cylindrospermum* and description of *Johanseniella tripurensis* gen. et sp. nov. from India. *FEMS Microbiology Letters* 369(1): fnac074.

Rippka R, Stanier RY, Deruelles J, Herdman M and Waterbury JB (1979) Generic assignments, strain histories and properties of pure cultures of cyanobacteria. *Microbiology* 111: 1–61.

Ronquist F, Teslenko M, van der Mark P, Ayres DL, Darling A, Höhna S, Larget B, Liu L, Suchard MA and Huelsenbeck JP (2012) MrBayes 3.2: Efficient Bayesian phylogenetic inference and model choice across a large model space. *Systematic Biology* 61: 539–542.

Saraf A, Dawda H, Suradkar A, Agre V and Singh P (2020) Fortiea necridiiformans sp. nov., a soil-dwelling cyanobacterium from Pachmarhi Biosphere Reserve, India. *International Journal of Systematic and Evolutionary Microbiology* 70(8): 4714–4724.

Saraf A, Suradkar A, Dawda HG, Gaysina LA, Gabidullin Y, Kumat A, Behere I, Kotulkar M, Batule P and Singh P (2019) Phylogenetic complexities of the members of Rivulariaceae with the re-creation of the family Calotrichaceae and description of *Dulcicalothrix necridiiformans* gen nov., sp nov., and reclassification of *Calothrix desertica. FEMS Microbiology Letters* 366(17): fnz219.

Schleifer KH and Kandler O (1989) Phylogenetic relationships of bacteria. In: The Hierarchy of Life (Eds. B. Fernholm, K. Bremer and H. Jornvall). Elsevier Science Publishers BV, Amsterdam. pp. 103–117.

Schmidt TM, Delong EF and Pace NR (1991) Analysis of a marine picoplankton community by 16S *rRNA* gene cloning and sequencing. *Journal of Bacteriology* 173(14): 4371–4378.

Sendall BC and McGregor GB (2018) Cryptic diversity within the *Scytonema* complex: Characterization of the paralytic shellfish toxin producer *Heterosyctonema crispum*, and the establishment of the family Heteroscytonemataceae (Cyanobacteria/Nostocales). *Harmful Algae* 80: 158–170.

Shalygin S, Shalygina R, Johansen JR, Pietrasiak N, Berrendero Gomez E, Bohunicka M, Mares J and Sheil CA (2017) *Cyanomargarita* gen. nov. (*Nostocales, Cyanobacteria*): Convergent evolution resulting in a cryptic genus. *Journal of Phycology* 53: 762–777.

Singh P, Fatma A and Mishra AK (2014) Molecular phylogeny and evogenomics of heterocystous cyanobacteria using *rbcl* gene sequence data. *Annals of Microbiology* 65(2): 799–807.

Singh P, Singh SS, Elster J and Mishra AK (2013) Molecular phylogeny, population genetics, and evolution of heterocystous cyanobacteria using *nifH* gene sequences. *Protoplasma* 250(3): 751–764.

Song G, Xiang X, Wang Z and Li R (2015) Polyphasic characterization of *Stigonema dinghuense*, sp. nov. (Cyanophyceae, Nostocophycidae, Stigonemaceae) from Dinghu Mountain, south China. *Phytotaxa* 213(3): 212–224.

Thiel T (2019) Organization and regulation of cyanobacterial *nif* gene clusters: Implications for nitrogenase expression in plant cells. *FEMS Microbiology Letters* 366(7): fnz077.

Thuret G (1875) Essai de classification des Nostochinées. *Ann Sci Nat Bot* 6(1): 372–382.

Tourova TP (2003) Copy number of ribosomal operons in prokaryotes and its effect on phylogenetic analyses. *Microbiology* 72: 389–402.

Trifinopoulos J, Nguyen LT, von Haeseler A and Minh BQ (2016) W-IQ-TREE: A fast online phylogenetic tool for maximum likelihood analysis. *Nucleic Acids Research* 44(W1): W232–W235.

Villanueva CD (2018) Molecular and Evolutionary Analysis of Cyanobacterial Taxonomic Methods. UNF Graduate Theses and Dissertations 810.

Waterbury J and Stanier R (1977) Two unicellular cyanobacteria which reproduce by budding. *Archives of Microbiology* 115(3): 249–257.

Wilmotte A and Herdman M (2001) Phylogenetic relationships among the cyanobacteria based on 16S *rRNA* sequences. In: Bergey's Manual of Systematic Bacteriology (Eds. D.R. Boone and R.W. Castenholz). 2nd ed. Vol. 1. Springer Verlag, New York. pp. 487–493.

Wilmotte A, Laughinghouse IV D, Capelli C, Rippka R and Salmaso N (2017) Taxonomic identification of cyanobacteria by a polyphasic approach. In: Molecular Tools for the Detection and Quantification of Toxigenic Cyanobacteria (Eds. R. Kumayer, K. Sivonen, A. Wilmotte and N. Salmaso). John Wiley & Sons, Hoboken, New Jersey. pp. 79–134.

Wilmotte A, Van der Auwera, G and De Wachter R (1993) Structure of the 16S ribosomal RNA of the thermophilic cyanobacterium *Chlorogloeopsis* HTF ("*Mastigocladus laminosus* HTF) strain PCC7518, and phylogenetic analysis. *FEBS Lett.* 317: 96–100.

Zuker M (2003) Mfold web server for nucleic acid folding and hybridization prediction. *Nucleic Acids Research* 31(13): 3406–3415.

2 Estimation of Growth and Photosynthetic Pigments in Cyanobacteria

Pankaj K. Maurya, Sapna Tiwari, Soumila Mondal, Anjali Gupta, and Shailendra P. Singh

2.1 INTRODUCTION

The first oxygenic photosynthesis on the Earth was performed by cyanobacteria during the Precambrian period. These Gram-negative bacteria are cosmopolitan in distribution and play an important role in any ecosystem because of their inherent ability to fix carbon dioxide and dinitrogen (Bryant 1994). Due to their photoautotrophic mode of nutrition, cyanobacteria have an obligatory requirement for the harvesting of solar radiation to produce adenosine triphosphate (ATP) and $NADPH_2$.

Cyanobacteria produce a variety of photosynthetic pigments such as chlorophylls, carotenoids and phycobiliproteins (PBPs) for the absorption of photons available in their habitats (Ho and Bryant 2019; Sanfilippo et al. 2019). The chemical energy produced during the light-dependent reaction, i.e., photochemistry, of photosynthesis is further utilized for the reduction of inorganic CO_2 in a light-independent process of photosynthesis called the Calvin-Benson-Bassham cycle (Bay and Pearlstein 1963; Tanaka et al. 2023).

Among the different photosynthetic pigments, carotenoids and chlorophylls are non-polar chemical molecules, while PBPs are proteinaceous water-soluble molecules with a chromophore attached to amino acid residues (Ho and Bryant 2019; Sanfilippo et al. 2019). These photosynthetic pigments constitute the light-harvesting complexes (LHCs) in cyanobacteria for the absorption of light energy to drive photochemistry. Also, the altered accumulation of pigments under different environmental conditions, including dynamic light conditions, gives an idea about abiotic stresses and acclimation to a light environment. For example, a high ratio of carotenoids to chlorophylls reflects oxidative stress, while the accumulation of different types and amounts of PBPs suggests acclimation to a prevalent light environment to maximize absorption of available light resources (Sanfilippo et al. 2019; Kumar et al. 2021; Maurya et al. 2023a,b). PBPs constitute a macromolecular complex called phycobilisome (PBS), which is the main LHC of the photosystem (PS) II in cyanobacteria, while chlorophylls and carotenoids are the constituents of the LHC of PSI. (Calzadilla and Kirilovsky 2020).

PBSs are hemidiscoidal structures where a core part is attached to the thylakoid membrane at the site of PSII and rods radiate from the core (Anderson and Toole 1998; Samsonoff and MacColl 2001; Calzadilla and Kirilovsky 2020). Acclimatization of cyanobacteria to diverse light conditions can modify or add complementary color to these organisms. The estimation of pigment concentration and ratios of chlorophyll *a*/carotenoids and different PBPs such as phycocyanin (PC, λ_{max} 620–638 nm), phycoerythrin (PE, λ_{max} 565 nm) and allophycocyanin (APC; λ_{max} 650 nm) gives an idea about cyanobacterial fitness in the corresponding light environment (Kehoe and Gutu 2006; Fatma 2009; Assunção et al. 2022). Also, changes in pigment composition under different light environments give a competitive advantage to cyanobacteria over other photosynthetic organisms and help them to flourish in that light environment (Whitton 1992; Sanfilippo et al. 2019).

DOI: 10.1201/9781003398387-2

Therefore, the estimation of growth together with photosynthetic pigments helps to understand cyanobacterial fitness under a particular light environment. The presence of yellowish-green cells with a high ratio of carotenoids to chlorophyll *a* generally indicates that the cells are growing under an intense light environment (Sanfilippo et al. 2019; Kumar et al. 2021; Maurya et al. 2023a; Maurya et al. 2023b). However, the dark bluish-green or reddish coloration of a cell culture reflects growth in a low-light environment associated with increased biosynthesis of chlorophylls and PBPs (Sanfilippo et al. 2019; Kumar et al. 2021; Maurya et al. 2023a; Maurya et al. 2023b). Generally, bacteria under any environmental condition take some time to acclimatize to a new environment in a batch culture and therefore exhibit slow growth (lag phase). After a few days of acclimatization, bacteria exhibit an exponential phase of growth where cells divide logarithmically (Figure 2.1).

After a long exponential phase, growth slows down because of nutritional and/or light limitation due to the high density of cells, which results in a stationary phase. Further, a death phase is observed after the stationary phase if fresh growth medium is not added to the culture (Figure 2.1). The growth of a cyanobacterium can be measured by recording the optical density (OD) of the cell culture at a wavelength that is not absorbed by cyanobacterial cells. Generally, a wavelength between 720 and 750 nm is used for this purpose, which reflects scattering of light as a proxy of growth, as it is dependent on the density/turbidity of cell culture (Janke et al. 1999). However, OD-based measurement of growth can be misleading, as the morphology of cells can have a significant impact on the scattering of light (Willey et al. 2011). A big spherical cell scatters more light and therefore gives more growth than in the case of smaller cells.

Therefore, several parameters such as dry cell biomass and amount of photosynthetic pigments, proteins or carbohydrates can be considered along with the OD-based method to get a better idea of growth under different experimental conditions (Whitman et al. 1998). In this chapter, we present methods for the estimation of growth and different photosynthetic pigments in cyanobacteria.

2.2 EQUIPMENT

1. 1000 μL Micropipette (Tarson, India).
2. Benchtop centrifuge machine (Remi, R-12C plus).

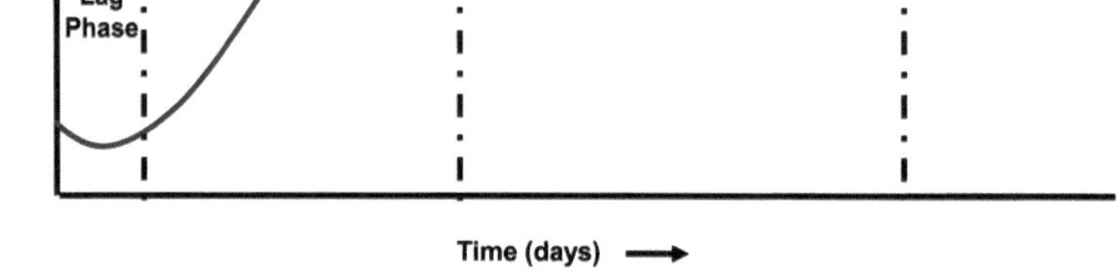

FIGURE 2.1 A typical growth curve diagram of a cyanobacterium growing in a batch culture. The diagram shows different stages of growth, i.e., lag phase, log or exponential phase, stationary phase and death or decline phase (color figure in e-book).

3. Benchtop cooling centrifuge (Remi, NEYA 16R).
4. Hot air oven (Shivam Instruments).
5. Laminar airflow (Shivam Instruments).
6. Magnetic stirrer plate (Tarson, Spinot, Model MC: 02).
7. pH meter (Eutech Instruments pH tutor).
8. Refrigerator (4°C) (Samsung, Korea).
9. Spectrophotometer (UV-1800, Shimadzu, Japan).
10. Vortex (Labquest, Borosil).
11. Weighing machine (Uniblog ATX224 Shimadzu, Japan).

2.3 MATERIALS

1. 0.22 μm syringe filter (Merck, catalogue no: GSWP02500).
2. 1.5 mL microcentrifuge tubes (Tarson, catalogue no: 500010).
3. 1000 mL beaker (Borosil).
4. 2.0 mL microcentrifuge tubes (Tarson, catalogue no: 500020).
5. $CaCl_2.2H_2O$ (calcium chloride dihydrate) (SRL, catalogue no: 70650).
6. Chloroform (Merck, catalogue no: DI7F67291).
7. Citric acid.$1H_2O$ (citric acid monohydrate) (SRL, catalogue no: 49887).
8. $CoCl_2.6H_2O$ (cobalt chloride hexahydrate) (SRL, catalogue no: 99412).
9. $CuSO_4.5H_2O$ (copper [II] sulfate pentahydrate) (SRL, catalogue no: 38869).
10. EDTA (SRL, catalogue no: 12070).
11. 50 mL Falcon tube (Tarson, catalogue no: P10404).
12. Ferric ammonium citrate (ammonium iron [III] citrate) (SRL, catalogue no: 35442).
13. 1000 mL glass bottle (Borosil).
14. 500 mL glass bottle (Borosil).
15. H_3BO_3 (boric acid) (SRL, catalogue no: 80266).
16. HEPES (free acid) (Himedia, catalogue no: RM380-500G).
17. $K_2HPO_4.3H_2O$ (potassium phosphate dibasic trihydrate) (SRL, catalogue no: 90654).
18. Lysozyme (Himedia, catalogue no: RM074).
19. Methanol (Merck, catalogue no: SK8F681128).
20. $MgSO_4.7H_2O$ (magnesium sulfate heptahydrate) (SRL, catalogue no: 85611).
21. $MnCl_2.4H_2O$ (manganese [II] chloride tetrahydrate) (SRL, catalogue no: 75113).
22. Na_2CO_3 (sodium carbonate) (SRL, catalogue no: 64079).
23. NaOH (sodium hydroxide pellets) (SRL, catalogue no: 96311).
24. 200–1000 μL pipette tips (Tarson, catalogue no: 521020).
25. Sodium chloride (SRL, catalogue no: 41721).
26. Spectrophotometric quartz VIS/UV-VIS cuvettes (Shimadzu, Japan).
27. Sucrose (SRL, catalogue no: 84973).
28. Syringe (DISPO-VAN).
29. Tris HCl (Himedia, catalogue no: 99438).
30. $ZnSO_4.7H_2O$ (zinc sulfate heptahydrate) (SRL, catalogue no: 76455).

2.4 RECIPES

2.4.1 BG-11 HEPES Medium Stock Solutions

All cyanobacterial cells used in this study were maintained in BG-11 growth medium (Allen 1968) supplemented with 20 mM HEPES (pH 8.0). The growth medium was prepared using different stock solutions described next.

2.4.1.1 Stock Solution A (100×)

Dissolve 0.1 g of Na_2EDTA (EDTA, disodium salt dihydrate or EDTA magnesium disodium salt), 0.6 g of ferric ammonium citrate, 0.6 g of citric acid (citric acid monohydrate) and 3.6 g of $CaCl_2$ (calcium chloride dihydrate) in 980 mL of Milli-Q water with the help of a magnetic stirrer hot plate. Adjust the final volume of the stock solution A to 1000 mL with Milli-Q water followed by sterilization by autoclave. Store the stock solution at 4°C after cooling.

2.4.1.2 Stock Solution B (100×)

Dissolve 7.5 g of $MgSO_4$ (magnesium sulfate heptahydrate) in 980 mL of Milli-Q water. Adjust the final volume of the stock solution B to 1000 mL with Milli-Q water and sterilize by autoclave. Store the stock solution at 4°C after cooling.

2.4.1.3 Stock Solution C (100×)

Dissolve 4 g of K_2HPO_4 (potassium phosphate dibasic trihydrate) in 980 mL Milli-Q water and adjust the final volume of the solution to 1000 mL using Milli-Q water. Sterilize stock solution C by autoclave and store at 4°C after cooling.

2.4.1.4 Stock Solution D (100×)

Dissolve 2 g of Na_2CO_3 (sodium carbonate) in 980 mL Milli-Q water. Adjust the final volume of the solution to 1000 mL using Milli-Q water. Sterilize stock solution D by autoclave and after cooling store at 4°C.

2.4.1.5 Stock Solution E (1000×)

Dissolve 0.286 g of H_3BO_3 (boric acid), 0.181 g of $MnCl_2$ (manganese [II] chloride tetrahydrate), 0.0222 g of $ZnSO_4$ (zinc sulfate heptahydrate), 0.0079 g of $CuSO_4$ (copper [II] sulfate pentahydrate), 0.0050 g of $CoCl_2$ (cobalt [II] chloride hexahydrate) and 0.039 g of Na_2MoO_4 (sodium molybdate, dihydrate) in 75 mL of Milli-Q water. Adjust the final volume of the solution to 100 mL with Milli-Q water. Sterilize stock solution E using a 0.22-µm membrane filter and store at 4°C.

2.4.1.6 1M HEPES pH 8.0

Dissolve 119.15 g HEPES in 350 mL Milli-Q water. Adjust the pH of the solution to 8.0 using a NaOH (sodium hydroxide) pellet. Adjust the final volume of the solution to 500 mL using Milli-Q water. Sterilize the solution using a 0.22-µm membrane filter and store at 4°C.

2.4.2 Preparation of BG-11 Liquid Growth Medium

Add 10 mL of stocks A, B, C and D and 1 ml of stock E solutions in 800 mL Milli-Q water. Thereafter, add 1.5 g $NaNO_3$ (sodium nitrate) to it and mix properly with the help of a magnetic stirrer on a hot plate. The pH of the solution is maintained at 7.1 using 1 M NaOH and HCl solutions. Adjust volume to 980 mL with Milli-Q water and sterilize by autoclave. After cooling, add 20 mL HEPES (1 M) and store BG-11 liquid medium at 4°C.

2.4.3 Methanol (90%)

Add 90 mL of absolute methanol to 10 mL Milli-Q water.

2.4.4 STES Buffer

1. Transfer 700 mL Milli-Q water to a 1000-mL beaker.
2. Add 2.93 g NaCl, 85.58 g sucrose, 50 mL 1 M Tris HCl (pH 8.0) and 20 mL 0.5 M EDTA (pH 8.0).
3. Maintain final volume at 1000 mL using Milli-Q water.
4. Sterilize using a 0.22-µm membrane filter and store at 4°C.
5. Before use, add 5 mg/mL lysozyme to the ice-cold STES buffer.

2.5 METHODS

2.5.1 Growth and Growth Rate Estimation (Campbell and Carpenter 1986; Kumar et al., 2021)

1. Take 500 µL of exponentially growing cyanobacterial cell culture from the stock. Add 500 µL BG-11 medium to it and mix by brief vortexing.
2. Calibrate the spectrophotometer using BG-11 growth medium as a blank at 750 nm.
3. Record the optical density of the culture at 750 nm (OD_{750}).
4. Start the cultures in BG-11 growth medium at 0.1–0.2 OD_{750} using information obtained from step 3 and grow them under desired experimental conditions.
5. Determine the OD_{750} of cell cultures every 24 hours. If OD_{750} comes to more than 1 arbitrary unit (AU) after a few days of growth, it is recommended to dilute the culture using BG-11 medium to bring the OD_{750} value within the range of one to avoid any instrumental error. Multiply the obtained value by the dilution factor and plot the growth curve as shown in Figure 2.2.
6. Calculate the growth rate by using the absorbance at different time intervals.

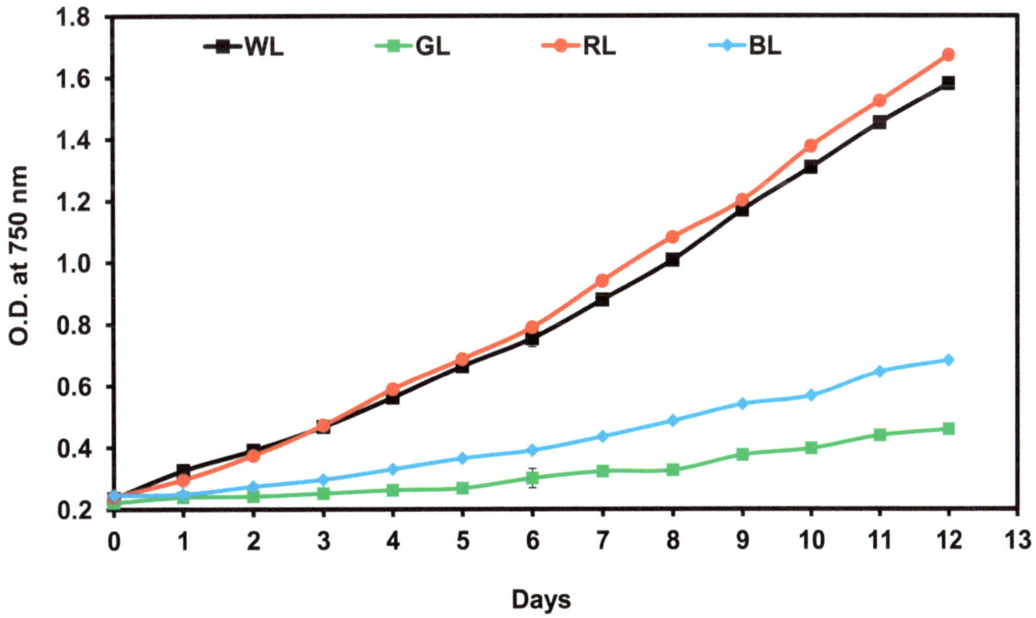

FIGURE 2.2 Growth curve of *Synechococcus elongatus* PCC 7942 grown under white light (WL), green light (GL), red light (RL) and blue light (BL) in BG-11 medium supplemented with 20 mM HEPES using 12 hours light:12 hours dark cycle (color figure in e-book).

The growth rate is calculated using the OD taken at two time intervals (Kumar et al. 2021). OD_{750} values taken at T1 and T2 time intervals are used for the calculation of growth rate using the following formula:

$$\text{Growth rate } (\mu) = [(\ln(\text{O.D.T2}) - \ln(\text{O.D.T1})]/(\text{T2} - \text{T1}) \tag{2.1}$$

2.5.2 Biomass Estimation

1. Start cyanobacterial cultures at 0.1–0.2 OD_{750} and transfer them under different experimental conditions. After growth for a defined period, collect 15 mL of culture in a 50-mL Falcon tube.
2. Centrifuge at 8000 × *g* for 10 minutes at 4°C.
3. Remove the supernatant and transfer the cell pellet to a new pre-weighed 2-mL microcentrifuge tube.
4. Wash the pellet using Milli-Q water and centrifuge at 10000 × *g* for 10 minutes at 4°C.
5. Remove the supernatant and dry the cell pellet in the hot air oven at 65°C for 3 hours until a constant weight is obtained. Do not over-dry the sample.
6. Calculate the biomass by using the following formula after measuring the final weight of the tube with a dried cell pellet.

$$\text{Biomass (mg/L)} = (\text{D2} - \text{D1}) \times 1000/15 \tag{2.2}$$

where D2 is the final weight of the microcentrifuge tube with the dried cell pellet and D1 is the initial empty weight of the tube without the cell pellet; 1000 is the conversion factor for gram (g) to milligram (mg), while 15 is the total volume of the cell culture taken for biomass estimation.

2.5.3 Estimation of Chlorophyll *A*, Carotenoids and Phycocyanin (De Marsac and Houmard 1988)

1. Start cyanobacterial cultures at 0.1–0.2 OD_{750} and transfer them under different experimental conditions. Collect 1 mL of cell culture in a 1.5-mL microcentrifuge tube for the extraction of chlorophyll *a* and carotenoids.
2. Centrifuge at 10000 × *g* for 5 minutes at room temperature. Discard the supernatant and add 1 mL of 90% methanol.
3. Vortex the tubes properly to suspend the pellet in the methanol.
4. Incubate the sample at 4°C for 12–16 hours in the dark.
5. After incubation, centrifuge at 10000 × *g* for 5 minutes at room temperature.
6. Weigh the empty 1.5-mL microcentrifuge tube.
7. Take out the supernatant in a pre-weighed 1.5-mL microcentrifuge tube and discard the pellet.
8. Record the weight of the methanol extract present in the microcentrifuge tube.
9. Calibrate the spectrophotometer using 90% methanol as a blank at 665 nm and 470 nm.
10. Take the absorbance of the methanolic extract at 665 nm (chlorophyll *a*) and 470 nm (carotenoids).
11. Calculate the concentration of chlorophyll *a* and carotenoids using the following equations:

$$\text{Chlorophyll } a \text{ (mg/mL)} = ((A_{665} \times 13.9 \times (\text{weight}/0.8)/1.5))/1000 \tag{2.3}$$

$$\text{Carotenoids } (\mu\text{g/mL}) = ((1000 \times A_{470}) - (1.91 \times \text{chl } a))/225 \tag{2.4}$$

12. Take the absorption spectrum of the methanolic extract as shown in Figure 2.3.

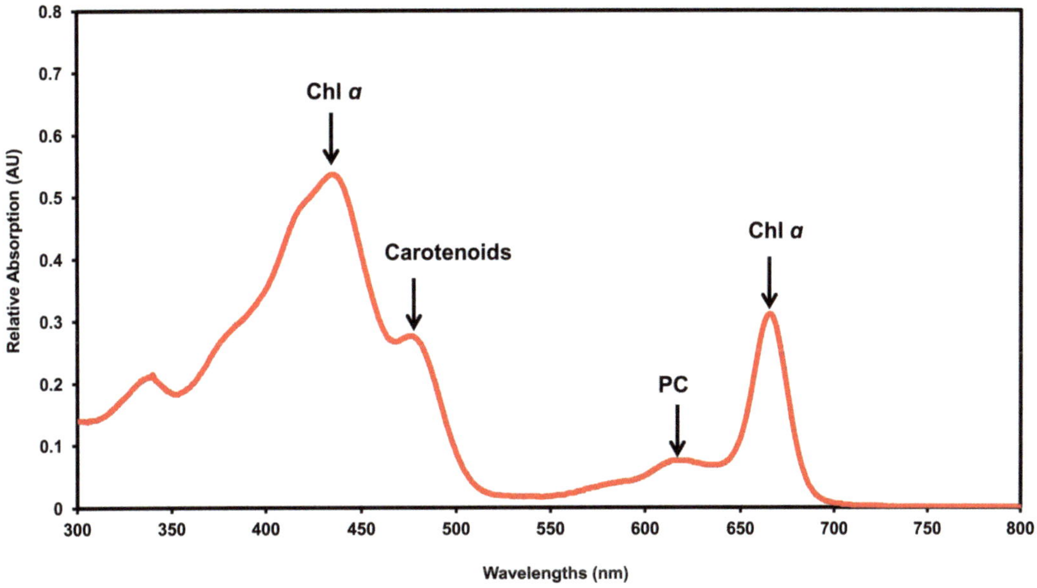

FIGURE 2.3 Absorption spectrum of methanolic extract of *Synechococcus elongatus* PCC 7942 grown under photosynthetic active radiation (PAR; 400–700 nm) in BG-11 medium supplemented with 20 mM HEPES. The spectrum shows the peak of chlorophyll *a* (chl *a*), carotenoids and phycocyanin (PC) (color figure in e-book).

2.5.4 Estimation of Phycocyanin by the Whole Cell-Absorbance Method (Schwarzkopf et al. 2014)

1. Collect 1 mL of cyanobacterial cell culture.
2. Take the absorbance at 625 nm (phycocyanin), 678 nm (allophycocyanin) and 750 nm (cell density).
3. Calculate the concentration of phycocyanin using the following equation:

$$\text{Phycocyanin (mg/mL/OD}_{750}) = 0.138445 \times (A_{625} - A_{750}) - 0.035483 \times (A_{678} - A_{750}) \quad (2.5)$$

2.5.5 Extraction of Phycobiliproteins (PE, PC and APC) (Kahn et al. 1997; Singh and Montgomery 2011)

1. Grow cyanobacterial cells under different experimental conditions.
2. Collect 1 mL of sample. (Optional: Maintain the OD_{750} of the sample at 0.6. If OD_{750} is higher than 0.6, than dilute the sample to 0.6. Similarly, if the OD_{750} is lower than 0.6, than concentrate the sample. This step is not required if biomass or other pigment concentration is used for the normalization.)
3. Incubate the sample on ice for 1 hour.
4. Centrifuge the sample at 10000 × *g* for 5 minutes at room temperature and discard the supernatant.
5. Resuspend the pellet in 1 mL of ice-cold STES buffer containing 5 mg/mL lysozyme (sonicate the sample, if required) and incubate in the dark at room temperature for 1 hour with shaking at 150 × *g*.
6. After incubation, centrifuge the sample at 13000 × *g* for 5 minutes at room temperature.
7. Collect supernatant in a new microcentrifuge tube.

8. Take the absorption spectrum between 400 nm and 700 nm and record absorbance values at 565 nm (PE), 620 nm (PC) and 650 nm (APC) using the STES buffer + lysozyme as a blank (Figures 2.4 and 2.5b).

$$\text{APC (mg/mL)} = (A_{650} - (0.19 \times A620))/5.65 \tag{2.6}$$

$$\text{PE (mg/mL)} = (A_{565} - (2.8 \times PC) - (1.34 \times APC))/12.7 \tag{2.7}$$

$$\text{PC (mg/mL)} = (A_{620} - (0.7 \times A650))/7.38 \tag{2.8}$$

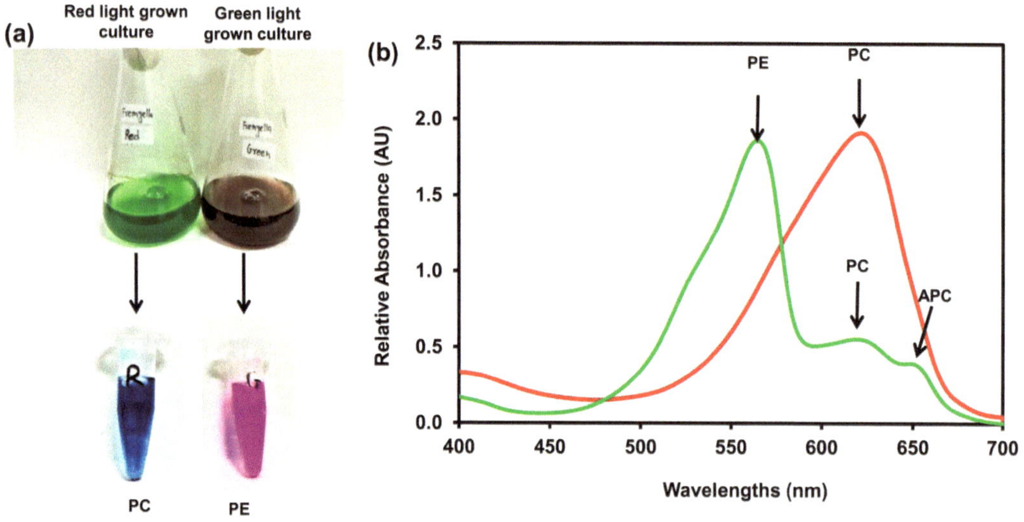

FIGURE 2.4 Extraction of phycobiliproteins (PBPS) from *Fremyella diplosiphon* SF33 (a) and absorption spectra of extracted PBPs (b). The cultures were grown under red light (~20 μmol photons m^{-2} s^{-1}) and green light (~20 μmol photons m^{-2} s^{-1}) conditions for 6 days in BG-11 medium supplemented with 20 mM HEPES. PC; Phycocyanin, PE; Phycoerythrin, APC; Allophycocyanin (color figure in e-book).

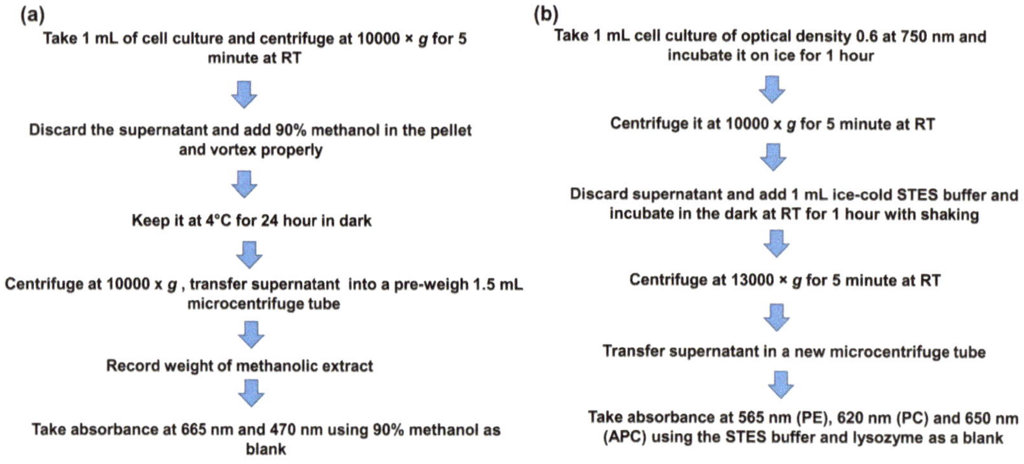

FIGURE 2.5 Steps involved in chlorophyll *a*, carotenoids (a) and phycobiliproteins (b) estimation in cyanobacteria (color figure in e-book).

2.6 DATA ANALYSIS

Figure 2.2 shows the growth curve of *Synechococcus elongatus* PCC 7942 grown under white light (WL), green light (GL), red light (RL) and blue light (BL) using a 12-hour light:12-hour dark cycle. However, unlike the bacterial growth curve shown in Figure 2.1, the growth curve of *S. elongatus* PCC 7942 does not show different growth phases because an exponential growing culture was used to start the experiment. Also, a longer period of growth under a low-light environment is required to reach a stationary phase in cyanobacteria. The growth obtained shows that *S. elongatus* PCC 7942 grows better under white and red lights, while it cannot efficiently use blue and green lights (Figure 2.2) to support its photoautotrophic growth, as reported earlier (Maurya et al. 2023b). The absorption spectrum of the methanolic extract of *S. elongatus* PCC 7942 shows absorption peaks of chl *a* (blue and red region peaks), carotenoids and phycocyanin (Figure 2.3). Also, absorption in the (ultraviolet radiation (UVR) region is seen, which is characteristic of water-soluble ultraviolet light (UV)–absorbing mycosporine-like amino acids (MAAs) in cyanobacteria. However, detailed high-performance liquid chromatography (HPLC) analysis suggested that *S. elongatus* PCC 7942 does not produce MAAs due to the absence of its biosynthetic pathway (Kumar et al. 2021). *Fremyella diplosiphon* is a filamentous cyanobacterium that exhibits a phenomenon of complementary chromatic acclimation (CCA), which permits the cyanobacterium to tailor its LHC PBSs in response to light signals, particularly under signals of red and green lights (Figure 2.4). RL-grown culture of *F. diplosiphon* exhibits a bluish-green color due to biosynthesis of PC, while GL-grown culture of *F. diplosiphon* exhibits a brick-red color due to the production of PE (Figure 2.4a). The absorption spectra of crude extracts of PBPs extracted from RL- and GL-grown cultures of *F. diplosiphon* shows absorption peaks of PE, PC and APC (Figure 2.4b).

2.7 NOTES/PRECAUTIONS

1. In OD-based growth measurements, the absorption value should be maintained within the detection limit of the spectrophotometer. To achieve this, growth medium can be used for the dilution of culture.
2. Changes in cell morphology can affect OD-based growth measurements.
3. Biomass should be properly dried at a fixed temperature; do not use temperatures above 65°C.
4. The minimum amount of light should be used during the extraction of photosynthetic pigments.
5. Dissolve lysozyme in STES buffer using a magnetic stirrer bar on the magnetic plate.

ACKNOWLEDGMENTS

This work was supported by funding from the Institute of Eminence Incentive Grant, Banaras Hindu University (R/Dev/D/IOE/Incentive/2021-2022/32399) and SERB, New Delhi, India, in the form of a Special Core Research Grant (SCP/2022/000201) to Shailendra P. Singh. Anjali Gupta and Sapna Tiwari are thankful to the ICMR, New Delhi, India (3/1/3/JRF -2019/HRD- LS) and University Grant Commission (UGC), New Delhi, India (NTA Ref. No.: 201610202339), respectively.

REFERENCES

Allen MM (1968) Simple conditions for growth of unicellular blue-green algae. *Journal of Phycology* 4: 1–4.

Anderson LK and Toole CM (1998) A model for early events in the assembly pathway of cyanobacterial phycobilisomes. *Molecular Microbiology* 30: 467–474.

Assunção J, Amaro HM, Malcata FX and Guedes AC (2022) Cyanobacterial pigments: Photosynthetic function and biotechnological purposes. In: The Pharmacological Potential of Cyanobacteria (Eds. G. Lopes, M. Silva and V. Vasconcelos). Academic Press. pp. 201–256. ISBN: 9780128214916

Bay Z and Pearlstein RM (1963) A theory of energy transfer in the photosynthetic unit. *Proceedings of the National Academy of Sciences USA* 50: 1071–1078.

Bryant DA (1994) The Molecular Biology of Cyanobacteria, First Ed. Springer, Salmon Tower Building, New York City, USA.

Calzadilla PI and Kirilovsky D (2020) Revisiting cyanobacterial state transitions. *Photochemical and Photobiological Sciences* 19: 585–603.

Campbell L and Carpenter EJ (1986) Diel patterns of cell division in marine *Synechococcus* sp. (Cyanobacteria): Use of the frequency of dividing cells technique to measure growth rate. *Marine Ecology Progress Series* 32: 139–148.

De Marsac NT and Houmard J (1988) Complementary chromatic adaptation: Physiological conditions and action spectra. *Methods in Enzymology* 167: 318–328.

Fatma T (2009) Screening of cyanobacteria for phycobiliproteins and effect of different environmental stress on its yield. *Bulletin of Environmental Contamination and Toxicology* 83: 509–515.

Ho MY and Bryant DA (2019) Global transcriptional profiling of the cyanobacterium *Chlorogloeopsis fritschii* PCC 9212 in far-red light: Insights into the regulation of chlorophyll *d* synthesis. *Frontiers in Microbiology* 10: 465.

Janke SA, Fortnagel P and Bergmann R (1999) Microbiological turbidimetry using standard photometers. *Biospektrum* 6: 501–502.

Kahn K, Mazel D, Houmard J, Tandeau de Marsac N and Schaefer MR (1997) A role for *cpeYZ* in cyanobacterial phycoerythrin biosynthesis. *Journal of Bacteriology* 179: 998–1006.

Kehoe DM and Gutu A (2006) Responding to color: The regulation of complementary chromatic adaptation. *Annual Review of Plant Biology* 57: 127–150.

Kumar V, Mondal S, Gupta A, Maurya PK, Sinha RP, Häder D-P and Singh SP (2021) Light-dependent impact of salinity on the ecophysiology of *Synechococcus elongatus* PCC 7942: Genetic and comparative protein structure analyses of UV-absorbing mycosporine-like amino acids (MAAs) biosynthesis. *Environmental and Experimental Botany* 191: 104620.

Maurya PK, Kumar V, Mondal S and Singh SP (2023a) Photoautotrophic black-colored cyanobacterial soil crust biosynthesizes photoprotective compounds and is capable of using blue, green, and red wavelengths of light for its growth. *Environmental Science and Pollution Research* 30(6): 16756–16769.

Maurya PK, Mondal S, Kumar V and Singh SP (2023b) Green and blue light-dependent morphogenesis, decoupling of phycobilisomes and higher accumulation of reactive oxygen species and lipid contents in *Synechococcus elongatus* PCC 7942. *Environmental and Experimental Botany* 205: 105105.

Samsonoff WA and MacColl R (2001) Biliproteins and phycobilisomes from cyanobacteria and red algae at the extremes of habitat. *Archives of Microbiology* 176: 400–405.

Sanfilippo JE, Garczarek L, Partensky F and Kehoe DM (2019) Chromatic acclimation in cyanobacteria: A diverse and widespread process for optimizing photosynthesis. *Annual Review of Microbiology* 73: 407–433.

Schwarzkopf M, Yoo YC, Hückelhoven R, Park YM and Proels RK (2014) Cyanobacterial phytochrome 2 regulates the heterotrophic metabolism and has a function in the heat and high-light stress response. *Plant Physiology* 164: 2157–2166.

Singh SP and Montgomery BL (2011) Temporal responses of wild-type pigmentation and RcaE-deficient strains of *Fremyella diplosiphon* during light transitions. *Communicative and Integrative Biology* 4: 503–510.

Tanaka K, Shirai T, Vavricka CJ, Matsuda M, Kondo A and Hasunuma T (2023) Dark accumulation of downstream glycolytic intermediates initiates robust photosynthesis in cyanobacteria. *Plant Physiology* 191(4): 2400–2413.

Whitman WB, Coleman DC and Wiebe WJ (1998) Prokaryotes: The unseen majority. *Proceedings of the National Academy of Sciences USA* 95: 6578–6583.

Whitton BA (1992) Diversity, ecology, and taxonomy of the cyanobacteria. In: Photosynthetic Prokaryotes (Eds. H.M. Nicholas and G.C. Noel). Springer, Inc, Boston, MA. pp. 1–51. ISBN: 9781475713329

Willey JM, Sherwood L and Woolverton CJ (2011) Prescott's Microbiology, Eighth Ed. McGraw-Hill, Singapore.

3 Quantitative Estimation of Total Carbohydrate, Protein and Lipid Content in Cyanobacteria

Pankaj K. Maurya, Sapna Tiwari, Soumila Mondal, Anjali Gupta, Vinod Kumar, and Shailendra P. Singh

3.1 INTRODUCTION

The continuous consumption of fossil fuel–based energy resources has threatened the sustainability of both the environment and energy supply (Houghton 2005). To overcome the negative impacts of fossil fuel burning on the energy supply and environment, biofuel production from the biomass of photoautotrophs has been advocated (Pinto et al. 2005; Rajneesh et al. 2017; Rodionova et al. 2017; Maurya et al. 2021). Depending on the source of the biomass, different generations (G) of biofuels such as 1G, 2G, 3G and 4G biofuels have been developed (Maurya et al. 2021). However, third and fourth generations of biofuel production from wild-type and genetically engineered strains, respectively, of microalgae and cyanobacteria have gained attention in the last two decades to avoid any competition for fertile land to produce food and energy (Pinto et al. 2005; Rodionova et al. 2017; Maurya et al. 2021).

Though the cyanobacterial biomass generally accumulates a lower percentage of lipid content in comparison to microalgae, these oxygen-evolving photosynthetic prokaryotic organisms have the potential to be utilized in 4G biofuel production using synthetic biology tools due to a simple growth requirement, high productivity and amenability to genetic modification (Rajneesh et al. 2017; Singh and Sinha 2020). The genetically engineered strains of these Gram-negative photoautotrophic bacteria absorb solar radiation and atmospheric CO_2 to produce energy-rich biomolecules (Berla et al. 2013; Rajneesh et al. 2017). The carbohydrate and lipid content of the biomass can be used for bioethanol and biodiesel production, respectively, while proteins are not used in biofuel production (Pinto et al. 2005). Thus, a microalgal and cyanobacterial biomass rich in carbohydrate and/or lipid content are of high value in the biofuel industry, and therefore, synthetic biology can play an important role in designing strains that have high carbohydrate and lipid content and accumulate minimal levels of proteins (Francisco et al. 2010; Singh and Sinha 2020; Maurya et al. 2021).

However, carbohydrate, lipid and protein content in the cyanobacterial biomass are a dynamic property that changes with growth as well as in the presence of abiotic stressors (Peramuna et al. 2015; Maurya et al. 2022). Therefore, it is important to monitor the changes in these cellular contents under different growth environments as well as in genetically engineered strains to select an ideal strain and growth condition that can support the maximum production of a carbohydrate- and lipid-enriched biomass. Also, the estimation of biomolecules helps to understand the allocation of cellular energy toward different metabolic pathways under different environmental conditions. In this chapter, we present biochemical assays

DOI: 10.1201/9781003398387-3

that are optimized for the estimation of carbohydrate, lipid and protein content in cyanobacterial biomass.

3.2 EQUIPMENT

1. Laminar airflow (Shivam instruments).
2. 100–1000 µL micropipette (Tarson).
3. 2–200 µL pipette tips (Tarson, catalogue No: 521010).
4. 200–1000 µL pipette tips (Tarson, catalogue No: 521020).
5. Glass test tubes (Borosil, catalogue No: 9820U04).
6. 1.5 mL microcentrifuge tubes (Tarson, catalogue no: 500010).
7. 2 mL microcentrifuge tubes (Tarson, catalogue no: 500020).
8. Spectrophotometer (Shimadzu, model: UV-1800).
9. Benchtop centrifuge (Remi, R-12C plus).
10. Benchtop cooling centrifuge (Remi, NEYA 16R).
11. Quartz UV-VIS cuvette (Shimadzu).
12. Refrigerator (4°C) (Samsung, model: RS55K5010S9/TL).
13. Weighing machine (Uniblog Shimadzu ATX224).
14. Vortex (Labquest).
15. Hot air oven (Shivam instruments).
16. 50 mL Falcon tubes (Abdos, catalogue No: P10404).
17. 15 mL Falcon tubes (Abdos, catalogue No: P10402).
18. Water bath (Labquest Borosil, India).
19. 250 mL conical flask (Borosil, catalogue No: 4980021).
20. Orbital incubator shaker (Shivam instruments).
21. 1000 mL glass bottle (Borosil).
22. Amber glass bottle (Borosil).

3.3 MATERIALS

1. Bovine serum albumin (BSA) (Himedia, catalogue No: GRM105-5G).
2. NaOH (SRL, catalogue No: 96311).
3. Na_2CO_3 (SRL, catalogue No: 64079).
4. $CuSO_4.5H_2O$ (Himedia, catalogue No: GRM630-500G).
5. Folin-Ciocalteau phenol reagent (FCP reagent) (Himedia, catalogue No: RM10822-500 mL).
6. Potassium sodium tartrate.$4H_2O$ (Merck, catalogue No: QI2Q622422).
7. Phenol (Merck, catalogue No: SE7RF67019).
8. Concentrated sulfuric acid (Merck, catalogue No: CA7C661076).
9. Pure methanol (Merck, catalogue No: SK8F681128).
10. Chloroform (Merck, catalogue No: DI7F67291).
11. Sucrose (SRL, catalogue No: 84973).
12. Nile red (Himedia, catalogue No: MB20056).
13. Dimethyl sulfoxide (DMSO) (SRL, catalogue No: 43404).
14. Immersion oil (Merck, catalogue No: DH6DF66567).

3.4 RECIPES

3.4.1 Reagent A (0.1 N NaOH and 2% Na_2CO_3)

Dissolve 0.4 g NaOH pellets and 2 g Na_2CO_3 in 100 mL Milli-Q (MQ) water in a glass bottle.

3.4.2 Reagent B (0.5% $CuSO_4.5H_2O$ + Potassium Sodium Tartrate.$4H_2O$)

Dissolve 1 g potassium sodium tartrate tetrahydrate in 100 ml MQ water, and afterward, add 0.5 g $CuSO_4.5H_2O$ to the same solution. This solution should be prepared in an amber glass bottle.

3.4.3 Reagent C (Reagent A + Reagent B)

Prepare reagent C by mixing reagent A and reagent B in a 50:1 ratio.

3.4.4 Reagent D

Mix the Folin-Ciocalteau reagent and MQ water in a 1:1 ratio. Always use freshly prepared reagent D.

3.4.5 BSA Stock Solution (1 mg/mL)

Dissolve 10 mg BSA in 10 mL MQ water. Vortex and store at –20°C.

3.4.6 Sucrose Stock Solution (1 mg/mL)

Dissolve 10 mg sucrose in 10 mL of MQ water. Vortex and store at 4°C.

3.4.7 80% Phenol

Mix 80 mL of phenol in 20 mL of MQ water.

3.4.8 Nile Red Stock Solution (1 mg/mL)

Dissolve 10 mg Nile red in 10 mL of DMSO. Vortex briefly and store at room temperature (~25°C) in the dark.

3.5 METHODS

3.5.1 Protein Estimation (Lowry et al. 1951)

The stock solution (1 mg/mL) of BSA in water is used to make 2 mL of different known (0–350 µg) concentrations of BSA for the development of a standard graph. Add 2 mL of reagent C and 200 µL of reagent D to 2 mL of different known (0–350 µg) concentrations of BSA and incubate samples for 30 minutes at room temperature for the development of a blue color. Thereafter, the intensity of the color is measured by recording the absorbance at 660 nm using 2 mL reagent C and 200 µL reagent D as a blank. Plot a standard graph for the known concentrations of BSA and intensity of the blue color to quantify the protein concentration in cyanobacterial samples as described next (Figure 3.1).

1. Collect 2 mL cyanobacterial culture and centrifuge at 10,000 × *g* for 10 minutes at 4°C to remove the medium.
2. Wash the cell pellet twice with MQ water.
3. Add 2 mL of alkaline reagent C to the cell pellet and vortex to suspend the cell pellet.
4. Incubate the sample for 15 minutes at room temperature (~25°C).
5. Add 200 µL of regent D and incubate in the dark for 30 minutes to develop the color.
6. Centrifuge the sample at 10,000 × *g* for 10 minutes at 4°C.
7. Transfer the supernatant to a new 2 mL microcentrifuge tube and measure absorbance at 660 nm using 2 mL reagent C and 200 µL reagent D mixture as a blank.
8. Calculate the concentration of protein in a given sample using the standard graph.

Collect 2 mL cyanobacterial culture and centrifuge at 10,000 × *g* at 4°C for 10 minute

↓

Discard the supernatant to obtain cell pellet

↓

Wash the cell pellet with autoclaved MQ water

↓

Add 2 mL of alkaline reagent C and vortex

↓

Incubate for 15 minute at room temperature

↓

Add 200 µL of reagent D and incubate sample in the dark for 30 minute to develop blue color

↓

Centrifuge at 10,000 × *g* for 10 minute at 4°C

↓

Collect supernatant and discard the pellet

↓

Measure absorbance at 660 nm using 2 mL of reagent C and 200 µL of reagent D as a blank

↓

Estimate concentration of protein using the standard graph

FIGURE 3.1 Flowchart depicting different steps involved in total protein estimation in a cyanobacterial cell biomass (color figure in e-book).

3.5.2 Carbohydrate Estimation (Dubois et al. 1956)

The stock solution of sucrose in water (1 mg/mL) is used to make 2 mL of different known concentrations of sucrose (0–120 µg). Add 50 µL of phenol (80%) and 5 mL of concentrated sulfuric acid to 2 mL of known concentrations of sucrose. Incubate the sample at room temperature for 10 minutes with intermittent mixing by a vortex. Thereafter, transfer the samples into a water bath and incubate at 30°C for 20 minutes to develop the yellow-orange color. Record the intensity of the color by measuring absorbance at 490 nm; use 50 µL of phenol (80%) and 5 m: of concentrated sulfuric acid as a blank. Plot a standard graph for the known concentrations of sucrose and intensity of the yellow-orange color to quantify the carbohydrate concentration in cyanobacterial samples as described next (Figure 3.2).

1. Collect a 5 mL cyanobacterial sample in a 15 mL Falcon tube and centrifuge at 10,000 × *g* for 10 minutes at 4°C.
2. Wash the cell pellet two times using MQ water.
3. Transfer the pellet to a glass test tube and add 50 µL of phenol (80%) and 5 mL of concentrated sulfuric acid directly to the cell pellet.
4. Incubate for 10 minutes with intermittent mixing by the vortex.
5. Transfer the sample to a water bath (30°C) for 20 minutes to develop the yellow-orange color.
6. Measure the absorbance at 490 nm using 50 µL of phenol (80%) and 5 mL of concentrated sulfuric acid as a blank.
7. Calculate the concentration of total carbohydrates in a given sample using a standard graph.

Collect 5 mL cyanobacterial culture and centrifuge at 10,000 × *g* at 4°C for 10 minute

↓

Wash the cell pellet with autoclaved MQ water

↓

Add 50 μL of 80% phenol and 5 mL of concentrated sulfuric acid to the cell pellet

↓

Incubate sample for 10 minute with intermittent mixing by the vortex

↓

Incubate sample in a water bath for 20 minute at 30°C to develop yellow-orange color

↓

Measure absorbance of yellow-orange colored product at 490 nm 50 μL of 80% phenol and 5 mL of concentrated sulfuric acid

↓

Estimate concentration of carbohydrates using the standard graph

FIGURE 3.2 Flowchart depicting different steps involved in total carbohydrate estimation in a cyanobacterial cell biomass (color figure in e-book).

3.5.3 Lipid Estimation by Gravimetric Method (Bligh and Dyer 1959)

1. Collect 25 mL cyanobacterial culture in a 50 mL Falcon tube and centrifuge at 8000 × *g* for 20 minutes at room temperature.
2. Discard the supernatant and add 2 mL of absolute methanol.
3. Vortex for 30 seconds to suspend the cell pellet.
4. Add 2 mL of chloroform and vortex to mix properly.
5. Incubate the sample for 12 hours at room temperature followed by the addition of 1 mL MQ water.
6. Vortex until a turbid solution is obtained.
7. Centrifuge at 1500 × *g* for 5 minutes at room temperature for phase separation.
8. Carefully collect the lower phase, which contains lipids in chloroform, using a Pasteur pipette in a glass test tube without disturbing the upper and middle phases. The upper phase contains water and methanol, while the middle phase contains cell debris.
9. Repeat steps 4–8 using 1 mL of chloroform and pool the lower phase.
10. Evaporate the pooled lower phase in a preweighed 2-mL microcentrifuge tube in the water bath at 65°C.
11. Dry the sample in a hot air oven at 65°C.
12. Record the weight of the microcentrifuge tube containing lipids and calculate the total lipid content using the following formula:

$$\text{Total lipids content}(\%\text{ of DCW}) = (W2 - W1)/DCW * 100$$

where, W1 is the weight of the empty tube, W2 is the weight of the tube containing total lipids, and DCW is the dry cell weight of the cyanobacterial sample (Figure 3.3).

3.5.4 Fluorescence-Based Lipid Estimation Using Nile Red Dye (Greenspan et al. 1985)

The stock solution (1 mg/mL) of Nile red in DMSO is used for the fluorescence microscope–based estimation of the lipid content in cyanobacteria. However, the optimal concentration of dye to be used

Collect 25 mL cyanobacterial culture in a 50 mL falcon tube and centrifuge at 8000 × *g* for 20 minute at room temperature

↓

Discard the supernatant, add 2 mL of methanol to the cell pellet and vortex for 30 s

↓

Add 2 mL of chloroform and vortex for 2 minute. Incubate overnight at room temperature

↓

Add 1 mL of autoclaved MQ water and vortex till turbid solution is obtained

↓

Centrifuge at 1500 × *g* for 5 minute at room temperature to develop three phases

↓

Collect lower chloroform phase containing lipids

↓

Repeat extraction procedure with chloroform and pool lower phase

↓

Evaporate lower phase in a preweighed 2 mL microcentrifuge tube in water bath at 60–65°C

↓

After complete evaporation, dry sample in a hot air oven at 60–65°C

↓

Record the weight of the microcentrifuge tube and calculate the total lipid contents as described in the method section

FIGURE 3.3 Flowchart depicting different steps involved in gravimetric total lipid estimation in a cyanobacterial cell biomass.

needs to be determined in advance before conducting the actual experiment. Collect exponentially growing cyanobacterial culture in a 15 mL Falcon tube and wash it using BG-11 growth medium. After washing, resuspend the cell pellet in an appropriate volume of BG-11 medium and adjust the optical density of the cell culture to 0.4 at 750 nm (OD_{750}). Make 1-mL aliquots of cyanobacterial culture in 1.5 mL microcentrifuge tubes having 0.4 OD_{750}. Add different concentrations of Nile red (1–20 µL of stock solution equivalent to 1–20 µg) to each sample and incubate with gentle shaking at room temperature for 20 minutes. After incubation, repeat the washing step to remove excess dye. After washing with BG-11 medium and centrifugation at 10,000 × *g* for 10 minutes, resuspend the cell pellet in 200 µL of fresh BG-11 medium and prepare a slide to find out the optimal concentration of dye using a fluorescence microscope. The optimal concentration of the dye will be a concentration that gives the maximum fluorescence signal and there is no further increase in signal with increasing dye concentration. Once the optimum concentration of the dye is determined, the estimation of lipid content in cyanobacterial samples can be done as described next (Figure 3.4).

1. Collect two sets of samples (1 mL) for each replicate and maintain OD_{750} of collected samples at 0.4 after washing with a fresh growth medium.
2. Treat one set of collected samples with an optimized concentration of Nile red dye (5 µL) using a stock solution of dye.
3. Treat another set of samples with 5 µL DMSO only, which is used as a control.
4. Incubate all samples for 20 minutes in the dark with gentle shaking at room temperature.
5. After incubation, wash samples with fresh BG-11 medium to remove unbound dye and subject cells to imaging by a fluorescence microscope in a dark room.
6. Visualize the samples in epifluorescence mode with a 100× objective lens using immersion oil.
7. Record the autofluorescence of photosynthetic pigments and Nile red fluorescence using the TRITC channel.

Collect two sets of samples (1 mL) for each replicate and maintain OD_{750} of collected samples at 0.4 after washing with a fresh growth medium

↓

Treat one set of collected samples with an optimized concentration of Nile red dye (5 μL) using a stock solution of dye (1 μg/μL in DMSO)

↓

Treat another set of samples with 5 μL DMSO only which is considered as a control

↓

Incubate all samples for 20 minute in the dark with gentle shaking at room temperature

↓

After incubation, wash samples with fresh BG-11 medium to remove unbound dye

↓

Visualize samples in epifluorescence mode with a 100 × objective lens using immersion oil

↓

Record autofluorescence of photosynthetic pigments and Nile red fluorescence using the TRITC channel

↓

Subtract DMSO fluorescence from the fluorescence of Nile red to estimate lipid content

FIGURE 3.4 Flowchart depicting different steps involved in fluorescence-based lipid estimation in cyanobacteria (color figure in e-book).

8. The fluorescence intensity of the DMSO-only treated sample is subtracted from the fluorescence intensity of the Nile red–treated samples to obtain the lipid content in a cyanobacterial sample.

3.6 DATA ANALYSIS

Figure 3.5 shows the concentration of protein accumulated in *Synechococcus elongatus* PCC 7942 after 12 days of growth under different light conditions. The protein was accumulated at the highest level in samples grown under blue light (BL), while samples grown under green light accumulated

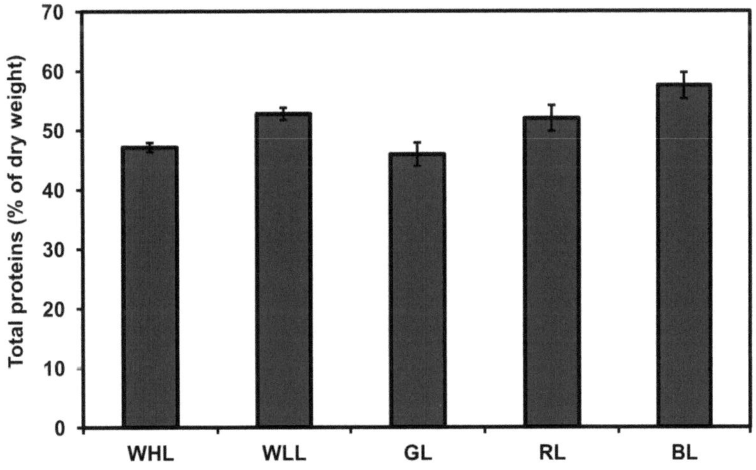

FIGURE 3.5 Percentage composition of protein (n = 3; ± SD) in the cell biomass of *Synechococcus elongatus* PCC 7942 under diurnal (12 hours light:12 hours dark) photoperiod of different light conditions. WHL, White high light; WLL, White low light; GL, Green light; RL, Red light; BL, Blue light (color figure in e-book).

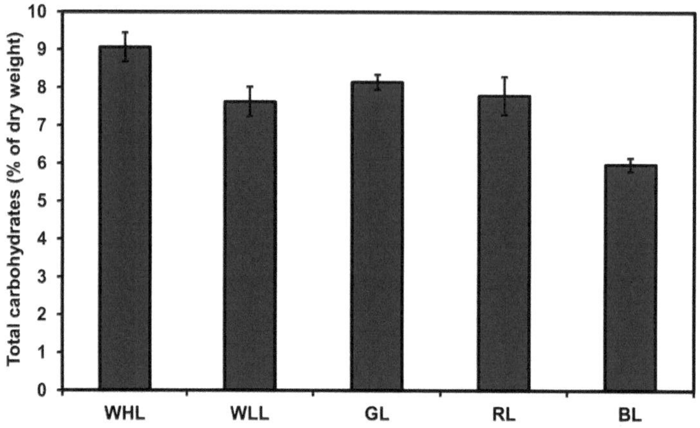

FIGURE 3.6 Percentage composition of carbohydrate (n = 3; ± SD) in the cell biomass of *Synechococcus elongatus* PCC 7942 under diurnal (12 hours light:12 hours dark) photoperiod of different light conditions. *Abbreviations*: BL, Blue light; GL, Green light; RL, Red light; WHL, White high light; WLL, White low light (color figure in e-book).

the lowest amount of protein. Figure 3.6 shows the concentration of carbohydrates accumulated in the *S. elongatus* PCC 7942 biomass. The highest amount of carbohydrate accumulated under white high light (WHL), while it accumulated to the lowest level under BL after 12 days of growth.

Figure 3.7 represents the gravimetric estimation of lipid content under different light conditions in *S. elongatus* PCC 7942. The highest amount of lipid accumulated under green and blue light conditions, while the lowest accumulation is observed under WHL after 12 days of growth. Similarly, fluorescence-based estimation of lipid is shown in Figure 3.8 where the TRITC channel of the fluorescence microscope was used to detect the fluorescence after staining cells with Nile red. It is evident that the intensity of fluorescence coming from the DMSO-only–treated sample is lower than the fluorescence coming from the Nile red–treated cells (Figure 3.8). The lipid content can be estimated by subtracting the fluorescence intensity of the DMSO-treated control sample from the fluorescence intensity of Nile red–treated samples (Greenspan et al. 1985).

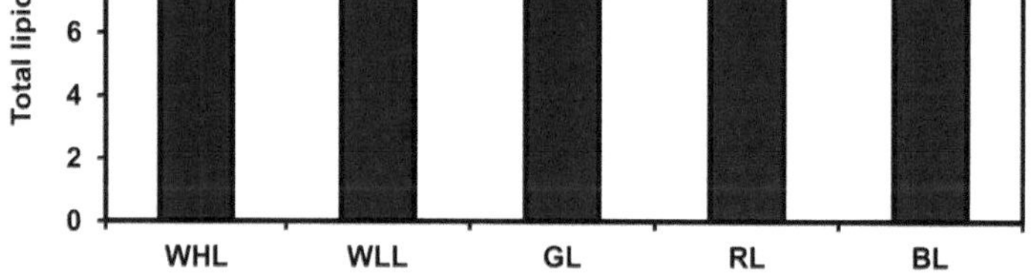

FIGURE 3.7 Percentage composition of lipid (n = 3; ± SD) in the cell biomass of *Synechococcus elongatus* PCC 7942 under diurnal (12 hours light:12 hours dark) photoperiod of different light conditions. *Abbreviations*: BL, Blue light; GL, Green light; RL, Red light; WHL, White high light; WLL, White low light (color figure in e-book).

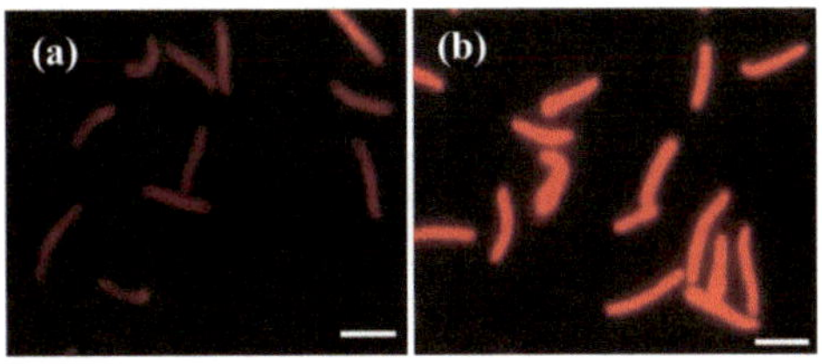

FIGURE 3.8 Nile red dye–based detection of lipid content in *Synechococcus elongatus* PCC 7942. Fluorescence microscopic images of *S. elongatus* PCC 7942 stained (a) without (DMSO only) and (b) with Nile red dye in DMSO. Images were acquired using a 100× objective with immersion oil. Scale bars, 5 µm (color figure in e-book).

3.7 NOTES/PRECAUTIONS

1. Always wear gloves during the handling of hazardous chemicals (sulfuric acid, phenol, etc.) and wear a mask while working around toxic fumes (phenol, chloroform, FCP reagent, etc.).
2. Light exposure to Nile red should be avoided.
3. Temperature should not exceed 70°C while evaporating chloroform using a water bath.

ACKNOWLEDGMENTS

This work was supported by funding from the Institute of Eminence Incentive Grant, Banaras Hindu University (R/Dev/D/IOE/Incentive/2021-2022/32399) and Science and Engineering Research Board (SERB), New Delhi (SCP/2022/000201) to Shailendra P. Singh. Anjali Gupta and Sapna Tiwari are thankful to the ICMR, New Delhi, India (3/1/3/JRF-2019/HRD-LS) and UGC, New Delhi, India (NTA Ref. No.: 201610202339), respectively.

REFERENCES

Berla BM, Saha R, Immethun CM, Maranas CD, Moon TS and Pakrasi HB (2013) Synthetic biology of cyanobacteria: Unique challenges and opportunities. *Frontiers in Microbiology* 4: 246.

Bligh EG and Dyer WJ (1959) A rapid method of total lipid extraction and purification. *Canadian Journal of Biochemistry and Physiology* 37: 911–917.

Dubois M, Gilles KA, Hamilton JK, Rebers PT and Smith F (1956) Colorimetric method for determination of sugars and related substances. *Analytical Chemistry* 28: 350–356.

Francisco EC, Neves DB, Jacob-Lopes E and Franco TT (2010) Microalgae as feedstock for biodiesel production: Carbon dioxide sequestration, lipid production and biofuel quality. *Journal of Chemical Technology and Biotechnology* 85: 395–403.

Greenspan P, Mayer EP and Fowler SD (1985) Nile red: A selective fluorescent stain for intracellular lipid droplets. *The Journal of Cell Biology* 100: 965–973.

Houghton J (2005) Global warming. *Reports on Progress in Physics* 68: 1343.

Lowry OH, Rosebrough NJ, Farr AL and Randall RJ (1951) Protein measurement with the Folin phenol reagent. *Journal of Biological Chemistry* 193: 265–275.

Maurya PK, Kumar V, Mondal S and Singh SP (2022) Chromatic acclimation in cyanobacteria: Photomorphogenesis in response to light quality. In: Ecophysiology and Biochemistry of Cyanobacteria (Ed. R.P. Rastogi). Springer, Singapore. pp. 209–223. ISBN: 978-981-16-4872-4

Maurya PK, Mondal S, Kumar V and Singh SP (2021) Roadmap to sustainable carbon-neutral energy and environment: Can we cross the barrier of biomass productivity? *Environmental Science and Pollution Research* 28: 49327–49342.

Peramuna A, Morton R and Summers ML (2015) Enhancing alkane production in cyanobacterial lipid droplets: A model platform for industrially relevant compound production. *Life* 5: 1111–1126.

Pinto AC, Guarieiro LL, Rezende MJ, Ribeiro NM, Torres EA, Lopes WA, Pereira PAD and Andrade JBD (2005) Biodiesel: An overview. *Journal of the Brazilian Chemical Society* 16: 1313–1330.

Rajneesh R, Singh SP, Pathak J and Sinha RP (2017) Cyanobacterial factories for the production of green energy and value-added products: An integrated approach for economic viability. *Renewable and Sustainable Energy Reviews* 69: 578–595.

Rodionova MV, Poudyal RS, Tiwari I, Voloshin RA, Zharmukhamedov SK, Nam HG, Zayadan BK, Bruce BD, Hou HJM and Allakhverdiev SI (2017) Biofuel production: Challenges and opportunities. *International Journal of Hydrogen Energy* 42: 8450–8461.

Singh SP and Sinha RP (2020) Marine photosynthetic microorganisms: Valuable compound production and bioenergy applications. *Encyclopedia of Marine Biotechnology* 4: 2229–2245.

4 Isolation and Purification of Phycobiliproteins

Jyoti Jaiswal, Sonal Mishra, Neha Kumari, Varsha K. Singh, and Rajeshwar P. Sinha

4.1 INTRODUCTION

Phycobiliproteins (PBPs) are the group of brilliantly colored, highly fluorescent, water-soluble proteins commonly present in cyanobacteria, red algae and cryptomonads possessing a wide range of applications (Singh and Montgomery 2011). These pigment-protein complexes are broadly categorized into four main groups based on their colors and absorption range in visible light *viz*. phycocyanin (PC, blue: λ_{max} = 610–625 nm), phycoerythrin (PE, red: λ_{max} = 490–570 nm), phycoerythrocyanin (PEC, purple: λ_{max} = 560–600 nm) and allophycocyanin (APC, bluish-green: λ_{max} = 650–660 nm) (Zuber 1986; Sinha et al. 1995; Bermejo et al. 2003, 2005; Kannaujiya and Sinha 2015, 2016). PBPs form up to 50% of the total soluble protein contents and about 24% of the dry weight of cyanobacterial cells (Bogorad 1975). Quantitatively, PC is the most abundant PBP in cyanobacteria followed by PE and APC (Wildman and Bowen 1974). PBPs play an indispensable role in the biotechnological industry due to their unique structural and functional properties.

However, the wide applications of PBPs are strongly dependent on their purity ratio. The extraction and purification of PBPs are extremely complex processes, which depend on the type of cyanobacterial strain. Several chemical and physical methods have been developed for the effective extraction of PBPs from any cyanobacteria (Sekar and Chandramohan 2008). However, among several techniques of extraction, freezing and thawing cycles of cyanobacterial biomass under –20 to 4°C (Sinha et al. 1995) is a more advantageous and effective technique for obtaining higher purity (Stewart and Farmer 1984).

Ammonium sulfate precipitation is a widely adopted method for partial purification of PBPs. The crude extract of PBPs is fractionally precipitated with 20–60% (w/v) ammonium sulfate (Kannaujiya and Sinha 2016). The precipitated solution is subjected to dialysis via dialysis tubing (<12 kDa) against the same buffer to remove the salt. A wide variety of column chromatography has been developed for subsequent purification of PBPs with a high purity index (Sonani et al. 2016).

4.2 EQUIPMENT

1. 50 mL centrifuge tube, RNase, DNase free (Thermo Fisher Scientific, Nunc™, catalog number: 339653).
2. Glass beakers, 10 mL and 1000 mL (Borosil, catalog number: 1000).
3. Measuring cylinder (Borosil, catalog number: 3021).
4. Glass rod (Fisher Scientific, catalog number: 11-380A).
5. Dialysis tubing cellulose membrane (MWCO: 12–14 kDa, Sigma-Aldrich, catalog number: D9277).
6. Aluminum foil.
7. Shaker.
8. Ultrasonicator (SONICS Vibra Cell, model: VCX 750).

DOI: 10.1201/9781003398387-4

9. Refrigerated centrifuge (Remi Electrotechnik, model: CM-12 PLUS).
10. Ultracentrifuge and tubes (Beckman Coulter Optima XPN-100, USA).
11. Ultrafiltration membrane filter (0.22 μm, MF-Millipore™, Merck, catalog number: GSWP04700).
12. Q-Sepharose Fast Flow column (GE Healthcare, Uppsala, Sweden, catalog number: 17051001).
13. Sephacryl S-100 HR column (GE Healthcare, Uppsala, Sweden, catalog number: 17061201).
14. Fast protein liquid chromatography system (AKTA™ Prime Plus, GE Healthcare, Uppsala, Sweden, model: 1338524).
15. Quartz cuvette (3.5 mL) (Cole-Parmer, JENWAY, catalog number: 035 028).
16. UV-VIS spectrophotometer (Shimadzu High-Technologies, model: UV-2600).
17. Fluorescence spectrophotometer (Agilent Technologies, model: Cary Eclipse).

4.3 MATERIALS

1. Cyanobacterial cells, e.g., *Spirulina* sp.
2. Phenylmethylsulfonyl fluoride (PMSF) (Sigma-Aldrich, catalog number: P7626).
3. Ethylenediamine tetra-acetic acid (EDTA) (Sigma-Aldrich, catalog number: E26282).
4. Sucrose (Sigma-Aldrich, catalog number: S8501).
5. Potassium dihydrogen phosphate (KH_2PO_4) (Sigma-Aldrich, catalog number: 104873).
6. Potassium phosphate dibasic anhydrous (K_2HPO_4) (Sigma-Aldrich, catalog number: 04248).
7. Tris-HCl (Sigma-Aldrich, catalog number: 1185531).
8. Ammonium sulfate (Sigma-Aldrich, catalog number: A4418).
9. Sodium chloride (Sigma-Aldrich, catalog number: S3014).
10. 100% ethanol (Sigma-Aldrich, catalog number: 459836).
11. Double distilled water (DDW).
12. 50 mM phosphate buffer (see Recipes).
13. 10 mM PMSF stock solution (see Recipes).

4.4 RECIPES

1. 50 mM phosphate buffer (400 mL).
   8.7 g K_2HPO_4.
   6.8 g KH_2PO_4.
   Adjust the pH to 7.0.
   Filter through 0.25-μm millipore membrane filter before use.
2. 10 mM (w/v) PMSF stock solution:
   Dissolve 0.00871 g of PMSF in 5 mL of absolute ethanol.

4.5 METHODS

1. A flowchart for isolation and purification of PBPs from cyanobacterium *Spirulina* sp. is shown in Figure 4.1.
2. Take 30 mL of cyanobacterial culture in a sterile centrifuge tube with the help of a measuring cylinder.
3. Cyanobacterial cells are harvested by centrifugation at 10,000 × *g*, 4°C for 10 minutes and the supernatant is discarded.
4. The pellet is washed twice with 50 mm phosphate buffer (pH 7.0) and resuspended in the same buffer together with 2% (w/v) EDTA, 1 mm PMSF and 5% sucrose.

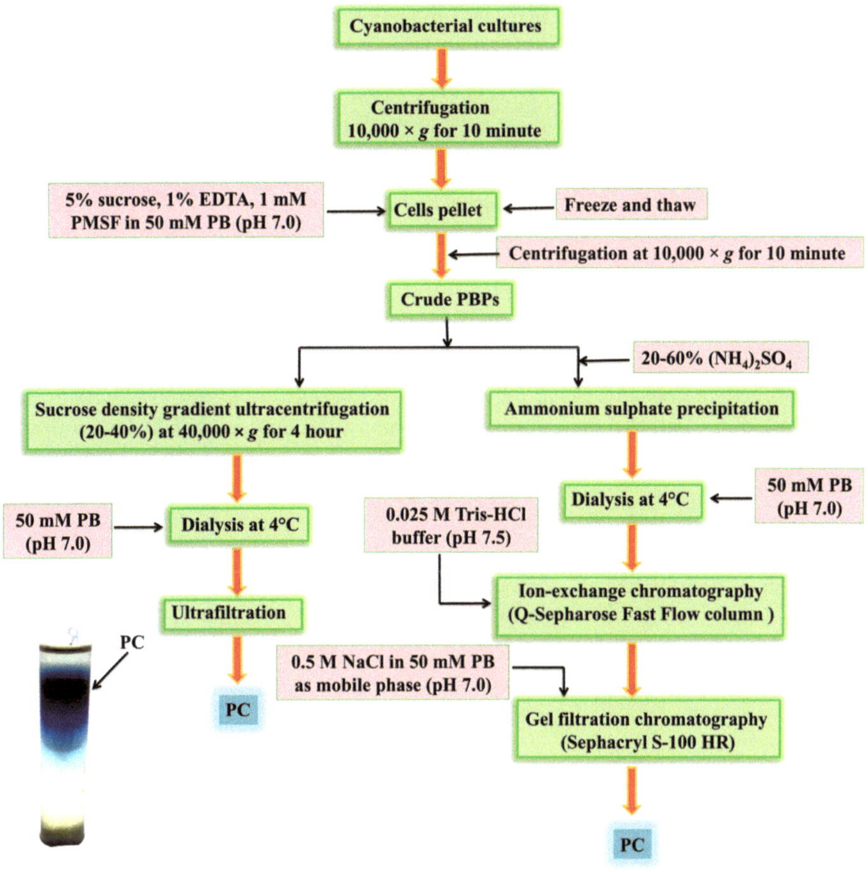

FIGURE 4.1 Flowchart for isolation and purification of PBPs from *Spirulina* sp. *Abbreviations*: PB, Phosphate buffer; PC, Phycocyanin (color figure in e-book).

5. Subsequently, cyanobacterial cells are broken using a sonicator (130 W, 20 kHz, 50% amplitude) for 5–10 minutes.
6. Suspended aliquots are subjected to repeated cycles of freeze and thaw at –20°C and 4°C, respectively.
7. The aliquots are centrifuged at 10,000 × *g* for 15 minutes to remove cell debris, and the remaining supernatant is considered partially purified PBPs.
8. To enrich the PBPs, the partially purified PBPs are fractionally precipitated with 20–60% (w/v) $(NH_4)_2SO_4$ in a stepwise manner. First, the crude extract is treated with 20% (w/v) $(NH_4)_2SO_4$ in 50 mm phosphate buffer (pH 7.0) for initial precipitation of large molecules. Thereafter, the subsequent supernatant is treated with 60% (*w/v*) $(NH_4)_2SO_4$ and left overnight at 4°C for complete precipitation of PBPs.
9. The pellet is collected by centrifugation at 10,000 × *g*, 4°C for 15 minutes and subsequently dissolved in 10 mL of 50 mm phosphate buffer (pH 7.0).
10. The precipitated solution is dialyzed against the same buffer overnight at 4°C using dialysis membranes to remove salts. The buffer is changed twice for the complete removal of $(NH_4)_2SO_4$.
11. For the sucrose density gradient ultracentrifugation purification method, crude PBPs are loaded on a continuous density gradient of 20–40% sucrose in 50 mM phosphate buffer (pH 7.0) using an ultracentrifuge tube and centrifuged at 40,000 × *g*, 4°C for 4 hours. The

blue-colored fraction of the PC layer is collected with the help of a syringe dialyzed against 50 mM phosphate buffer (pH 7.0) to remove the sucrose and concentrated by ultrafiltration using a 0.22-μm membrane filter.

12. For ion-exchange chromatography, Q-Sepharose Fast Flow column is used, which is pre-equilibrated with the 0.025 M Tris-HCl buffer (pH 7.5) at 40 cm/hour. The dialyzed sample is loaded onto the column. The non-adsorbed proteins are removed by washing with the same buffer. Elution is carried out using a linear gradient of 0–1 M NaCl with a gradient volume of 50 mL in 0.025 M Tris-HCl buffer (pH 7.5) with a flow rate of 1 mL min^{-1}.
13. Further purification of PC is done using a gel filtration column of Sephacryl S-100 HR. The column is pre-equilibrated with the 50 mM phosphate buffer (pH 7.0) at 0.8 mL min^{-1}. The sample is not loaded more than 5–10% of the total column volume and eluted with a mobile phase of 50 mM phosphate buffer and 0.5 M NaCl at a flow rate of 0.5 mL min^{-1}.
14. Blue-colored fractions eluted from the column and collected after ultracentrifugation, rich in PCs, are collected, and their absorbance is measured at 615 and 280 nm using a spectrophotometer for the purity ratio analyses.

4.6 DATA ANALYSIS

The purity index of the purified PC was analyzed based on the absorption ratio at 280 and 615 nm as follows: $A_{615}/A_{280} \geq 3.5$ for PC. Absorbance of PBPs was recorded at 200–700 nm against the phosphate buffer (50 mM, pH 7.0) as a blank by using a UV-VIS double beam spectrophotometer (UV-2600, Shimadzu, Tokyo, Japan) with 1-cm path length. Purified PCs absorb at 615 nm (Figure 4.2). The concentration of the purified PC in the sample was calculated using an equation (Bryant et al. 1979) as follows:

$$\text{PC}\left(\text{mg mL}^{-1}\right) = \left[A_{615} - 0.474\left(A_{652}\right)\right]/5.34 \tag{4.1}$$

The fluorescence of PBPs was measured by using a fluorescence spectrophotometer at an excitation of 615 nm. The FL solution software was used for data analysis provided by the manufacturer. The purified PC shows a sharp peak at the emission wavelengths of 642 nm (Figure 4.3).

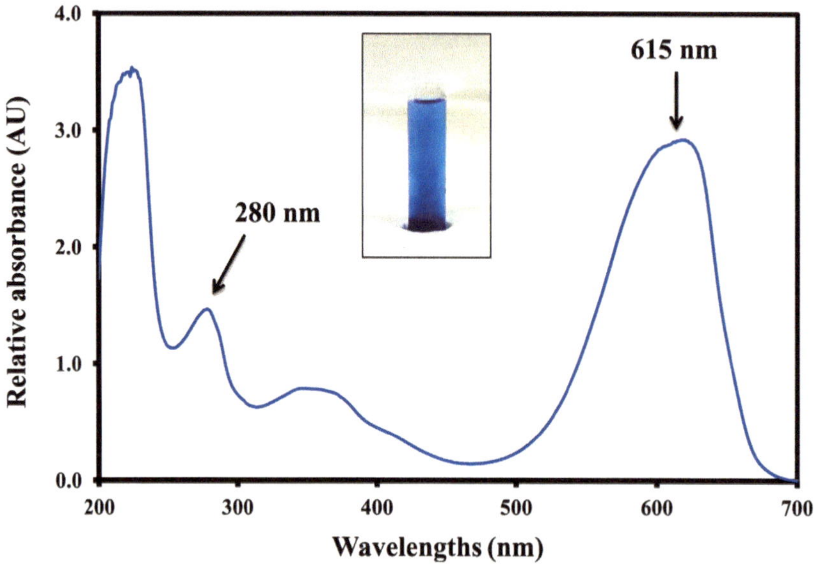

FIGURE 4.2 Absorption spectrum of purified PC from *Spirulina* sp. The purity of PC was monitored by measuring the absorption of total protein at 280 nm (color figure in e-book).

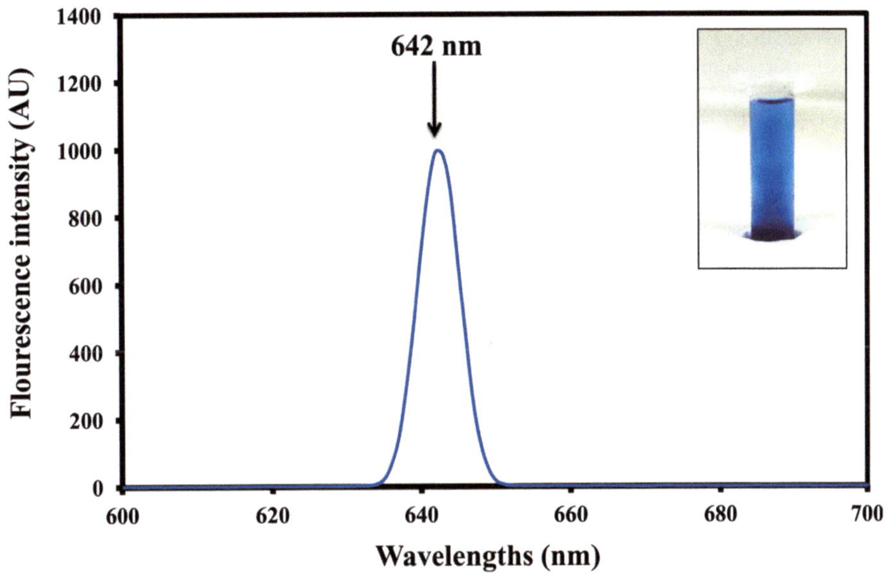

FIGURE 4.3 Fluorescence emission spectrum of purified PC from *Spirulina* sp. after excitation at 615 nm showing characteristic emission peak at 642 nm (color figure in e-book).

4.7 NOTES/PRECAUTIONS

1. All solutions were filtered through a 0.25-µm millipore membrane filter before use.
2. The pH of the buffer solution should be maintained properly before use.

ACKNOWLEDGMENTS

J. Jaiswal and S. Mishra are thankful to the University Grants Commission (UGC), New Delhi, India, for the financial assistance in the form of a fellowship as junior research fellows. N. Kumari is grateful to CSIR, New Delhi, India, for providing financial support in the form of a senior research fellowship. V.K. Singh is thankful to CSIR, New Delhi, India, for providing financial support in the form of a junior research fellowship.

REFERENCES

Bermejo R, Acien FG, Ibanez MJ, Fernandez JM, Molina E and Alvarez-Pez JM (2003) Preparative purification of B-phycoerythrin from the microalga *Porphyridium cruentum* by expanded-bed adsorption chromatography. *Journal of Chromatography B: Analytical Technologies in the Biomedical and Life Sciences* 790: 317–325.

Bogorad L (1975) Phycobiliproteins and complementary chromatic adaptation. *Annual Review of Plant Physiology* 26: 369–401.

Bryant DA, Guglielmi G, Tandeau de Marsac N, Castlets AM and Cohen-Bazire G (1979) The structure of cyanobacterial phycobilisomes: A model. *Archives in Microbiology* 123: 113–127.

Kannaujiya VK and Sinha RP (2015) Impacts of varying light regimes on phycobiliproteins of *Nostoc* sp. HKAR-2 and *Nostoc* sp. HKAR-11 isolated from diverse habitats. *Protoplasma* 252: 1551–1561.

Kannaujiya VK and Sinha RP (2016) An efficient method for the separation and purification of phycobiliproteins from a rice-field cyanobacterium *Nostoc* sp. strain HKAR-11. *Chromatographia* 79: 335–343.

Sekar S and Chandramohan M (2008) Phycobiliproteins as a commodity: Trends in applied research, patents and commercialization. *Journal of Applied Phycology* 20: 113–136.

Singh SP and Montgomery BL (2011) Determining cell shape: Adaptive regulation of cyanobacterial cellular differentiation and morphology. *Trends in Microbiology* 19: 278–285.

Sinha RP, Kumar A, Tyagi MB and Häder D-P (2005) Ultraviolet-B-induced destruction of phycobiliproteins in cyanobacteria. *Physiology and Molecular Biology in Plants* 11: 313.
Sinha RP, Lebert M, Kumar A, Kumar HD and Häder D-P (1995) Spectroscopic and biochemical analyses of UV effect on phycobiliproteins of *Anabena* sp. and *Nostoc carmium*. *Botanica Acta* 108: 87–92.
Sonani RR, Rastogi RP, Patel R and Madamwar D (2016) Recent advances in production, purification and applications of phycobiliproteins. *World Journal of Biological Chemistry* 26: 100–109.
Stewart DE and Farmer FH (1984) Extraction, identification and quantitation of phycobiliprotein pigments from phototropic plankton. *Limnology and Oceanography* 29: 392–397.
Wildman RB and Bowen CC (1974) Phycobilisomes in blue green algae. *Journal of Bacteriology* 117: 866–881.
Zuber H (1986) Structure of light-harvesting antenna complexes of photosynthetic bacteria, cyanobacteria and red algae. *Trends in Biochemical Sciences* 11: 414–419.

5 Methods for Estimation of Oxidative Stress Indices in Cyanobacteria

Vinod Kumar, Anjali Gupta, Soumila Mondal, Sapna Tiwari, and Shailendra P. Singh

5.1 INTRODUCTION

Cyanobacteria are a primitive group of photosynthetic microorganisms that perform oxygenic photosynthesis in the presence of sunlight. These prokaryotes have emerged as model organisms in the last decade for the production of bioenergy and industrially important chemicals in a carbon-neutral manner (Pathak et al. 2018; Rajneesh et al. 2017). Cyanobacteria are also good sources of natural antioxidants that are produced as a protective mechanism against toxic reactive oxygen species (ROS) (Jerez-Martel et al. 2017; Guerreiro et al. 2020). Due to their photoautotrophic mode of nutrition, cyanobacteria experience seasonal and diurnal changes in quality and quantity of light, temperature, nutrient availability, CO_2 levels, salinity and pH (Singh and Montgomery 2013; Kumar et al. 2021).

These fluctuations in environmental conditions compromise the growth and fitness of photosynthetic organisms by disturbing the electron transport chain of both photosynthesis and respiration. The disturbance in electron transport results in the production of ROS such as superoxide anion radicals ($O_2\cdot^-$ and $O_2\cdot^{2-}$), hydroxyl radicals (OH·), hydrogen peroxide (H_2O_2), hypochlorous acid (HClO) and hydroperoxyl radicals. ROS are even produced at a minimal level during normal metabolism (Gorman and Rodgers 1992; Foyer 1997; Apel and Hirt 2004; Latifi et al. 2009). These radicals can cause oxidative damage and possibly even cell death when they interact with other biological molecules such as lipids, pigments, proteins and nucleic acids (Kasai 1997; Wagner et al. 2004). Consequently, the cellular machinery and its functions are severely harmed by excessive amounts of ROS (Mittler 2002; Hu et al. 2008). In addition to their deleterious effects, ROS can also serve as a crucial developmental signal by integrating with other signaling networks (Foyer and Noctor 2009; Mittler et al. 2011).

However, to combat the negative effects of ROS produced under a range of stressors, cyanobacteria have evolved both enzymatic and non-enzymatic antioxidant defense mechanisms (Shao et al. 2008; Castenholz and Garcia-Pichel 2000; He et al. 2002). Several enzymes such as glutaredoxins, thioredoxins, superoxide dismutase, catalase, ascorbate peroxidase and peroxidase (POD) can convert ROS into non-toxic compounds and eliminate extra free oxygen radicals from the biological system (Figure 5.1). Cyanobacteria also have a highly evolved non-enzymatic antioxidant system with phytonutrients, pigments, phycobiliproteins, carotenoids, phenolic compounds, glutathione (GSH), ascorbate and other antioxidative substances (Ismaiel et al. 2014). These non-enzymatic antioxidants can reduce cellular oxidative stress by inhibiting the formation of superoxide anions and by detoxifying ROS/reactive nitrogen species (Noctor et al. 2002). In this chapter, we present some of the important enzymatic and non-enzymatic assays that are used to assess the fitness and functioning of cyanobacterial cells in the presence of different abiotic stressors.

DOI: 10.1201/9781003398387-5

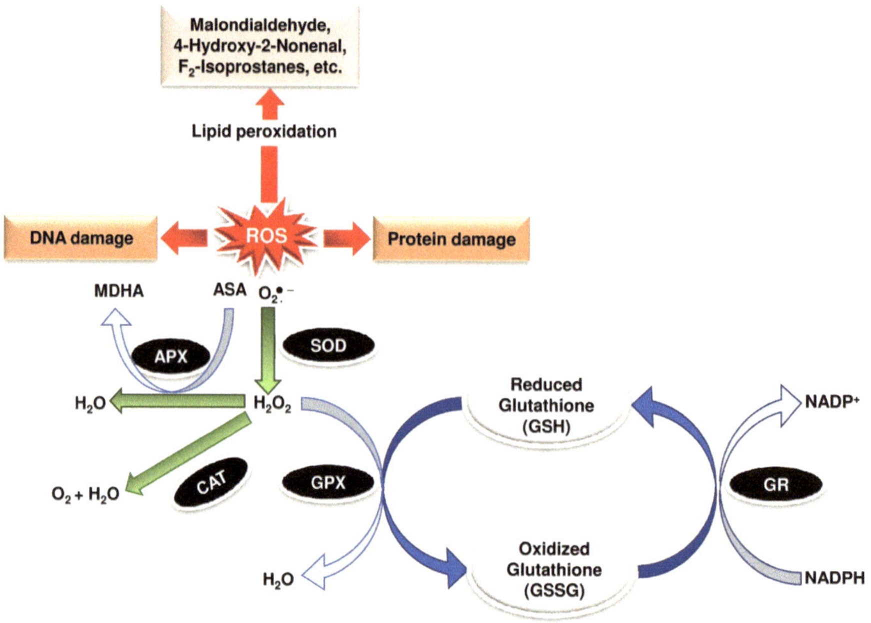

FIGURE 5.1 Enzymatic and non-enzymatic stress tolerance mechanisms in cyanobacteria. The reactive oxygen radical (O_2–) is the form of chemically ROS. Because of their intrinsic chemical properties, ROS react with different biological targets. The effect of ROS is shown in boxes. The O2 is disproportionated to hydrogen peroxide (H_2O_2) and O_2 by superoxide dismutase (SOD), and the H_2O_2 is reduced to water by catalases and guaiacol peroxidase (GPX). Ascorbate (ASA) is oxidized by ROS and converted into monodehydroascorbate (MDHA) by ascorbate peroxidase (APX). APX, CAT, and GPX act as the main H_2O_2 detoxifying enzymes. ASA and GSH are antioxidants. GSH participates in ROS scavenging and is converted into GSSG by the enzyme GPX. GSSG can be reconverted/recycled again into GSH by the activity of glutathione reductase (GR). (color figure in e-book).

5.2 EQUIPMENT

1. Ultrasonic probe sonicator (Sonics and Materials, USA).
2. Icebox (Cello, India).
3. Autoclave (NSW-227, NSW-India).
4. Refrigerated centrifuge (NEYA16R, Remi, India).
5. Water bath (LABQUEST, Borosil, India).
6. Stirring plate with heat source (Tarson, India).
7. –20°C and 4°C side-by-side refrigerator (Samsung, Korea).
8. –80°C freezer (Haier, China).
9. Vortex (LABQUEST, Borosil, Model: VMS, India).
10. UV-Visible (UV-VIS) Spectrophotometer 1800 (Shimadzu, catalog number: 80626, Japan).
11. 1-cm path-length quartz cuvette (Quartz cuvette, catalog number: MCQ-254, India).
12. pH meter (Eutech Instruments, India).

5.3 MATERIALS

1. 1.5 mL microcentrifuge amber tubes (Tarson, catalog number: 500010X, India).
2. 1.5 mL microcentrifuge tubes (Tarson, catalogue no: 500010, India).
3. 2.0 mL microcentrifuge tubes (Tarson, catalogue no: 500020, India).

4. 1 mL micropipette tips (Abdos, catalog number: P10109, India).
5. 200 μL micropipette tips (Abdos, catalog number: P10140, India).
6. 10 μL micropipette tips (Abdos, catalog number: P10116, India).
7. Micropipettes (Eppendorf, Germany).
8. 15 mL polypropylene tubes (Abdos, catalog number: P10430, India).
9. 50 mL plastic polypropylene tubes (Abdos, catalog number: P10404, India).
10. 250 mL flask (Borosil, catalog number: 4980021, India).
11. 100 mL graduated cylinder (Borosil, catalog number: 3021016, India).
12. 50 mL graduated cylinder (Borosil, catalog number: 3012, India).
13. Absolute ethanol (Merck, catalog number: K48191683632, India).
14. Trichloroacetic acid (TCA) (Merck, catalog number:1.94971.0521, India).
15. 2-Thiobarbituric acid (TBA) (Merck, catalog number:1.08180.0025, India).
16. 1 M dipotassium phosphate (K_2HPO_4) (SRL, catalog number: 90654, India).
17. 1 M monopotassium phosphate (KH_2PO_4) (SRL, catalog number:52403, India).
18. 1 M potassium iodide (KI) (SRL, catalog number: 27874, India).
19. 0.5 M disodium ethylenediaminetetra acetic acid (Na_2EDTA) (SRL, catalog number: 12070, India).
20. 1% polyvinyl pyrrolidone (PVP) (Himedia, catalog number: RM854, India).
21. 500 μM nitro blue tetrazolium (NBT) stock (Merck, catalog number: 104186, India).
22. 50 mM methionine stock (Merck, catalog number: 8.16010.0025, India).
23. 1 mM riboflavin stock (Merck, catalog number: 555682, India).
24. 0.05 M sodium carbonate (Na_2CO_3) (SRL, catalog number: 64079, India).
25. Disodium hydrogen phosphate dihydrate ($Na_2HPO_4 \cdot 2H_2O$) (SRL, catalog number: 32080, India).
26. 50% H_2O_2 stock (Merck, catalog number: 1.93408.0521, India).
27. Tris hydrochloride (Tris-HCl) (SRL, catalog number: 99438, India).
28. Reduced nicotinamide adenine dinucleotide phosphate (NADPH) (SRL, catalog number: 10107824001, India).
29. Oxidized glutathione (GSSG) (SRL, catalog number: RM550, India).
30. Magnesium chloride ($MgCl_2$) (Merck, catalog number: MK7M561509, India).
31. L-ascorbic acid (AsA) (SRL, catalog number: 23006, India).
32. 5-Sulfosalicylic acid (Himdia, catalog number: GRM1076, India).
33. 5,5′-Dithiobis (2-nitrobenzoic acid) (DTNB) (Sigma, catalog number: D8130, India).
34. Reduced GSH (SRL, catalog number: G4251, India).
35. Glutathione reductase (GR) (0.5 U) (Sigma, catalog number: G3664, India).
36. Methanol (Merck, catalog number: 1.94516.2521, India).
37. 2,2-Diphenyl-1-picrylhydrazyl (DPPH) (SRL, catalog number: D9132-1G, India).
38. Autoclaved double distilled water or Milli-Q water (MQ water, India).
39. Exponentially growing cultures (0.6 $O.D._{750}$) of *Synechococcus elongatus* PCC 7942.

Note: *S. elongatus* PCC 7942 were routinely subcultured in a sterilized BG-11 medium supplemented with 20 mM HEPES buffer pH 7.1 (Allen 1968) in a culture room at 25 ± 2°C under white fluorescent light (~80 μmol photons m^{-2} s^{-1}).

5.4 RECIPES

5.4.1 0.1% TCA

Dissolve 0.1 g of TCA in 80 mL of MQ water. Adjust the final volume to 100 mL using MQ water to obtain 0.1% (w/v) TCA solution. Store the 0.1% (w/v) TCA in the dark at 4–8°C.

5.4.2 20% TCA

Dissolve 20 g of TCA in 80 mL of MQ water. Adjust the final volume to 100 mL using MQ water to obtain 20% (w/v) TCA solution. Store the solution in the dark at 4–8°C.

5.4.3 0.5% TBA

Dissolve 0.5 g of TBA in 20% TCA. Adjust the final volume to 100 mL using 20% TCA to obtain 0.5% TBA in 20% (w/v) TCA. Store at 4–8°C.

5.4.4 50 mM Potassium Phosphate Buffer (pH 7.0)

Mix 30.75 mL of 1 M K_2HPO_4 with 19.25 mL of 1 M KH_2PO_4 in a 250-mL glass bottle. Adjust the final volume to 100 mL using MQ water. Store the solution at room temperature (RT; 25°C).

5.4.5 1 M KI Solution

Dissolve 16.60 g of KI in 80 mL of MQ water. Stir properly on a magnetic stirrer. Adjust the final volume to 100 mL using MQ water. Store the solution at RT.

5.4.6 0.5 M Na_2EDTA

Dissolve 181.6 g of $Na_2EDTA{\cdot}2H_2O$ in 800 ml of MQ water. Stir properly on a magnetic stirrer. Adjust pH to 8.0 using pellets of NaOH (~20 g of NaOH is required). Adjust the final volume to 1 L using MQ water. Sterilize by autoclaving and store at RT.

5.4.7 100 mM Potassium Phosphate Buffer (pH 7.0)

Mix 61.50 mL of 1 M K_2HPO_4 with 38.5 mL of 1 M KH_2PO_4 to prepare 100 mM potassium phosphate buffer pH 7.0 at 25°C. Store at RT.

5.4.8 1% PVP Lysis Buffer

Mix 500 µL of 0.5 M Na_2EDTA stock solution (pH 8.0) with 99.50 mL of 50 mM potassium phosphate buffer (pH 7.0) to obtain 100 mL solution. Add 1 g of PVP to 100 mL of lysis buffer before use and store at RT.

5.4.9 500 mM NBT Stock

Dissolve 41 mg of NBT in 100 mL of MQ water. Stir it properly on the stirrer. Store the solution in the dark at 4°C.

5.4.10 50 mM Methionine Stock Solution

Dissolve 193.973 mg of methionine in 100 mL of MQ water. Mix it properly to make a clear solution and store it at 4°C.

5.4.11 100 mM Potassium Phosphate Buffer (pH 7.8)

Mix 90.8 mL of 1 M K_2HPO_4 and 9.2 mL of 1 M KH_2PO_4 into a 250-mL glass bottle to make the final solution of 100 mL of pH 7.8. Store at RT.

5.4.12 1 mM Riboflavin Stock Solution

Dissolve 37.64 mg of riboflavin in 100 mL of MQ water. Store the stock solution at 4°C in the dark.

5.4.13 0.05 M Na_2CO_3 (pH 10.2)

Dissolve 1.325 g of sodium carbonate in 200 mL of MQ water in a glass bottle. Maintain the pH of the solution at 10.2 using NaOH and adjust the final volume to 250 mL using MQ water. Store at RT.

5.4.14 50 mM Sodium Phosphate Buffer (pH 7.0)

Dissolve 6.81 g KH_2PO_4 in 1000 mL of MQ water to prepare solution A and 8.90 g of $Na_2HPO_4{\cdot}2H_2O$ in 1000 mL of MQ water to prepare solution B. Mix solution A and B in a proportion of 1:1.5 (v/v) when needed. Store at RT.

5.4.15 100 mM H_2O_2

Add 0.284 mL of 50% H_2O_2 to a final volume of 50 mL MQ water before use.

5.4.16 3 mM Na_2EDTA

Dilute 0.5 M Na_2EDTA stock solution with MQ water to make 3 mM Na_2EDTA solution. Store at RT.

5.4.17 0.05 M Tris-HCl pH 7.5

Dissolve 6.06 g of tris-HCl and 1.39 g of tris base in 800 mL of MQ water in a glass bottle. Stir until it is completely dissolved. Adjust the pH to 7.5 and adjust the final volume to 1000 mL using MQ water. Store at RT.

5.4.18 10 mM NADPH Stock

Dissolve 20.83 mg of NADPH in 5 mL of MQ water. Mix it properly and store it at 4°C.

5.4.19 0.5 mM GSSG

Dissolve 30.632 mg of GSSG in 80 mL of MQ water. Adjust the final volume to 100 mL using MQ water. Store the 1-mL aliquots at 4°C.

5.4.20 1 M $MgCl_2$

Dissolve 203.3 g of $MgCl_2.6H_2O$ in 800 mL of MQ water. Adjust the final volume to 1000 mL using MQ water. Sterilize by autoclaving and store at RT.

5.4.21 0.1 mM EDTA

Dilute 0.5 M stock solution of ETDA using MQ water to obtain 0.1 mM EDTA solution. Store at RT.

5.4.22 0.5 mM L-Ascorbic Acid

Dissolve 8.8 mg of L-ascorbic acid in 100 mL of MQ water. Store it at 4°C in the dark.

5.4.23 0.1 mM H_2O_2

Dilute 100 mM stock of H_2O_2 using MQ water to obtain 0.1 mM H_2O_2 solution. Store at RT.

5.4.24 5% Sulfosalicylic Acid

Dissolve 5 g of sulfosalicylic acid in 75 mL of MQ water. Stir it to mix properly. Adjust the final volume to 100 mL using MQ water. Store at 4°C.

5.4.25 0.15% DTNB

Dissolve 0.15 g of DTNB in 100 mL of methanol. Mix it properly and store it at 4°C.

5.4.26 0.4% NADPH

Dissolve 0.40 g of NADPH in 100 mL of MQ water. Mix it properly and store it at 4°C.

5.4.27 0.1 mM DPPH

Dissolve 39.40 mg of DPPH in 1000 mL of methanol. Mix it properly and store it at 4°C.

5.5 METHODS

5.5.1 Quantification of Lipid Peroxidation

Lipid peroxidation is a sign of oxidative stress in cells, and it is a recognized mechanism of cellular damage in all organisms (Latifi et al. 2009). Lipids that contain unsaturated fatty acids are particularly susceptible to free radicals. Lipid peroxidation damages biological membranes, and therefore, it is very harmful to living organisms (Figure 5.1). Several by-products are formed during the process of lipid peroxidation that can be estimated using different assays to monitor lipid peroxidation (Ayala et al. 2014). The most commonly used method involves the estimation of aldehydic compounds due to their propensity to combine with TBA to produce a TBA–reactive substance (RS) complex, which can be easily detected by UV-VIS spectrophotometry (Cakmak and Horst 1991).

1. Transfer 10 mL of cyanobacterial cell culture to 15-mL polypropylene tubes.
2. Centrifuge at 10,000 × *g* for 10 minutes at 4°C and pour off as much supernatant as possible.
3. Wash the cell pellet two to three times with MQ water.
4. Transfer the pellet to a 2-mL microcentrifuge tube.
5. Add 1 mL of 0.1% (w/v) ice-cold TCA to the pellet.
6. Sonicate the sample for 5–7 minutes (10-second pulse, 750 watt, 20 kHz; Sonics and Materials, USA) in the ice bath. After sonication, centrifuge at 13,000 × *g* for 15 minutes.
7. Transfer 500 µL of supernatant to a new 2-mL microcentrifuge tube.
8. Add 1.5 mL of TBA (0.5% in 20% TCA) to the supernatant. Incubate the mixture at 95°C for 30–40 minutes until the solution becomes pink.
9. Place the tubes in the ice bath for 10 minutes to stop the reaction.
10. Centrifuge at 10,000 × *g* for 10 minutes at 4°C. Transfer the supernatant to a 2-mL microcentrifuge tube.
11. Measure the absorbance at two wavelengths, i.e., 532 and 600 nm, using 0.1% (w/v) TCA and TBA (0.5% in 20% TCA) as a blank.
12. The extinction coefficient (E_c) (absorbance of one molar solution) of the malondialdehyde (MDA)–TBA complex is 155 mM^{-1} cm^{-1} and the degree of lipid peroxidation is expressed in nM MDA mg^{-1} biomass.

5.5.1.1 Principle of the Method

The total amount of TBA-RS in an extract is used to assess the oxidative damage to lipids. The TBA-RS assay quantifies the quantity of aldehyde generated by lipid peroxidation. The total amount of TBA-RS, expressed as the equivalent of MDA, is used to quantify the oxidative damage to lipids. The absorbance at 532 nm is used to estimate the MDA-TBA complex and is corrected for non-specific turbidity by subtracting the absorbance at 600 nm. The MDA content can be calculated using the following formula:

$$\text{MDA content (nM MDA mg}^{-1}\text{ biomass)} = \frac{\Delta A}{Ec \times l} \times \frac{V1 \times V2}{V3 \times m} \times f \qquad (5.1)$$

where

ΔA = (A_{532} A_{600})
A = Absorbance
E_c = Extinction coefficient
l = Optical path length, i.e., 1 cm
V1 = Total volume of reaction system
V2 = Volume of 0.1% TCA for preparing the homogenate
V3 = Volume of cyanobacterial extract added to the reaction mixture, i.e., 500 µL
f = Dilution factor of sample
m = Dry weight of biomass

5.5.2 Estimation of H_2O_2 Content

H_2O_2 is commonly produced in a cell during normal physiological processes and in response to stress conditions (Foyer and Shigeoka 2011). It is generally produced at higher levels during oxidative stress, but it also acts as a signaling molecule at lower levels in several cellular processes (Neill et al. 2002; Apel and Hirt 2004). Considering the biological relevance of H_2O_2, spectrophotometer-, chemiluminescence-, enzymatic-, and electrochemistry-based methods have been developed for the estimation of the H_2O_2 level in a cell (Liu and Zweier 2001; Lu et al. 2009; Nagaraja et al. 2009). Here, we present an optimized colorimetric method developed by Velikova et al. (2000) for determining the level of H_2O_2 in cyanobacteria.

1. Grow the cyanobacterial cell culture starting from 0.2 OD_{750}.
2. Add the required concentrations of the oxidizing agent, like H_2O_2, to one set of the cell culture (OD_{750} 0.2) and take the other set without H_2O_2 as a control.
3. Collect 10 mL of the cyanobacterial sample and centrifuge at 8000 × *g* for 5 minutes. Cyanobacteria can be grown with and without known concentrations of H_2O_2 to optimize the protocol in the laboratory or as a positive control.
4. Wash the pellet two to three times with MQ water.
5. Add 1 mL of 0.1% TCA to the washed pellet and incubate on an ice bath for 10 minutes.
6. Centrifuge the sample at 8000 × *g* for 10 minutes at 4°C.
7. Transfer 1 mL of supernatant to a glass test tube and add 1 mL of potassium phosphate buffer (pH 7.0) and 1.5 mL of 1 M KI solution. Cover the tube to avoid light exposure and shake properly.
8. Take the absorbance of the solution at 390 nm against a blank lacking the cyanobacterial extract.
9. Prepare a standard curve for known concentrations of H_2O_2 by using the following steps.
10. Dilute 100 µL of 100 mM H_2O_2 stock solution with 9.90 mL water to form 1 mM H_2O_2 solution in a total volume of 10 mL.
11. Afterward, prepare different dilutions (0–1000 µM) using 1 mM H_2O_2 and water.

12. Add 1 mL potassium phosphate buffer (pH 7.0) and 1.5 mL 1 M KI solution to 1 mL of known concentrations of H_2O_2 (step 11) in a labeled glass test tube and mix properly by shaking.
13. Take the absorbance of the solution at 390 nm against a blank that contains all the reagents except the H_2O_2 solution.
14. The standard curve can be plotted as OD_{390} values on the y-axis and the concentration on the x-axis. Calculate the H_2O_2 content from a linear equation based on the calibration curve, which is used for quantification of H_2O_2 and expressed as μM H_2O_2 as given below.

$$y = (mx + c)$$

where y represents OD at 390 nm, x is the known concentration of H_2O_2 in (μM), m is the slope and c is a constant or y-intercept.

5.5.2.1 Principle of the Method

H_2O_2 reacts with potassium iodide in an acidic environment to produce iodine (I_2) and water molecules. In a classic illustration of an oxidation-reduction reaction, when H_2O_2 is added to a colorless solution of KI, the iodide ions (I^-) are slowly oxidized in iodine. In the presence of iodide, iodine reacts to form triiodide (I^{-3}), resulting in a blue color. Thus, the level of H_2O_2 in a cyanobacterial sample can be quantified by measuring the intensity of the blue color by spectrophotometer.

5.5.3 SOD, CAT, GR and APX Activity

1. Collect 10 mL of the cyanobacterial sample in a 15-mL polypropylene tube and centrifuge at 10,000 × *g* for 10 minutes at 4°C.
2. Wash the pellet two to three times with MQ water.
3. Add 1 mL of lysis buffer containing 100 mM potassium phosphate buffer (pH 7.0); add 1 mM EDTA and 1% PVP to the washed pellet and sonicate for 5–7 minutes (10-second pulse, 750 watts, 20 kHz) in the ice bath.
4. Centrifuge at 13,000 × *g* for 30 minutes at 4°C.
5. Transfer the supernatant (enzymatic extracts) in a 1.5-mL microcentrifuge tube, and, if required, the supernatant can be stored at –20°C for later analysis of SOD, CAT, GR, and APX activity.

5.5.3.1 Estimation of SOD Activity

SOD is a crucial component of an antioxidant defense system in almost all living organisms that are exposed to oxygen. SOD acts by catalyzing the dismutation (or partitioning) of the superoxide ($O_2^{\cdot-}$) radical into typical molecular oxygen (O_2) and H_2O_2 (Figure 5.1). Superoxide is a by-product of the oxygen metabolism that can lead to a variety of cell damage (Hayyan et al. 2016). Here, we present a colorimetric method for measuring the activity of superoxide dismutase (Giannopolitis and Ries 1977).

1. Take two sets of 3 mL reaction mixture containing 50 mM potassium phosphate buffer (pH 7.8), 2.0 μM riboflavin, 13 mM methionine, 0.05 M Na_2CO_3 (pH 10.2), 50 μM NBT, 100 mM EDTA and 100 μL of cyanobacterial enzymatic extract. Add riboflavin in the last in reaction mixture and perform all procedures in the dark.
2. Shake tubes properly and place one set under 80–100 μmol photons m^{-2} s^{-1} of white fluorescent light until a blue color develops (higher light intensity can also be used that will decrease the incubation time, but care should be taken to avoid damage to the enzyme due to excess heat). Keep another set under dark conditions for the same duration.
3. Prepare the third set containing only the reaction mixture, and keep it under the same light condition and duration. Use the reaction mixture only in the dark as a blank.
4. Measure the absorbance of all samples at 560 nm and calculate SOD activity in the cyanobacterial extract by using the following formula (5.2).

5. SOD activity is expressed in terms of units of SOD mg^{-1} biomass.
6. One unit of enzyme is defined as the amount that causes 50% inhibition of the photochemical reduction of NBT and chromogen formation.

5.5.3.1.1 Principle of the Method

SOD activity measures the inhibition of the photochemical reduction of NBT salt to purple chromogen. Illumination of riboflavin in the presence of O_2 and electron donor methionine generates superoxide anions, which reduce the NBT to purple chromogen.

$$\% \text{ Inhibition of NBT reduction by SOD (X)} = \left(\frac{Ao-(A-Ab)}{Ao}\right)\times 100 \tag{5.2}$$

Ao is the absorbance at 560 nm for a reaction mixture without enzymatic extract kept in light, ***A*** is the absorbance at 560 nm for a reaction mixture with enzymatic extract kept in light, and ***Ab*** is the absorbance of the reaction mixture with enzymatic extract kept in the dark.

Calculate the SOD activity (unit of SOD mg^{-1} biomass) in a cyanobacterial extract using % inhibition of NBT reduction by SOD (X) as given next.

$$\text{Units of SOD} = \frac{1}{50}\times X \tag{5.3}$$

5.5.3.2 Estimation of Catalase Activity

Catalase (CAT) helps in the breakdown of H_2O_2 into water (H_2O) and oxygen (O_2) (Chelikani et al. 2004) (Figure 5.1). A catalase molecule may convert millions of H_2O_2 molecules into water and oxygen per second, making it one of the enzymes with the fastest turnover rates (Goodsell 2004). In the following section, we present a method for determining the activity of CAT enzyme in the cyanobacterial extract (Aebi 1984).

1. Mix 100 μL of enzymatic extract, 50 mM Na-K phosphate buffer (pH 7.0) and 10 mM of H_2O_2 in a total volume of 1 mL.
2. Adjust the final volume of the reaction mixture to 1 mL using the Na-K phosphate buffer (pH 7.0).
3. Use a reaction mixture without cyanobacterial enzymatic extract as a blank.
4. Measure CAT activity by recording a decrease in the absorbance at 240 nm for 3 minutes, using a kinetics setting in a spectrophotometer.
5. One unit of CAT is defined as one μM of H_2O_2 utilized min^{-1}.

5.5.3.2.1 Principle of the Method

The CAT activity is estimated by measuring the decrease in absorbance at 240 nm due to the breakdown of H_2O_2 by CAT into H_2O and O_2. Thus, a decrease in absorbance at 240 nm can be directly related to the breakdown of H_2O_2, and therefore, the activity of CAT in a reaction mixture.

$$\text{Catalase activity } (\mu\text{M } H_2O_2 \text{ decomposed min}^{-1}\text{ mg}^{-1}\text{ biomass}) = \frac{\Delta A}{\text{Ec}\times 1}\div \text{T}\times\frac{\text{V1}\times\text{V2}}{\text{V3}\times\text{Cpr or m}}\times \text{f} \tag{5.4}$$

where

ΔA = (Initial absorbance at 240 nm – final absorbance at 240 nm)
T = Time interval, i.e., 3 minutes
E_c = Extinction coefficient of H_2O_2, i.e., 39.4 mM^{-1} cm^{-1}
l = Optical path length, i.e., 1 cm
V1 = Total reaction volume, i.e., 1 mL
V2 = Volume of lysis buffer for preparing homogenate or total volume of enzymatic extract

V3 = Volume of enzymatic extract added to the reaction mixture, i.e., 100 μL
f = Dilution factor of the sample before being tested
C_{pr} = Protein concentration in the cyanobacterial sample
m = Dry weight of cyanobacterial cell biomass

5.5.3.3 Estimation of Glutathione Reductase Activity

Glutathione reductase is a ubiquitous enzyme that maintains a high ratio of reduced and oxidized glutathione (GSH/GSSG) and plays an important role in the regulation of cellular metabolism and signaling. The enzyme glutathione reductase, also referred to as glutathione-disulfide reductase (GSR), catalyzes the conversion of glutathione disulfide (GSSG) to glutathione sulfhydryl form (GSH) (Figure 5.1). GSH is a crucial molecule for protecting the cells from oxidative stress and preserving the reducing environment of the cell (Mannervik 1987; Meister 1988; Deponte 2013). The following section describes the steps involved in measuring glutathione reductase activity in a cyanobacterial enzymatic extract (Schaedle and Bassham 1977).

1. Mix 200 μL of enzymatic extract with a reaction mixture containing 50 mM potassium phosphate buffer (pH 7.8), 2 mM Na_2EDTA, 0.15 mM NADPH, 0.5 mM GSSG and 3 mM $MgCl_2$ in a total volume of 1 mL.
2. Measure glutathione reductase activity by recording a decrease in the absorbance at 340 nm for 3 minutes using a kinetics setting in a spectrophotometer.
3. One enzyme unit corresponds to one μM NADPH oxidized min^{-1}.

5.5.3.3.1 Principle of the Method

GR is a highly specific enzyme that utilizes NADPH to reduce oxidized glutathione. The GR activity is estimated by monitoring the decrease in the absorbance at 340 nm due to the oxidation of NADPH. The formula to calculate GR activity is given next.

$$\text{GR activity } (\mu\text{m of NADPH oxidized min}^{-1}\text{ mg}^{-1}\text{ biomass}) = \frac{\Delta A}{\text{Ec} \times 1} \div \text{T} \times \frac{\text{V1} \times \text{V2}}{\text{V3} \times \text{Cpr or m}} \times \text{f} \tag{5.5}$$

where

ΔA = (Initial absorbance at 340 nm – final absorbance at 340 nm)
T = Time interval, i.e., 3 minutes
E_c = Extinction coefficient of NADPH, i.e., 6.22 mM^{-1} cm^{-1}
l = Optical path length, i.e., 1 cm
V1 = Total volume of reaction mixture, i.e., 1 mL
V2 = Volume of lysis buffer used for preparing homogenate
V3 = Volume of cyanobacterial extract added to the reaction mixture, i.e., 200 μL
f = Dilution factor, if any
C_{pr} = Protein concentration in cyanobacterial sample
m = Dry weight biomass

5.5.3.4 Estimation of Ascorbate Peroxidase Activity

APX utilizes ascorbate as its specific electron donor to neutralize H_2O_2. In the following section, we present a method for determining APX activity in a cyanobacterial enzymatic extract (Nakano and Asada 1981).

1. Mix 200 μL of enzymatic extract with 800 μL of 50 mm potassium phosphate (pH 7.0) containing 0.5 mm ascorbic acid and 0.1 mm EDTA.
2. Start the reaction by adding 0.1 mm H_2O_2 and record a decrease in the absorbance at 290 nm for 3 minutes against a blank lacking cyanobacterial extract.

3. One enzyme unit corresponds to the amount of enzyme required to oxidize one μM of ascorbate $min^{-1.}$

5.5.3.4.1 *Principle of the Method*

APX reduces H_2O_2 to water using ascorbic acid as a reductant. During this reaction ascorbate is converted to its oxidized form, i.e., mono-dehydroascorbic acid (MDHA) (Figure 5.1). The decrease in the absorbance at 290 nm depicts the oxidation of ascorbate due to APX activity in the enzymatic extract. APX activity can be calculated by using the following formula:

$$\text{APX activity } (\mu\text{M ascorbate oxidized min}^{-1}\ \text{mg}^{-1}\ \text{biomass}) = \frac{\Delta A}{\text{Ec} \times 1} \div \text{T} \times \frac{\text{V1} \times \text{V2}}{\text{V3} \times \text{Cpr or m}} \times \text{f} \tag{5.6}$$

where

ΔA = (initial absorbance at 290 nm – final absorbance at 290 nm)
T = Time interval, i.e., 3 minutes
E_c = Extinction coefficient of ascorbate, i.e., 2.8 $mM^{-1}\ cm^{-1}$
l = Optical path of quartz cuvette, i.e., 1 cm
V1 = Total volume of reaction mixture, i.e., 1 mL
V2 = Volume of lysis buffer used for preparing homogenate
V3 = Volume of cyanobacterial extract added to the reaction mixture, i.e., 200 μL
f = Dilution factor, if any
C_{pr} = Protein concentration in the cyanobacterial sample
m = Dry weight of cyanobacterial biomass

5.5.4 Quantification of Reduced, Oxidized and Total Glutathione Contents

In cyanobacteria and plant cells, the ratio between the reduced and oxidized forms of glutathione (GSH/GSSG) plays a significant part in signaling and the activation of several defense mechanisms (Figure 5.1). Here, we present a method for determining glutathione content in cyanobacteria based on the protocol developed by Anderson (1985).

1. Collect 10 mL of cyanobacterial culture sample in a 15-mL polypropylene tube and centrifuge at 13,000 × *g* for 10 minutes.
2. Discard the supernatant and wash the pellet with MQ water.
3. Add 2 mL of 5% sulfosalicylic acid to the washed pellet and break the cells by sonication (10-second pulse, 750 watt, 20 kHz).
4. Centrifuge at 10,000 × *g* for 10 minutes. Collect the supernatant in a new tube and discard the pellet. Keep the supernatant on ice before use.
5. Mix 500 μL of supernatant with 600 μL of 100 mM phosphate buffer (pH 7.0) and 40 μL of DTNB. Incubate the sample at RT (~25°C) for 2 minutes.
6. After incubation, record absorbance at 412 nm to determine the level of reduced glutathione content.
7. Afterward, add 50 μL of 0.4% NADPH and 1 μL of glutathione reductase (0.5 U) to the same reaction mixture and incubate at room temperature (~25°C) for 30 minutes.
8. Record absorbance at 412 nm against the blank to determine total glutathione content.
   The blank solution contains all reagents except the enzymatic extract. Determine glutathione content from a standard curve of GSH.
9. Prepare a standard curve for known concentrations of GSH by using the following steps.
10. Prepare a stock solution of GSH (1 mM) by dissolving 0.0307 g of GSH in a final volume of 100 mL of 0.2 M EDTA solution. Different known concentrations of GSH (25, 50, 100, 150 and 250 μM) are made using 1 mM GSH stock solution and 0.2 M EDTA solution.

11. Mix 500 μL of GSH solution from each dilution with 600 μL of 100 mM phosphate buffer (pH 7.0) and 40 μL of DTNB in a labeled 1.5-mL microcentrifuge tube. Incubate samples at room temperature (~25°C) for 2 minutes.
12. After incubation, record absorbance at 412 nm to determine the level of reduced glutathione content.
13. Afterward, add 50 μL of 0.4% NADPH and 1 μL of glutathione reductase (0.5 U) to the same reaction mixture and incubate at room temperature (~25°C) for 30 minutes.
14. Record absorbance at 412 nm against the blank to determine the total glutathione content. The blank solution contains all reagents except the GSH solution.
15. The standard curve can be plotted as OD_{412} values on the y-axis and concentration on the x-axis.
16. Calculate the linear trendline and the equation of the line based on calibration curve data.
17. Calculate the reduced glutathione and total glutathione concentration by using the following equations:

$$y = (mx + c)$$

where y represents OD at 412 nm, x is the known concentration of GSH, m is the slope and c is a constant or y-intercept.

$$\text{Reduced glutathione (GSH)} = \frac{\text{OD at 412} - \text{(y intercept)]}}{\textit{Slope}} \times \text{Sample dilution} \tag{5.7}$$

$$\text{Total glutathione (GSSG +GSH)} = \frac{\text{OD at 412} - \text{(y intercept)]}}{\textit{Slope}} \times \text{Sample dilution} \times 2 \tag{5.8}$$

The equation is multiplied by 2 as $1\,\text{GSSG} = 2\,\text{GSH}$.

Calculate the oxidized glutathione by subtracting the level of reduced glutathione from total glutathione as given next.

$$\begin{aligned}&[\text{Oxidized glutathione (GSSG) } (\mu\text{m mg}^{-1} \text{ biomass})\\&= (\text{Total glutathione} - \text{Reduced glutathione})]\end{aligned} \tag{5.9}$$

5.5.4.1 Principle of the Method

The assay is based on the interaction of GSH with DTNB (also known as Ellman's reagent), which results in the production of oxidized glutathione-TNB adduct (GS-TNB). GS-TNB has a maximum absorbance at 412 nm. The quantity of GSH in the sample directly correlates with the rate of TNB adduct production that results in the absorbance at 412 nm. GSH is then recycled back into the process when glutathione reductase reduces the disulfide product (GS-TNB) in the presence of NADPH.

5.5.5 DPPH Radical Scavenging Assay

The DPPH assay is used to investigate the antioxidant properties of a cell extract. It is an inexpensive and easy method to assess the antioxidant potential of both hydrophilic and lipophilic antioxidants present in a cell (Bae et al. 2012). This assay has been successfully used to measure the antioxidant activities of carotenoids, polysaccharides and methanol extracts of microorganisms (Kusmita et al. 2017; Petruk et al. 2019). Here, we present an effective method for comprehensively screening the antioxidant effects of cyanobacterial extracts containing various molecules.

1. Collect 10 mL of a cyanobacterial culture sample in a 15-mL polypropylene tube and centrifuge at 13,000 × *g* for 10 minutes.

2. Discard the supernatant and wash the pellet with MQ water.
3. Add 1 mL of 90% methanol to the washed pellet and incubate the sample overnight at 4°C in the dark for extraction.
4. After extraction, centrifuge the sample at 12,000 × *g* for 20 minutes at 4°C.
5. Discard the pellet and take 50 μL of supernatant in two sets in 2-mL microcentrifuge tubes. Add 450 μL of 0.1 mM DPPH to one set (A) while adding the same volume of 90% methanol to the second set (A_b) which acts as a control.
6. Mix the reaction mixture using a vortex and incubate at RT for 30 minutes in the dark.
7. Record the absorbance of the first set (A), second set (A_b) and DPPH solution at 517 nm using a UV-VIS spectrophotometer and 90% methanol as a blank.
8. Calculate the DPPH scavenging activity in a cell extract by using the following formula.

5.5.5.1 Principle of the Method

DPPH is a stable nitrogen-centered radical that is widely used to test the free radical scavenging activity of samples. Antioxidants present in the sample reduce the DPPH radical (violet color) to DPPH-H (yellowish color) (Shirwaikar et al. 2006).

$$\text{DPPH scavenging activity } (\%) = \frac{[\text{Ao} - (\text{A} - \text{Ab})]}{\text{Ao}} \times 100 \qquad (5.10)$$

where $\mathbf{A_0}$ is OD_{517} of the DPPH solution, **A** is OD_{517} of the cell extract with DPPH solution and $\mathbf{A_b}$ is OD_{517} of the cell extract with 90% methanol.

5.6 DATA ANALYSIS

To demonstrate the efficiency of methods to detect oxidative damage, we calculated lipid peroxidation and H_2O_2 content in cyanobacterial samples treated with 0.5 mM H_2O_2. The results are shown in Table 5.1. The light blue and light pink colors were obtained for H_2O_2 and MDA (lipid peroxidation) content estimation, respectively, in cyanobacterial samples without H_2O_2 treatment (Figure 5.2). However, 5.0 mM H_2O_2–treated samples showed dark blue and pink colors during estimation of H_2O_2 and MDA content, respectively. These observations were supported by the quantitative estimation of H_2O_2 and MDA content, which were higher in the 5.0 mM H_2O_2–treated samples (Table 5.1).

Figure 5.3 shows the colorimetric assay for SOD enzyme activity in a cyanobacterial cell extract; the reaction mixture developed a more intense blue color (Tube 1) without the enzymatic extract. However, the addition of cyanobacterial enzymatic extract in a reaction mixture resulted in the decreased intensity of the blue color due to a decreased level of chromogen formation in the presence of the SOD enzyme (Tube 2). The absorbance was recorded by UV-VIS spectrophotometer at 560 nm. The enzyme activity was calculated and presented in Table 5.2 using the formula given by Giannopolitis and Ries (1977).

TABLE 5.1
H_2O_2 and MDA Contents in Cyanobacterial Samples Treated with and without 5.0 mm H_2O_2

Biochemical Assays	Results	
	0 mM H_2O_2	5.0 mM H_2O_2
MDA content (nM MDA mg^{-1} biomass)	0.054	0.156
H_2O_2 content (μM mg^{-1} biomass)	0.21	0.456

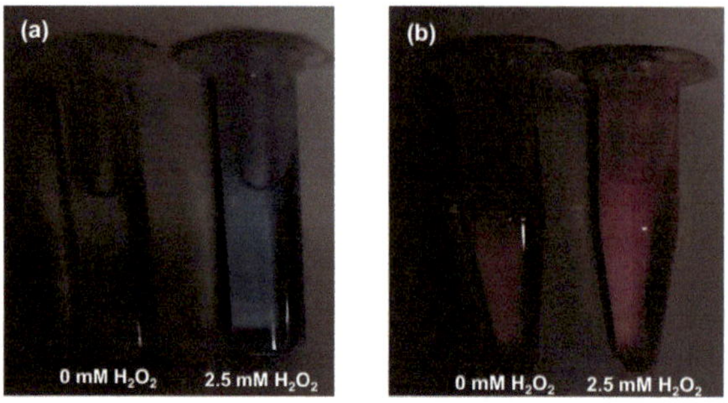

FIGURE 5.2 The colorimetric estimation of H_2O_2 (a) and MDA (b) contents in cyanobacterial cells treated with 0 mM and 5.0 mM H_2O_2 (color figure in e-book).

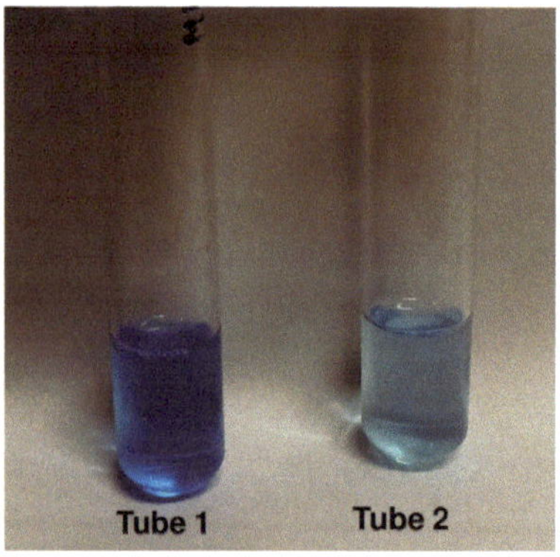

FIGURE 5.3 Superoxide dismutase (SOD) activity in cyanobacterial cells. Tube 1 shows chromogen formation in the absence of the enzymatic extract during the illumination of the reaction mixture, while tube 2 shows the inhibition of chromogen formation in the presence of the enzymatic extract during the illumination of the reaction mixture (color figure in e-book).

TABLE 5.2
Results of Antioxidants Assay Carried with the Cyanobacterial Extract

Enzymatic and Non-Enzymatic Assays	Results
Superoxide dismutase (unit mg^{-1} biomass)	0.124
Catalase (μm H_2O_2 decomposed min^{-1} mg^{-1} biomass)	0.017
Glutathione reductase (μm of NADPH oxidized min^{-1} mg^{-1} biomass)	0.015
Ascorbate peroxidase (μm of ascorbate oxidized min^{-1} mg^{-1} biomass)	0.390
Oxidized glutathione (μm mg^{-1} biomass)	15.160
Reduced glutathione (μm mg^{-1} biomass)	10.552
Total glutathione (μm mg^{-1} biomass)	25.712
DPPH radical scavenging activity (%)	19.402

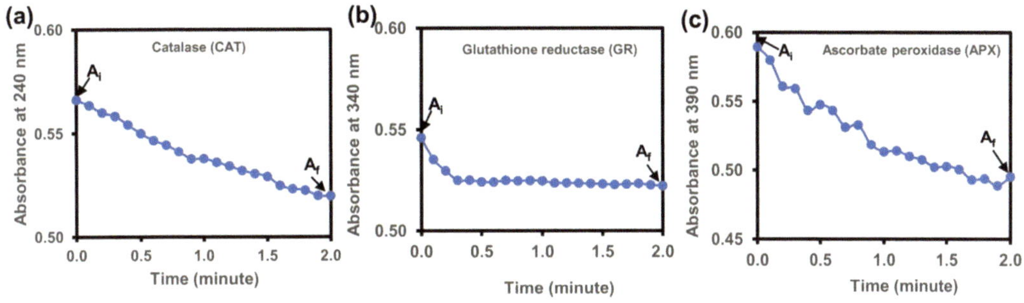

FIGURE 5.4 A view of the kinetic curve reflecting the course of enzymatic reaction in determining the activity of catalase (CAT) (a), glutathione reductase (GR) (b) and ascorbate peroxidase (APX) (c) in the cyanobacterial sample. *Note*: A_i represents the initial absorbance value, and **af** represents the final absorbance at their respective wavelengths (color figure in e-book).

Figure 5.4 represents kinetic graphs for estimating the enzymatic activity in the cyanobacterial extract. The reaction proceeds with a decrease in the absorbance, as the detection of enzymatic activity is based on determining the concentration of substrates by measuring the optical absorbance at a respective wavelength (Figures 5.4a–5.4c). A decrease in the concentration of substrates per unit of time is directly proportional to the activity of the enzyme in the analyzed samples (Table 5.2).

To determine the redox state of cyanobacterial cells, reduced glutathione and total glutathione contents (Table 5.2) are estimated. The oxidized glutathione content was estimated (Table 5.2) by subtracting the reduced glutathione content from the total glutathione pool (Anderson 1985). The results of antioxidant activity suggested the presence of scavenging activity in 90% methanol extract of the cyanobacterium (Table 5.2).

5.7 NOTES/PRECAUTIONS

1. Do not use TCA directly because contact with the skin causes corrosion. This results in irreversible damage to the skin.
2. TBA has a strong mercaptan odor. Do not breathe the dust. Avoid contact with skin and eyes. Do not leave the reagent bottles open. Replace the caps as soon as the desired volume is removed.
3. KI is yellow in air as it reacts with oxygen and free iodine is released. Replace the caps as soon as the desired volume is removed. Wash skin thoroughly after handling.
4. The Na_2EDTA will not go into the solution until the pH of the solution is adjusted to approximately 8.0 by the addition of NaOH.
5. Always prepare fresh NBT solution before use. In some cases, a precipitate may occur, which is easily brought into solution by briefly warming the substrate at 37°C.
6. H_2O_2 solution is light sensitive, so keep it in the dark during estimation.
7. Store NADPH stock solution for no longer than 2 weeks; store it frozen at –20°C and thaw it on ice before use. NADPH is light sensitive; therefore, store in amber vials to protect it from light.
8. The cell extract for glutathione content determination needs to be used within 2–4 hours after preparation. After 2 hours and after freezing, a drastic reduction in GSH content is commonly observed.
9. $MgCl_2$ is extremely hygroscopic. Do not store opened bottles for long periods. Once the crystals become saturated with H_2O, dispose of the chemical properly.
10. Optimize all the methods and sonication times for different types of cyanobacterial samples before following the methods described earlier.

11. A new calibration curve must be analyzed for new experiments.
12. Estimation through the colorimetric method is light sensitive, so it is very important to protect samples and solutions from light during estimation.
13. During MDA content estimation, do not place the sample in a hot water bath for more than 1 hour for pink color development.
14. Use freshly prepared stock solutions of buffers during the extraction of enzymes from cyanobacterial cells.
15. Prepare aliquots of enzyme extract after extraction from cells to avoid multiple freezes and thaws. Thaw samples on ice for use in assays and store them at –20°C for further use.
16. It is highly recommended to perform the measurements immediately after obtaining the extract of the sample. Freezing the samples is not recommended because after thawing a drastic reduction in the antioxidant content is commonly observed.

ACKNOWLEDGMENT

This work was supported by funding from the Institute of Eminence Incentive Grant, Banaras Hindu University (R/Dev/D/IOE/Incentive/2021-2022/32399) and SERB, New Delhi, India, in the form of a core research grant (SCP/2022/000201) to Shailendra P. Singh. Anjali Gupta and Sapna Tiwari are thankful to the ICMR, New Delhi, India (3/1/3/SRF -2019/HRD-LS) and UGC, New Delhi, India (NTA Ref. No.: 201610202339) for a senior research fellowship and a junior research fellowship, respectively.

REFERENCES

Aebi H (1984) Catalase in vitro. *Methods in Enzymology* 105: 121–126.

Allen MM (1968) Simple conditions for growth of unicellular blue-green algae. *Journal of Phycology* 4: 1–4.

Anderson ME (1985) Determination of glutathione and glutathione disulfide in biological samples. In: Methods in Enzymology (Ed. A. Meister). Academic Press. pp. 548–555.

Apel K and Hirt H (2004) Reactive oxygen species: Metabolism, oxidative stress, and signal transduction. *Annual Review of Plant Biology* 55: 373–399.

Ayala A, Muñoz MF and Argüelles S (2014) Lipid peroxidation: Production, metabolism, and signaling mechanisms of malondialdehyde and 4-hydroxy-2-nonenal. *Oxidative Medicine and Cellular Longevity* 2014: 360438.

Bae H, Jayaprakasha G, Jifon J and Patil BS (2012) Variation of antioxidant activity and the levels of bioactive compounds in lipophilic and hydrophilic extracts from hot pepper (*Capsicum* spp.) cultivars. *Food Chemistry* 134: 1912–1918.

Cakmak I and Horst WJ (1991) Effect of aluminium on lipid peroxidation, superoxide dismutase, catalase, and peroxidase activities in root tips of soybean (*Glycine max*). *Physiologia Plantarum* 83(3): 463–468.

Castenholz RW and Garcia-Pichel F (2000) The ecology of cyanobacteria: Their diversity in time and space. In: The Ecology of Cyanobacteria II (Eds. A.B. Whitton and M. Potts). Kluwer Academic Publishers, Springer, Dordrecht, pp. 591–611. ISBN: 978-94-007-3854-6

Chelikani P, Fita I and Loewen PC (2004) Diversity of structures and properties among catalases. *Cellular and Molecular Life Sciences* 61(2): 192–208.

Deponte M (2013) Glutathione catalysis and the reaction mechanisms of glutathione-dependent enzymes. *Biochimca et Biophysica Acta* 1830(5): 3217–3226.

Foyer CH (1997) Oxygen metabolism and electron transport in photosynthesis. In: Molecular Biology of Free Radical Scavenging Systems (Ed. J. Scandalios). Cold Spring Harbor Laboratory Press, New York, NY, USA. pp. 587–621.

Foyer CH and Noctor G (2009) Redox regulation in photosynthetic organisms: Signaling, acclimation, and practical implications. *Antioxidants and Redox Signaling* 11(4): 861–905.

Foyer CH and Shigeoka S (2011) Understanding oxidative stress and antioxidant functions to enhance photosynthesis. *Plant Physiology* 155: 93–100.

Giannopolitis CN and Ries SK (1977) Superoxide dismutases: I. Occurrence in higher plants. *Plant Physiology 59*(2): 309–314.

Goodsell DS (2004) Catalase. In: Molecule of the Month, RCSB Protein Data Bank.

Gorman AA and Rodgers MA (1992) Current perspectives of singlet oxygen detection in biological environments. *Journal of Photochemistry and Photobiology B* 14: 159–176.

Guerreiro A, Andrade MA, Menezes C, Vilarinho F and Dias E (2020) Antioxidant and cytoprotective properties of cyanobacteria: Potential for biotechnological applications. *Toxins 12*(9): 548.

Hayyan M, Hashim MA and AlNashef IM (2016) Superoxide ion: Generation and chemical implications. *Chemical Reviews* 116(5): 3029–3085.

He YY, Klisch M and Häder D-P (2002) Adaptation of cyanobacteria to UV-B stress correlated with oxidative stress and oxidative damage. *Photochemistry and Photobiology* 76(2): 188–196.

Hu WH, Song XS, Shi K, Xia XJ, Zhou YH and Yu JQ (2008) Changes in electron transport, superoxide dismutase and ascorbate peroxidase isoenzymes in chloroplasts and mitochondria of cucumber leaves as influenced by chilling. *Photosynthetica* 46(4): 581–588.

Ismaiel M, El-Ayouty YM and Piercey-Normore MD (2014) Antioxidants characterization in selected cyanobacteria. *Annals of Microbiology* 64(3): 1223–1230.

Jerez-Martel I, García-Poza S, Rodríguez-Martel G, Rico M, Afonso-Olivares C and Gómez-Pinchetti JL (2017) Phenolic profile and antioxidant activity of crude extracts from microalgae and cyanobacteria strains. *Journal of Food Quality* 2017: 2924508.

Kasai H (1997) Analysis of a form of oxidative DNA damage, 8-hydroxy-2′-deoxyguanosine, as a marker of cellular oxidative stress during carcinogenesis. *Mutation Research/Reviews in Mutation Research* 387(3): 147–163.

Kumar V, Mondal S, Gupta A, Maurya PK, Sinha RP, Häder D-P and Singh SP (2021) Light-dependent impact of salinity on the ecophysiology of *Synechococcus elongatus* PCC 7942: Genetic and comparative protein structure analyses of UV-absorbing mycosporine-like amino acids (MAAs) biosynthesis. *Environmental and Experimental Botany* 191: 104620.

Kusmita L, Mutiara EV, Nuryadi H, Pratama PA, Wiguna AS and Radjasa OK (2017) Characterization of carotenoid pigments from bacterial symbionts of soft-coral *Sarcophyton* sp. from North Java Sea. *International Aquatic Research* 9: 61–69.

Latifi A, Ruiz M and Zhang CC (2009) Oxidative stress in cyanobacteria. *FEMS Microbiology Reviews* 33(2): 258–278.

Liu X and Zweier JL (2001) A real-time electrochemical technique for measurement of cellular hydrogen peroxide generation and consumption: Evaluation in human polymorphonuclear leukocytes. *Free Radical Biology and Medicine* 31: 894–901.

Lu S, Song J and Campbell-Palmer L (2009) A modified chemiluminescence method for hydrogen peroxide determination in apple fruit tissues. *Scientia Horticulturae* 120: 336–341.

Mannervik B (1987) The enzymes of glutathione metabolism: An overview. *Biochemical Society Transactions* 15(4): 717–718.

Meister A (1988) Glutathione metabolism and its selective modification. *Journal of Biological Chemistry* 263(33): 17205–17208.

Mittler R (2002) Oxidative stress, antioxidants and stress tolerance. *Trends in Plant Science* 7(9): 405–410.

Mittler R, Vanderauwera S, Suzuki N, Miller GAD, Tognetti VB, Vandepoele K, Gollery M, Shulaev V and Van Breusegem F (2011) ROS signaling: The new wave? *Trends in Plant Science* 16(6): 300–309.

Nagaraja P, Shivakumar A and Shrestha AK (2009) Quantification of hydrogen peroxide and glucose using 3-methyl-2-benzothiazolinonehydrazone hydrochloride with 10,11-dihydro-5H-benz(b,f)azepine as chromogenic probe. *Analytical Biochemistry* 395: 231–236.

Nakano Y and Asada K (1981) Hydrogen peroxide is scavenged by ascorbate-specific peroxidase in spinach chloroplasts. *Plant and Cell Physiology* 22(5): 867–880.

Neill S, Desikan R and Hancock J (2002) Hydrogen peroxide signaling. *Current Opinion in Plant Biology* 5: 388–395.

Noctor G, Gomez L, Vanacker H and Foyer CH (2002) Interactions between biosynthesis, compartmentation and transport in the control of glutathione homeostasis and signaling. *Journal of Experimental Botany* 53: 1283–1304.

Pathak J, Rajneesh, Maurya PK, Singh SP, Häder D-P and Sinha RP (2018) Cyanobacterial farming for environment friendly sustainable agriculture practices: Innovations and perspectives. *Frontiers in Environmental Science* 6: 7.

Petruk G, Roxo M, De Lise F, Mensitieri F, Notomista E, Wink M, Izzo V and Monti DM (2019) The marine Gram-negative bacterium *Novosphingobium* sp. PP1Y as a potential source of novel metabolites with antioxidant activity. *Biotechnology Letters* 41: 273–281.

Schaedle M and Bassham JA (1977) Chloroplast glutathione reductase. *Plant Physiology* 59(5): 1011–1012.

Shao H-B, Ye Chu L, Hua Lu Z and Min Kang C (2008) Primary antioxidant free radical scavenging and redox signaling pathways in higher plant cells. *International Journal of Biological Science* 4: 8–14.
Shirwaikar A, Prabhu KS and Punitha ISR (2006) *In vitro* antioxidant studies of *Sphaeranthus indicus* (Linn). *Indian Journal of Experimental Biology* 44(12): 993–996.
Singh SP and Montgomery BL (2013) Distinct salt-dependent effects impair *Fremyella diplosiphon* pigmentation and cellular shape. *Plant Signaling and Behavior* 8(7): e24713.
Rajneesh, Singh SP, Pathak J and Sinha RP (2017) Cyanobacterial factories for the production of green energy and value-added products: An integrated approach for economic viability. *Renewable and Sustainable Energy Reviews* 69: 578–595.
Velikova V, Yordanov I and Edreva A (2000) Oxidative stress and some antioxidant systems in acid rain-treated bean plants: Protective role of exogenous polyamines. *Plant Science* 151: 59–66.
Wagner D, Przybyla D, op den Camp R, Kim C, Landgraf F, Lee KP, Würsch M, Laloi C, Nater M, Hideg E and Apel K (2004) The genetic basis of singlet oxygen-induced stress responses of *Arabidopsis thaliana. Science* 306(5699): 1183–1185.

6 Estimation of Glutamine Synthetase Enzyme

Sonal Mishra, Neha Kumari, Varsha K. Singh, and Rajeshwar P. Sinha

6.1 INTRODUCTION

Glutamine synthetase (EC 6.3.1.2) is a significant enzyme involved in assimilation of NH_4^+ in N_2-fixing cyanobacteria confirmed by evidence based on enzyme studies using ^{15}N tracer and analogues (Dharmawardene et al. 1973; Stewart and Rowell 1975; Rowell et al. 1977; Ladha et al. 1978). Detailed studies on the regulation of glutamine synthetase are needed to isolate it, which was previously facilitated by the purification of this enzyme from two cyanobacterial species, *Anabaena cylindrica* and *Nostoc* sp. (Sampaio et al. 1979). Glutamine synthetase catalyzes glutamine synthesis from glutamate and ammonia at the expense of adenosine triphosphate (ATP). Thereafter, glutamine is utilized by the enzyme glutamate synthase or glutamine oxoglutarate aminotransferase (GOGAT) to synthesize glutamate by shifting an amide group to 2-oxoglutarate. The total glutamate molecule generated in every cycle of glutamine synthetase is the means to incorporate nitrogen into several organic compounds like amino acids, amino sugars, purines and pyrimidines.

Two different types of glutamine synthetase have been found in cyanobacteria to be encoded by the gene *glnA*, known as glutamine synthetase type I and glutamine synthetase type II. Many strains have only type I (Domínguez-Martín 2016). Cyanobacterial glutamine synthetase type I is composed of 12 identical subunits (50 kDa each) (Valentine et al. 1968; Yamashita et al. 1989; Mérida et al. 1991). Glutamine synthetase type III coexists with glutamine synthetase type I in several cyanobacterial strains, where glutamine synthetase type III requires nitrogen-deficient conditions for its induction (Reyes and Florencio 1994; García-Domínguez et al. 1997; Sauer et al. 2000). To date, a broadly applicable method is reported here for the estimation of glutamine synthetase enzymes in cyanobacteria.

6.2 EQUIPMENT

1. Ultraviolet cutoff filter (Ultraphan, Digefra, Munich, Germany) to facilitate the desired wavelength of ultraviolet radiation (UV-B; 280–315 nm).
2. Ultraviolet treatment chamber fixed with UV-B (Philips TL 40 W: 12, Philips Lighting, model: TL 40W/12 RS SLV/25).
3. Glass Petri dishes (Corning, catalogue number: 3160-102).
4. Light microscope (OLYMPUS, model number: CX21i-TR).
5. Measuring cylinder.
6. Microcentrifuge tubes (Tarsons, Cliklok Micro Centrifuge Tube).
7. Refrigerated centrifuge (Remi Electrotechnik, Model: CM-12 PLUS).
8. Shaker.
9. UV-VIS spectrophotometer (Shimadzu, Japan, model: U-2600, Double beam spectrophotometer).
10. Cuvette (quartz, 3.5 mL) (Cole-Parmer, JENWAY, catalogue number: 035028).

DOI: 10.1201/9781003398387-6

11. Ultrasonicator (SONICS Vibra Cell, model: VC X 750; power: 750 watts, frequency: 20 KHz).
12. 100 mL glass beaker (Borosil, model number 1000D16).
13. 500 mL glass flask (Borosil, model number 4980024).

6.3 MATERIALS

1. Toluene (Merck, CAS number: 108-88-3).
2. Imidazole (Merck, CAS number: 288-32-4).
3. Sodium glutamate solution (Merck, CAS number: 142-47-2).
4. Magnesium chloride (Merck, CAS number: 7791-18-6).
5. Glutamine solution (Merck, Catalogue number: 25030-081).
6. Potassium arsenate (Merck, CAS number: 7784-41-0).
7. Cyanobacterial cells.
   Note: Cyanobacterium was grown under axenic conditions in BG-11$^-$ (nitrate negative) or BG-11$^+$ (nitrate positive) medium (Safferman and Morris 1964) based on their capacity to fix nitrogen. The temperature was maintained at 20 ± 2°C under continuous white light (12 ± 2 W m^{-2}) until the OD_{750} of culture reaches between 0.7 and 0.8.
8. Adenosine 5′-triphosphate solution (Merck, CAS number: 34369-07-8).
9. Adenosine 5′-diphosphate (Merck, CAS number: 16178-48-6).
10. Hydroxylamine.
11. Manganese chloride (Merck, CAS number: 7773-01-5).
12. Ferric chloride (Merck, CAS number: 7705-08-0).
13. Hydrochloric acid (Merck, CAS number: 7647-01-0).
14. TCA.

6.4 RECIPES

6.4.1 50 mM Imidazole HCl Buffer (100 mL)

Imidazole in deionized water; maintain pH 7 at 37°C with 1 M HCl.

6.4.2 Sodium Glutamate Solution (3 M; 10 mL)

L-Glutamic acid in deionized water using monosodium salt.

6.4.3 Glutamine Solution

L-Glutamine in cold deionized water (freshly prepared solution is preferred).

6.4.4 Adenosine 5′-triphosphate Solution (250 mM; 5 mL)

Adenosine 5′-triphosphate disodium salt in deionized water (freshly prepared solution is preferred).

6.4.5 Magnesium Chloride Solution (900 mM; 10 mL)

Magnesium chloride hexahydrate in deionized water.

6.5 METHODS

The method for glutamine synthetase enzyme activity is adapted from (Sinha et al. 1995). A flow-chart for the estimation of glutamine synthetase enzyme is shown in Figure 6.1, while the step-by-step method is given next.

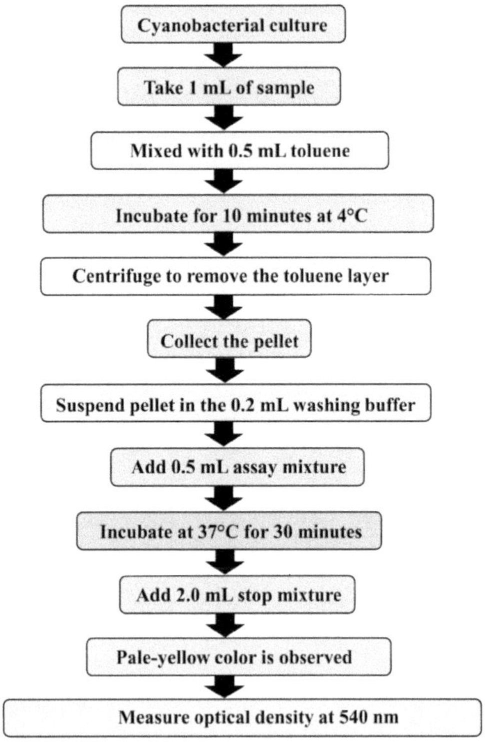

FIGURE 6.1 Flowchart showing the different steps involved in the protocol.

1. Transfer 10 mL of cyanobacterial culture to sterile glass Petri dish using a measuring cylinder. These Petri dishes are covered with a UV cutoff filter of 295 nm; transfer the culture to the UV light treatment chamber.
2. Cyanobacteria are irradiated with UV-B; maintain the temperature of the chamber to avert a heating effect. Stir the culture with a glass rod at regular time intervals to avoid clump formation.
3. After the desired time interval, take 1 mL of the aliquot and mix it with 0.5 mL toluene.
4. Thereafter, incubate it for 10 minutes at 4°C.
5. Centrifuge the cells to remove the toluene layer and collect the pellet.
6. Suspend the pellet in 0.2 mL washing buffer containing 50 mM imidazole, 5 mM magnesium chloride and 10 mM L-glutamic acid (sodium glutamate solution).
7. After that add 0.5 mL of the assay mixture composed of 0.1 M imidazole HCl buffer, 1.0 M potassium arsenate, 2.0 M hydroxylamine, 0.1 M glutamine, 0.1 M manganese chloride and 0.01 M sodium adenosine 5′-diphosphate.
8. Incubate it at 37°C for 30 minutes.
9. Add 2.0 mL of the stop mixture containing 10% ferric chloride; 6 N hydrochloric acid and 24% TCA to terminate the reactions.
10. The absorbance of the pale-yellow color is measured at 540 nm in a double-beam spectrophotometer (Shimadzu, Japan, model: U-2600) (Figure 6.2).

6.6 DATA ANALYSIS

The effect of different periods of UV-B irradiation on the *in vivo* glutamine synthetase activity is shown in Figure 6.2. A gradual decrease in glutamine synthetase activity was shown with increasing UV-B exposure time up to 120 minutes, and there was no complete inactivation even after 120 minutes of UV-B exposure (Sinha et al. 1995).

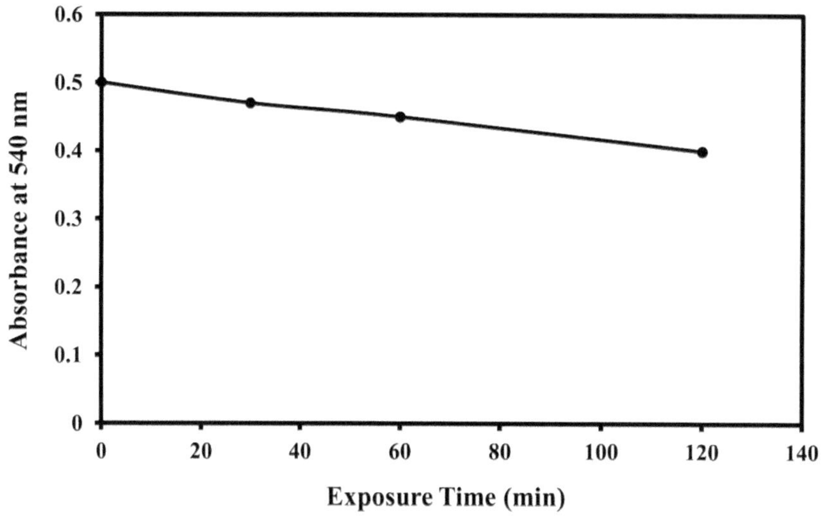

FIGURE 6.2 *In vivo* glutamine synthetase enzyme activity of *Nostoc commune* after increasing exposure to UV-B (adapted from Sinha et al. 1995).

6.7 NOTES/PRECAUTIONS

1. Maintain the pH of the 50 mM imidazole HCl buffer (100 mL) at 7 with 1 M HCl.
2. Maintain the temperature of the 50 mM imidazole HCl buffer (100 mL) at 37°C.
3. Freshly prepared solutions are preferred for glutamine synthetase enzyme activity.
4. Where Merck product or CAS numbers are specified, equivalent reagents may be substituted.

ACKNOWLEDGMENTS

Sonal Mishra and Varsha K. Singh thank UGC (Joint CSIR-UGC JRF-2019/NTA Ref. No.: 191620046790) and CSIR (09/0013(12862)/2021-EMR-I), New Delhi, India, respectively, for the financial assistance in the form of junior research fellowships. Neha Kumari thanks the Council of Scientific and Industrial Research (09/013(0819)/2018-EMR-I), New Delhi, India, for the financial support in the form of a senior research fellowship.

REFERENCES

Dharmawardene MWN, Haystead A and Stewart WD (1973) Glutamine synthetase of the nitrogen-fixing alga *Anabaena cylindrica*. *Archives of Microbiology* 90: 281–295.

Domínguez-Martín MA, Díez J and García-Fernández JM (2016) Physiological studies of glutamine synthetases I and III from *Synechococcus* sp. WH7803 reveal differential regulation. *Frontiers in Microbiology* 7: 969.

García-Domínguez M, Reyes JC and Florencio FJ (1997) Purification and characterization of a new type of glutamine synthetase from cyanobacteria. *European Journal of Biochemistry* 244: 258–264.

Ladha JK, Rowell P and Stewart WD (1978) Effects of 5-hydroxylisine on acetylene reduction and NH_4^+ assimilation in the cyanobacterium *Anabaena cylindrica*. *Biochemical and Biophysical Research Communications* 83: 688–696.

Mérida AN, Candau PE and Florencio FJ (1991) Regulation of glutamine synthetase activity in the unicellular cyanobacterium *Synechocystis* sp. strain PCC 6803 by the nitrogen source: Effect of ammonium. *Journal of Bacteriology* 173: 4095–4100.

Reyes JC and Florencio FJ (1994) A new type of glutamine synthetase in cyanobacteria: The protein encoded by the *glnN* gene supports nitrogen assimilation in *Synechocystis* sp. strain PCC 6803. *Journal of Bacteriology* 176: 1260–1267.

Rowell P, Enticott S and Stewart WD (1977) Glutamine synthetase and nitrogenase activity in the blue-green alga *Anabaena cylindrica*. *New Phytologist* 79: 41–54.
Safferman RS and Morris ME (1964) Growth characteristics of the blue-green algal virus LPP1. *Journal of Bacteriology* 88: 771–775.
Sampaio MJ, Rowell P and Stewart WD (1979) Purification and some properties of glutamine synthetase from the nitrogen-fixing cyanobacteria *Anabaena cylindrica* and a *Nostoc* sp. *Microbiology* 111: 181–191.
Sauer J, Dirmeier U and Forchhammer K (2000) The *Synechococcus* strain PCC 7942 glnN product (glutamine synthetase III) helps recovery from prolonged nitrogen chlorosis. *Journal of Bacteriology* 182: 5615–5619.
Sinha RP, Kumar HD, Kumar A and Häder D-P (1995) Effects of UV-B irradiation on growth, survival, pigmentation and nitrogen metabolism enzymes in cyanobacteria. *Acta Protozoologica* 34: 187.
Stewart WD and Rowell P (1975) Effects of L-methionine-DL-sulphoximine on the assimilation of newly fixed NH_3, acetylene reduction and heterocyst production in *Anabaena cylindrica*. *Biochemical and Biophysical Research Communications* 65: 846–856.
Valentine RC, Shapiro BM and Stadtman ER (1968) Regulation of glutamine synthetase. XII. Electron microscopy of the enzyme from *Escherichia coli*. *Biochemistry* 7: 2143–2152.
Yamashita MM, Almassy RJ, Janson CA, Cascio D and Eisenberg D (1989) Refined atomic model of glutamine synthetase at 3.5 Å resolution. *Journal of Biological Chemistry* 264: 17681–17690.

7 Estimation of Nitrate Reductase Enzyme

Sonal Mishra, Prashant R. Singh, Niharika Sahu, and Rajeshwar P. Sinha

7.1 INTRODUCTION

Ferredoxin-dependent cyanobacteria nitrate reductase (EC 1.7.7.2) catalyzes the reduction of nitrate (NO^{3-}) to nitrite (NO^{2-}) (Manzano et al. 1976; Losada and Guerrero 1979; Herrero et al. 1981). The final product of the nitrate-reducing pathway, i.e., ammonium, acts as a potent antagonist of nitrate assimilation in both prokaryotes and eukaryotes. Various repressive effects of this compound on nitrate uptake, activity of nitrate reductase and the synthesis of nitrate reductase have already been described (Syrett and Leftley 1976; Losada and Guerrero 1979; Herrero et al. 1981). The regulation of nitrate reductase synthesis in eukaryotic algae has been investigated in some detail, in which in most instances it has been found that ammonium behaves as a repressor (Morris and Syrett 1963; Vega et al. 1971; Herrera et al. 1972).

However, the synthesis of NAD(P)H-dependent nitrate reductase (EC 1.7.1.4) plays a role in the activity of nitrate in both green and red algae in nitrogen-containing (amino acid) or nitrogen-free medium (Vega et al. 1971; Herrera et al. 1972; Rigano and Violante 1973; Syrett and Hipkin 1973; Sosa et al. 1978). In cyanobacteria, information related to the control of nitrate reductase synthesis is insufficiently available. Nitrate induction has been observed in both *Agmenellum quadruplicatum* (unicellular) and *Anabaena cylindrical* (filamentous nitrogen fixer) (Hattori 1962; Ohmori and Hattori 1970; Stevens and Van Baalen 1974). Ammonium repression has been observed in two unicellular strains, *Anacystis nidulans* and *A. quadruplicatum* (Stevens and Van Baalen 1974; Manzano et al. 1976). Here, we present a broadly applicable method for the estimation of nitrate reductase enzyme activity in cyanobacteria.

7.2 EQUIPMENT

1. Ultraviolet cutoff filter (Ultraphan, Digefra, Munich, Germany) to facilitate the desired wavelength of ultraviolet radiation (UV-B; 280–315 nm).
2. Ultraviolet treatment chamber fixed with UV-B tubes (Philips TL 40 W: 12, Philips Lighting, model: TL 40W/12 RS SLV/25).
3. Glass Petri dishes (Corning, catalogue number: 3160-102).
4. Light microscope (OLYMPUS, model number: CX21i-TR).
5. Measuring cylinder.
6. Microcentrifuge tubes (Tarsons, Cliklok microcentrifuge tube).
7. Refrigerated centrifuge (Remi Electrotechnik, model: CM-12 PLUS).
8. Shaker.
9. UV-VIS spectrophotometer (Shimadzu, Japan, model: U-2600, Double beam spectrophotometer).
10. 3.5 mL quartz cuvette (Cole-Parmer, JENWAY, catalogue number: 035028).
11. 100 mL glass beaker (Borosil, model number 1000D16).
12. 500 mL glass flask (Borosil, model number 4980024).

 DOI: 10.1201/9781003398387-7

7.3 MATERIALS

1. Sulfanilamide (Merck, CAS number 63-74-1).
2. Nitrate reductase dye or *N*-(1-naphthyl) ethylenediamine dihydrochloride (Merck, CAS number 1465-25-4).
3. K_2HPO_4 (Merck, CAS number 7758-11-4).
4. KH_2PO_4 (Merck, CAS number 7778-77-0).
5. KNO_3 (Merck, CAS number 7757-79-1).
6. Cyanobacterial cell culture.

Note: Cyanobacteria can be grown under axenic conditions in BG-11$^-$ (nitrate negative) or BG-11$^+$ (nitrate positive) medium (Safferman and Morris 1964,) depending on its ability to fix atmospheric dinitrogen. The culture was grown at 20 ± 2°C under continuous white light (12 ± 2 W m^{-2}) until the OD_{750} reached between 0.7 and 0.8.

7.4 RECIPES

7.4.1 Nitrate Reductase Dye (100 mL)

200 mg in 100 mL of double distilled water (DDW).

7.4.2 50 mM Potassium Phosphate Buffer (200 mL)

$$\left.\begin{array}{l} K_2HPO_4\,(8.7\text{g})\text{ in }100\text{ mL} \\ KH_2PO_4\,(6.8\text{g})\text{ in }100\text{ mL} \end{array}\right\} 20\text{ mL} + 180\text{ mL double distilled water (DDW)}$$

7.4.3 Sulfanilamide (100 mL)

1 g in 100 mL of 1: 4 HCl (20 mL DDW + 80 mL HCl)

7.4.4 KNO3 (0.02 M; 100 mL)

0.2 g in 100 mL

7.5 METHODS

The method for nitrate reductase enzyme activity is adapted from (Sinha et al. 1995). A flowchart for the estimation of nitrate reductase enzyme is shown in Figure 7.1.

1. Transfer 10 mL of cyanobacterial culture to a sterile glass Petri dish using a measuring cylinder. These Petri dishes are covered with a 295-nm cutoff filter and transferred to the UV light treatment chamber.
2. The temperature of the chamber is kept constant during irradiation of cyanobacteria with UV-B to avoid any damage by heat. Culture is mixed using a glass rod at regular intervals to avoid clump formation.
3. After light or any other treatment, collect 1 mL of cell culture and mix it with 2 mL of sulfanilamide (see Recipes).
4. Thereafter, incubate the sample for 10 minutes at 30°C.
5. After incubation, add 2 mL nitrate reductase dye or N-(1-naphthyl) ethylenediamine dihydrochloride (see Recipes).

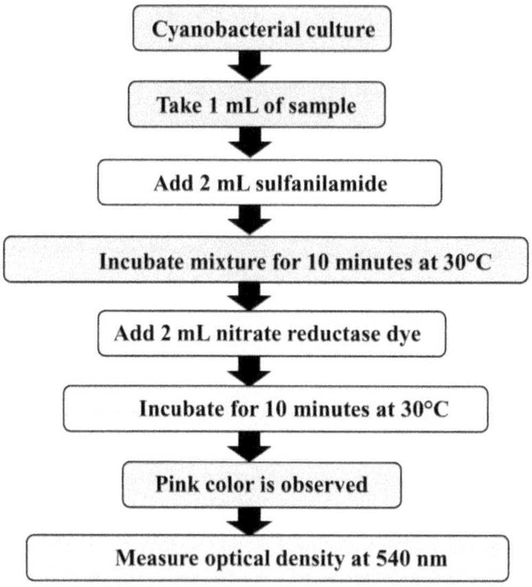

FIGURE 7.1 Flowchart showing different steps involved in the protocol.

6. Incubate the reaction mixture for 10 minutes at 30°C.
7. The absorbance of the pink color is measured at 540 nm using a double-beam spectrophotometer (Figure 7.2).

7.6 DATA ANALYSIS

The effect of different periods of UV-B irradiation on the *in vivo* nitrate reductase activity is shown in Figure 7.2. A gradual increase in nitrate reductase activity was shown with increasing UV-B exposure time up to 120 minutes, and there was no negative effect on nitrate reductase activity even after 120 minutes of exposure to UV-B radiation (Sinha et al. 1995).

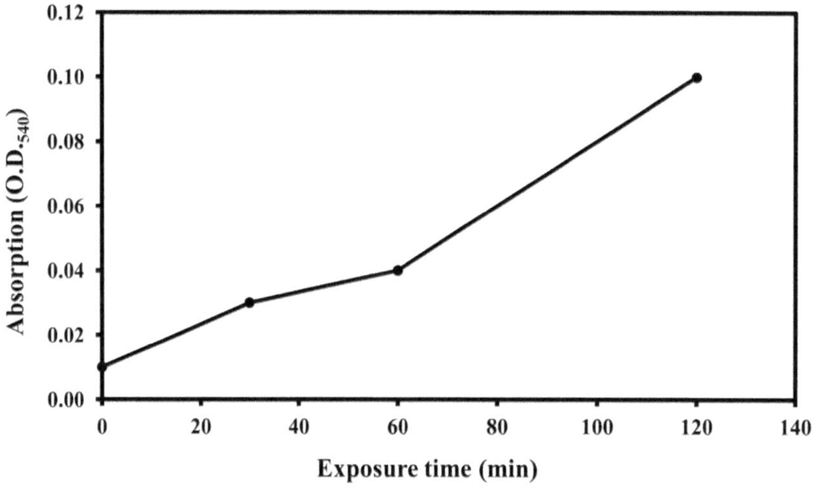

FIGURE 7.2 *In vivo* nitrate reductase enzyme activity of *Nostoc commune* after increasing exposure to UV-B radiation (adapted from Sinha et al. 1995).

7.7 NOTES/PRECAUTIONS

1. Freshly prepared solutions are preferred for nitrate reductase enzyme activity.
2. Where Merck products or CAS numbers are specified, equivalent reagents may be substituted.
3. Other conditions for this assay are as follows: Temperature 30°C, pH 7.3 and light path length 1 cm.

ACKNOWLEDGMENTS

Sonal Mishra is thankful to the University Grants Commission (Joint CSIR-UGC JRF-2019/NTA Ref. No.: 191620046790), New Delhi, India, for the financial assistance in the form of a junior research fellowship. Prashant R. Singh (09/013(0795)/2018-EMR-I) is grateful to the Council of Scientific and Industrial Research (CSIR), New Delhi, India, for providing financial support in the form of a senior research fellowship. Niharika Sahu (09/013(0927)/2020-EMR-I) is thankful to CSIR, New Delhi, India, for the financial support in the form of fellowships.

REFERENCES

Hattori A (1962) Adaptive formation of nitrate reducing system in *Anabaena cylindrica. Plant and Cell Physiology* 3: 371–377.

Herrera J, Paneque A, Maldonado JM, Barea JL and Losada M (1972) Regulation by ammonia of nitrate reductase synthesis and activity in *Chlamydomonas reinhardi. Biochemical and Biophysical Research Communications* 48: 996–1003.

Herrero AN, Flores EN and Guerrero MG (1981) Regulation of nitrate reductase levels in the cyanobacteria *Anacystis nidulans, Anabaena* sp. strain 7119, and *Nostoc* sp. strain 6719. *Journal of Bacteriology* 145: 175–180.

Losada MA and Guerrero MG (1979) The photosynthetic reduction of nitrate and its regulation. In: Photosynthesis in Relation to Model Systems (Ed. J. Barber). Elsevier/North Holland Biomedical Press, Amsterdam. pp. 365–408. ISBN: 9780444415967

Manzano C, Candau P, Gomez-Moreno C, Relimpio AM and Losada M (1976) Ferredoxin-dependent photosynthetic reduction of nitrate and nitrite by particles of *Anacystis nidulans. Molecular and Cellular Biochemistry* 10: 161–169.

Morris I and Syrett PJ (1963) The development of nitrate reductase in *Chlorella* and its repression by ammonium. *Archives of Microbiology* 47: 32–41.

Ohmori K and Hattori A (1970) Induction of nitrate and nitrite reductases in *Anabaena cylindrica. Plant and Cell Physiology* 11: 873–878.

Rigano C and Violante U (1973) Effect of nitrate, ammonia and nitrogen starvation on the regulation of nitrate reductase in *Cyanidium caldarium. Archives of Microbiology* 90: 27–33.

Safferman RS and Morris ME (1964) Growth characteristics of the blue-green algal virus LPP1. *Journal of Bacteriology* 88: 771–775.

Sinha RP, Kumar HD, Kumar A and Häder D-P (1995) Effects of UV-B irradiation on growth, survival, pigmentation and nitrogen metabolism enzymes in cyanobacteria. *Acta Protozoologica* 34: 187.

Sosa FM, Ortega T and Barea JL (1978) Mutants from *Chlamydomonas reinhardii* affected in their nitrate assimilation capability. *Plant Science Letters* 11: 51–58.

Stevens SE Jr and Van Baalen C (1974) Control of nitrate reductase in a blue-green alga: The effects of inhibitors, blue light, and ammonia. *Archives of Biochemistry and Biophysics* 161: 146–152.

Syrett PJ and Hipkin CR (1973) The appearance of nitrate reductase activity in nitrogen-starved cells of *Ankistrodesmus braunii. Planta* 111: 57–64.

Syrett PJ and Leftley JW (1976) Nitrate and urea assimilation by algae. In: Perspectives in Experimental Biology (Ed. N. Sunderland). Pergamon Press, Oxford. pp. 221–234. ISBN: 0-08-019868-6

Vega JM, Herrera J, Aparicio PJ, Paneque A and Losada M (1971) Role of molybdenum in nitrate reduction by *chlorella. Plant Physiology* 48: 294–299.

8 Estimation of Nitrogenase Enzyme Activity in Cyanobacteria

Sonal Mishra, Amit Gupta, Ashish P. Singh, and Rajeshwar P. Sinha

8.1 INTRODUCTION

Diazotrophic microorganisms are proficient at fixing atmospheric nitrogen (N_2) through the activity of biological nitrogen fixation (Alam et al. 2003; Dixon and Kahn 2004; Raja et al. 2006; Mia and Shamsuddin 2010). This process profits the agriculture sector especially for reducing expenses of artificial inorganic nitrogen fertilizers. It is due to the potential of microorganisms to turn atmospheric nitrogen (N_2) into ammonia (NH_3), which is useful for the plants. Accordingly, several studies in this area of research were conducted to optimize the N_2 fixation activities (Rebah et al. 2007). The significant factor for an effective biological nitrogen fixation process is the presence of the nitrogenase enzyme, which catalyzes the fixation process (Sur et al. 2010). Nitrogen fixation is an essential process for all life forms, as fixed nitrogen is essential for the biosynthesis of nucleotides and amino acids. The nitrogenase enzymes are encoded by the *Nif* genes and their homologs. Due to the O_2-sensitive nature of nitrogenase, several diazotrophic bacteria have evolved multiple O_2-resistance mechanisms (Sabra et al. 2000).

Nitrogenases are enzymes (EC 1.18.6.1/EC 1.19.6.1) that perform a key step in the process of nitrogen fixation. Nitrogenase is produced by certain bacteria such as cyanobacteria and rhizobacteria. The cyanobacterium *Aulosira fertilissima* is an efficient biofertilizer, especially in rice (paddy) fields, due to its ability to fix atmospheric nitrogen (Stewart 1980). In addition to playing a significant role in the reduction of nitrogen (N_2) to ammonia (NH_3), nitrogenases are also responsible for the reduction of acetylene (C_2H_2) to ethylene (C_2H_4) (Dilworth 1966).

The ^{15}N incorporation assay is used to measure the activity of nitrogenase enzymes, but it is not cost-effective. An alternative to the ^{15}N incorporation assay, a highly sensitive and inexpensive assay, i.e., acetylene reduction assay, can be used to measure the activity of nitrogenase enzymes (Stewart et al. 1968). This acetylene reduction assay depends on the use of gas chromatography to observe the reduction of acetylene to ethylene (Hardy et al. 1968). Observation of this alternative assay gives a broadly applicable method for the estimation of nitrogenase enzyme activity.

8.2 EQUIPMENT

1. Ultraviolet cutoff filter (Ultraphan, Digefra, Munich, Germany) to select a wavelength of ultraviolet radiation (UV-B; 280–315 nm).
2. Ultraviolet treatment chamber equipped with UV-B tubes (Philips TL 40 W: 12, Philips Lighting, model: TL 40W/12 RS SLV/25).
3. Glass Petri dishes (Corning, catalogue number 3160-102).
4. Light microscope (OLYMPUS, model number CX21i-TR).

DOI: 10.1201/9781003398387-8

5. Measuring cylinder.
6. Microcentrifuge tubes (Tarsons, Cliklok tube).
7. Refrigerated centrifuge (Remi Electrotechnik, model CM-12 PLUS).
8. Shaker.
9. Vacutainer tubes (Beeton Dickinson, Rutherford, NJ).
10. Hypodermic syringe.
11. Gas chromatograph (Thermo Fisher Scientific).
12. Porapak-R column.
13. Flame ionization detector.

8.3 MATERIALS

1. Calcium carbide (CaC_2) (Merck, CAS number 75-20-7).
2. Double distilled water (DDW).
3. Twenty-four-inch section of Tygon tubing with needle fitting.
4. 1 L volumetric flask with armhole and appropriately sized rubber stopper #8.
5. Tedlar gas collection bag.
6. PVC coring cylinders (5 cm) with yellow caps.
7. Canning jar (500 mL) with a hole drilled in the lid.
8. Number 4 one-hole rubber stopper fitted with glass tubing.
9. 1 mL pipette (Abdos).
10. Gas-tight syringes (60 mL) with two-way stopcocks.
11. Twenty-two-gauge syringe needles.
12. Labco Exetainer sample tubes (pre-evacuated).
13. Cyanobacterial cell culture.

Note: The cyanobacterium can be grown under axenic conditions in BG-11$^-$ (nitrate negative) or BG-11$^+$ (nitrate positive) medium (Safferman and Morris 1964) depending on its ability to fix atmospheric dinitrogen. The culture is grown at 20 ± 2°C under continuous white light (12 ± 2 W m^{-2}) until the OD_{750} reaches between 0.7 and 0.8.

8.4 RECIPES

8.4.1 For Making Acetylene Gas (C_2H_2)

Incorporate a suitable number of CaC_2 rocks and half a cup of water in a flask. Thereafter, quickly close the flask with a rubber stopper for some seconds before injecting the syringe needle into the collection bag. Ensure that the needle does not rupture the gas bag. Let the bag fill with C_2H_2 and after that remove the needle from the bag and flask. Put the flask in the hood until reaction shown next is complete.

$$\mathbf{CaC_2 + 2H_2O \rightarrow Ca(OH)_2 + C_2H_2}$$

8.5 METHODS

The method to quantify nitrogenase enzyme activity is adapted from Sinha et al. (1996). A flowchart for the estimation of nitrogenase enzyme activity is shown in Figure 8.1.

1. Transfer 10 mL of cyanobacterial culture to a sterile glass Petri dish using a measuring cylinder. The petri dishes are covered with a UV cutoff filter of 295 nm and transferred to the UV light treatment chamber.

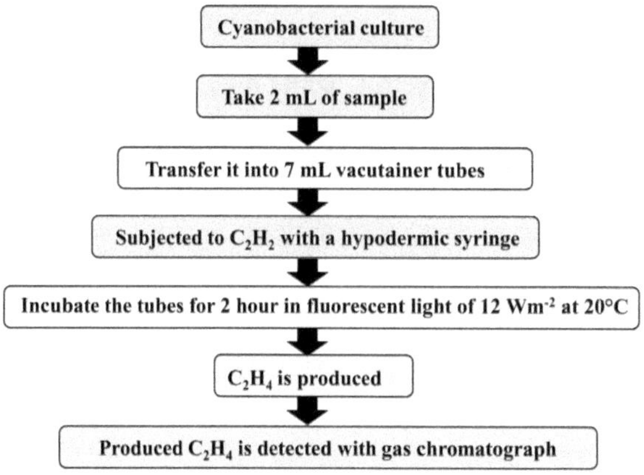

FIGURE 8.1 Flowchart showing the different steps involved in the protocol.

2. Cyanobacteria are irradiated with UV radiation; maintain the temperature of the chamber at a constant rate to avert a heating effect. Stir the culture with a glass rod at regular intervals to avoid clump formation.
3. After exposure, collect 2 mL of cell culture in a 7-mL vacutainer tube (Beeton Dickinson, Rutherford, NJ).
4. Thereafter, the sample is subjected to C_2H_2 with a hypodermic syringe to accomplish a final concentration of 10%.
5. After that incubate the tubes in the fluorescent light (12 Wm^{-2}) at 20°C.
6. The formation of C_2H_4 is determined after 2 hours of incubation using a gas chromatograph fitted with a Porapak-R column and flame ionization detector.

8.6 DATA ANALYSIS

The effect of different periods of UV irradiation on *in vivo* nitrogenase enzyme activity is shown in Figure 8.2. A gradual decrease in nitrogenase activity was achieved with increasing UV exposure time. A 50% reduction in nitrogenase activity was observed after 25 minutes of UV exposure, whereas complete inhibition was observed after 45 minutes of UV exposure (Sinha et al. 1996).

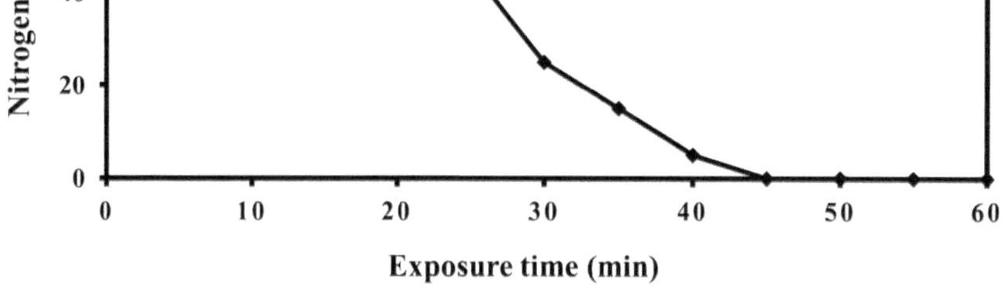

FIGURE 8.2 *In vivo* nitrogenase enzyme activity of *Anabaena* sp. after increasing exposure to UV (5 W m^{-2}) for regular intervals of 5 minutes, it is subjected to a C_2H_2 reduction assay (adapted from Sinha et al. 1996).

8.7 NOTES/PRECAUTIONS

1. Remove needle from the bag and flask when acetylene production is finished.
2. Do not overinflate the bags.

ACKNOWLEDGMENTS

Sonal Mishra thanks the University Grants Commission (Joint CSIR-UGC JRF-2019/NTA Ref. No.: 191620046790), New Delhi, India, for the financial assistance in the form of a junior research fellowship. Amit Gupta (09/013(0912)/2019-EMR-I) thanks the Council of Scientific & Industrial Research (CSIR), New Delhi, India, for a junior research fellowship. Ashish P. Singh thanks the University Grant Commission (UGC), New Delhi, India, (NTA Ref. No. 191620014505) for providing a junior research fellowship.

REFERENCES

Alam R, Karim MZ, Hasan MF, Hossain MA and Mafizur Rahman AB (2003) Evaluation of sugarcane clones with biological nitrogen fixation endowment. *Asian Journal of Plant Sciences* 2: 216–219.

Dilworth MJ (1966) Acetylene reduction by nitrogen fixing preparations from *Clostridium pasteurianum. Biochimica et Biophysica Acta (BBA) General Subjects* 127: 285–294.

Dixon R and Kahn D (2004) Genetic regulation of biological nitrogen fixation. *Nature Reviews Microbiology* 2: 621–631.

Hardy RW, Holsten RD, Jackson EK and Burns R (1968) The acetylene-ethylene assay for N_2 fixation: Laboratory and field evaluation. *Plant Physiology* 43: 1185–1207.

Mia MB and Shamsuddin ZH (2010) Nitrogen fixation and transportation by rhizobacteria. *International Journal of Botany* 6: 235–242.

Raja P, Uma S, Gopal H and Govindarajan K (2006) Impact of bio inoculants consortium on rice root exudates, biological nitrogen fixation and plant growth. *Journal of Biological Sciences* 6: 815–823.

Rebah FB, Prévost D, Yezza A and Tyagi RD (2007) Agro-industrial waste materials and wastewater sludge for rhizobial inoculant production: A review. *Bioresource Technology* 98: 3535–3546.

Sabra W, Zeng AP, Lünsdorf H and Deckwer WD (2000) Effect of oxygen on formation and structure of *Azotobacter vinelandii* alginate and its role in protecting nitrogenase. *Applied and Environmental Microbiology* 66: 4037–4044.

Safferman RS and Morris ME (1964) Growth characteristics of the blue-green algal virus LPP1. *Journal of Bacteriology* 88: 771–775.

Sinha RP, Singh N, Kumar A, Kumar HD, Häder M and Häder D-P (1996) Effects of UV irradiation on certain physiological and biochemical processes in cyanobacteria. *Journal of Photochemistry and Photobiology B: Biology* 32: 107–113.

Stewart WD (1980) Some aspects of structure and function in nitrogen fixing cyanobacteria. *Annual Reviews in Microbiology* 34: 497–536.

Stewart WD, Fitzgerald GP and Burris RH (1968) Acetylene reduction by nitrogen-fixing blue-green algae. *Archives of Microbiology* 62: 336–348.

Sur S, Bothra AK and Sen A (2010) Symbiotic nitrogen fixation-a bioinformatics perspective. *Biotechnology* 9: 257–273.

9 Isolation and Purification of Cyanobacterial Toxins

Sonal Mishra and Rajeshwar P. Sinha

9.1 INTRODUCTION

Cyanobacteria are photosynthetic microorganisms with ~3500 million years of existence on Earth, producing a diverse group of bioactive compounds (Mishra et al. 2022a,b,c). The most commonly investigated bloom-forming, toxin-producing cyanobacterial strains belong to the genera *Microcystis*, *Merismopedia*, *Dolichospermum* or *Anabaena*, *Aphanizomenon*, *Anabaenopsis* and *Oscillatoria* (Spoof and Catherine 2016; Svirčev et al. 2019; Mishra et al. 2019a, 2022a) (Figure 9.1). Global changes emerging from human consequences such as eutrophication, CO_2 increases and global warming are prime driving forces for the intensification of cyanobacterial blooms. Cyanobacterial blooms produce highly potent cyanobacterial toxins, where microcystins (MCs) are the most frequently recognized cyanobacterial toxins worldwide, followed by cylindrospermopsins (CYNs), nodularins (NODs), anatoxins (ATXs) and saxitoxins (STXs) (Svirčev et al. 2019).

Cyanobacterial toxins endanger ecosystem functioning and deteriorate the quality of water used for drinking, dialysis and food items, which exert toxic effects on human health and the economy (Mishra et al. 2022a). Furthermore, according to World Health Organization (WHO) standards, the level of cyanobacterial toxins in drinking water should not surpass 1 μg/L, while the Environmental Protection Agency (EPA) has stated that the tolerable level for children under 6 years of age is ≤0.3 μg/L and ≤1.6 μg/L for adults (World Health Organization 2011; US EPA 2015a,b). Cyanobacterial toxins are chemical compounds that cause several diseases associated with digestion, liver and neurological disorders when consumed by birds and mammals, including humans (Metcalf and Codd 2012; Merel et al. 2013). To date, a broadly applicable method for the isolation and purification of cyanobacterial toxins based on solid-phase extraction (SPE), high-performance liquid chromatography (HPLC) and liquid chromatography-mass spectrometry (LC-MS) have been reported (Zervou et al. 2017; El-Nahhal et al. 2021).

9.2 EQUIPMENT

1. Analytical balance (LABCON, LAB-104, catalogue number: 9502104584).
2. Reverse-phase HPLC (Photodiode Array Detector, Waters 2998, 515 HPLC pump, Nova-Pak® C18 reverse-phase guard column; 4.6 mm × 150 mm inside diameter).
3. Liquid chromatography (model LC 1220 Infinity, Agilent Technology, USA, dual-channel gradient pump, reverse-phase C18 analytical column (3.5 μm, 4.6 mm × 100 mm inside diameter, XBridge, Waters).
4. Refrigerated centrifuge (Remi Electrotechnik, Model: CM-12 PLUS).
5. UV-VIS spectrophotometer (Shimadzu, Japan, model: U-2600, Double beam spectrophotometer).
6. Ultrasonicator (SONICS Vibra Cell, model: VC X 750; power: 750 watts, frequency: 20 KHz).

9.3 MATERIALS

1. Glass Petri dishes (Corning, catalogue number: 3160-102).
2. Measuring cylinder (Borosil).

 DOI: 10.1201/9781003398387-9

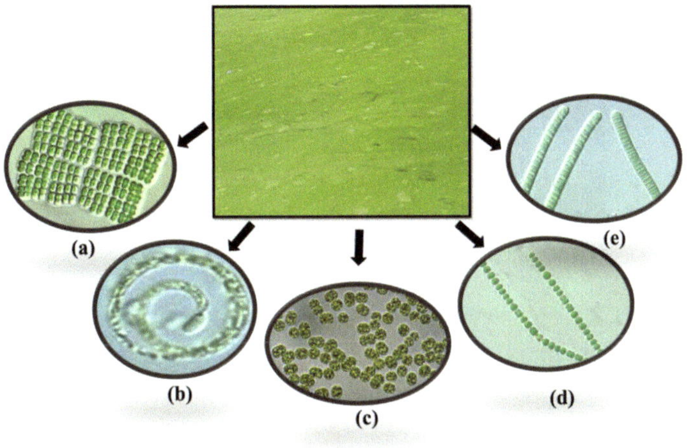

FIGURE 9.1 Bloom-forming, toxin-producing strains. (a) *Merismopedia* sp., (b) *Anabaenopsis* sp., (c) *Microcystis* sp., (d) *Anabaena* sp. and (e) *Oscillatoria* sp. (color figure in e-book).

3. Microcentrifuge tubes (Tarsons, Cliklok Microcentrifuge Tube).
4. Glass rod (Fisher Scientific, catalogue number: 11-380A).
5. Cuvette (quartz, 3.5 mL) (Cole-Parmer, JENWAY, catalogue number: 035028).
6. Methanol, HPLC grade (Merck, catalogue number: 106009).
7. Acetonitrile (Merck, catalogue number: 100003).
8. Trifluoroacetic acid (Merck, catalogue number: 302031).
9. HPLC-grade ultra-pure water (A10 Milli-Q).
10. Nitrogen or argon gas, >99.99%.
11. Borosilicate glass vials/tubes.
12. Syringe-driven filter (before analysis by LC-MS), filter type: 0.2 µm pore-sized, 13 mm (Pall Life Sciences, Ann Arbor, MI).
13. Nitrogen evaporation unit (rotary evaporator or centrifugal evaporator).
14. Glass beaker, 100 mL (Borosil, Model No: 1000D16).

9.4 RECIPES

9.4.1 Isolation Solvent for Cyanobacterial Toxins

Dissolve the dried cyanobacteria in methanol on a volumetric basis (e.g., 65% methanol: 65 mL methanol + 35 mL ultra-pure water).

9.4.2 Purification Solvent Mixture for Cyanobacterial Toxins

Mobile phase A (0.05% trifluoroacetic acid in acetonitrile).
Mobile phase B (0.05% trifluoroacetic acid in ultra-pure water).

9.5 METHODS

A flowchart for the isolation and purification of cyanobacterial toxins is shown in Figure 9.2.

1. Dry cyanobacteria by lyophilization and dissolve in 5 mL methanol on a volumetric basis.
2. Stock solutions are prepared by evaporating methanol under a nitrogen stream, and the extracted toxins are redissolved in 100 µL 'Gradient A10 Milli-Q'–produced ultra-pure water.

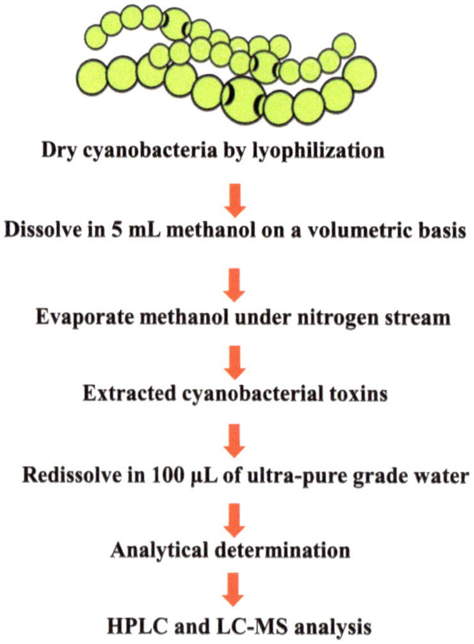

FIGURE 9.2 Flowchart showing the different steps involved in the protocol (color figure in e-book).

3. The protocol is improved by an additional rinse step, where 4.5 mL of methanol out of 5 mL is evaporated and the remaining 0.5 mL transferred to a vial. The remaining 0.5 mL is dissolved with 0.5 mL methanol, and this volume is added to the vial. The solution in the vial is evaporated to dryness and the extracted cyanobacterial toxin dissolved in 100 µL of ultra-pure water.
4. Linearity plots are formulated for the HPLC-photodiode array detector by analysis of standard solutions at eight different concentrations in the range of 20–5000 µg/L.
5. Thereafter, an analytical determination using HPLC and LC-MS analysis is performed.

9.6 DATA ANALYSIS

HPLC analysis is performed with the isolated cyanobacterial toxins. The recorded retention time (RT) corresponding to MCs in ultra-pure water are 8.247 minutes and 9.225 minutes, respectively, for MC-RR and MC-LR at 500 µg/L each. Analytes are determined using LC-MS analysis (Thuret-Benoist et al. 2019).

9.7 NOTES/PRECAUTIONS

1. All solutions are filtered through a 0.2-µm pore-size sterilized microcentrifuge syringe-driven filter before use.
2. Once prepared for HPLC and LC-MS analysis, the solvent is kept in an axenic environment to avoid bacterial contamination (freshly prepared solvent is preferred).
3. Safety equipment such as gloves and face masks should be used to avoid the harmful effects of the cyanobacterial bloom, and lyophilized cyanobacterial material must be handled with extreme care to avoid a hazard.

ACKNOWLEDGMENTS

Sonal Mishra is thankful to the University Grants Commission (UGC), New Delhi, India, for financial assistance in the form of a fellowship as a senior research fellow.

REFERENCES

El-Nahhal H, Yassin M, Alzaharna M, El-Nahhal I and El-Nahhal Y (2021) Extraction methods of cyanotoxins aqueous media and sediments. *American Journal of Analytical Chemistry* 12: 311–323.

Merel S, Walker D, Chicana R, Snyder S, Baurès E and Thomas O (2013) State of knowledge and concerns on cyanobacterial blooms and cyanotoxins. *Environment International* 59: 303–327.

Metcalf JS and Codd GA (2012) Cyanotoxins. In: Ecology of Cyanobacteria II: Their Diversity in Space and Time (Ed. B.A. Whitton). Springer, Dordrecht. pp. 651–675. ISBN: 978-94-007-3855-3

Mishra S, Kumari N, Häder D-P and Sinha RP (2022a) Cyanobacterial blooms and their implications in the changing environment. *Advances in Environmental and Engineering Research* 3: 1–41.

Mishra S, Pandey A, Ahmed H, Kumar D, Singh V and Sinha RP (2019a) Mycosporine-like amino acids (MAAs) profile of cyanobacteria from different historical kunds of Varanasi, India. *International Journal of Applied Sciences and Biotechnology* 7: 317–326.

Mishra S, Pandey A and Sinha RP (2019b) Characterization of photoprotective and pharmaceutically important compounds from cyanobacteria and algal assemblages from historical kunds of Varanasi, India. *Journal of Chemical and Pharmaceutical Research* 11: 24–36.

Mishra S, Richa and Sinha RP (2022b) Biotechnological exploitation of cyanobacterial photoprotective metabolites. *Vegetos* 35: 281–297.

Mishra S, Sahu N, Kumari N, Richa and Sinha RP (2022c) Identification and characterization of mycosporine-like amino acid in the cyanobacterium *Scytonema* sp. HKAR-16 by HPLC, Raman spectroscopy, and ESI-MS. *Songklanakarin Journal of Science and Technology* 44: 836–840.

Spoof L and Catherine A (2016) Appendix 3: Tables of microcystins and nodularins. In: Handbook of Cyanobacterial Monitoring and Cyanotoxin Analysis (Eds. J. Meriluoto, L. Spoof and G.A. Codd). John Wiley & Sons, Ltd, Chichester, UK. pp. 526–537. ISBN: 9781119068761

Svirčev Z, Lalić D, Bojadžija Savić G, Tokodi N, Drobac Backović D, Chen L, Meriluoto J and Codd GA (2019) Global geographical and historical overview of cyanotoxin distribution and cyanobacterial poisonings. *Archives of Toxicology* 93: 2429–2481.

Thuret-Benoist H, Pallier V and Feuillade-Cathalifaud G (2019) Quantification of microcystins in natural waters by HPLC-UV after a pre-concentration step: Validation of the analytical performances and study of the interferences. *Environmental Toxicology and Pharmacology* 72: 103223.

US EPA (2015a) Drinking water health advisory for the cyanobacterial microcystin toxins. https://www.epa.gov/sites/default/files/2017-06/documents/microcystins-report-2015.pdf

US EPA (2015b) Drinking water health advisory for the cyanobacterial toxin *Cylindrospermopsin*. https://www.epa.gov/sites/default/files/2017-06/documents/cylindrospermopsin-report-2015.pdf

World Health Organization (2011) Guidelines for drinking-water quality. http://apps.who.int/iris/bitstream/10665/44584/1/9789241548151_eng.pdf

Zervou SK, Christophoridis C, Kaloudis T, Triantis TM and Hiskia A (2017) New SPE-LC-MS/MS method for simultaneous determination of multi-class cyanobacterial and algal toxins. *Journal of Hazardous Materials* 323: 56–66.

10 Isolation and Purification of Scytonemin

Neha Kumari, Sonal Mishra, Jyoti Jaiswal, Prashant R. Singh, and Rajeshwar P. Sinha

10.1 INTRODUCTION

Cyanobacteria are photoautotrophic prokaryotes that can convert CO_2 directly into organic compounds using solar energy. Cyanobacteria are naturally found in various extreme environments such as oceans, alkaline waters, freshwater, deserts and hot springs. To withstand these diverse habitats, cyanobacteria have developed a variety of adaptive strategies. Excessive ultraviolet (UV) radiation has harmful effects on cyanobacteria. They cause cell damage by both direct effects on proteins and nucleic acids and indirect effects via the production of reactive oxygen species (Rastogi et al. 2014). To counteract the adverse effects of UV radiation, cyanobacteria produce photoprotective compounds like scytonemin and mycosporine-like amino acids (MAAs) (Gao and Garcia-Pichel 2011; Gao et al. 2021).

Scytonemin is naturally present in the extracellular sheath of cyanobacteria and has great potential in biomedical and cosmetic applications. Scytonemin is a yellow-brown pigment with a molecular weight of 546 Da, and it is soluble in lipid. It is a dimeric compound composed of indolic and phenolic subunits. Scytonemin exists in two forms: Oxidized and reduced. Three more derivatives of scytonemin have been found, i.e., dimethoxyscytonemin, tetramethoxyscytonemin and scytonin (Bultel-Poncé et al. 2004). Scytonemin mainly absorbs UV-A radiation and can prevent about 90% of UV-A from entering cells (Garcia-Pichel and Castenholz 1991). It also absorbs UV-B and UV-C radiation.

Scytonemin has an *in vivo* maximum absorption at 370 nm and an *in vitro* maximum absorption at 384 nm (Garcia-Pichel and Castenholz 1991). Scytonemin can be extracted from samples with solvents such as 100% acetone or 100% ethyl acetate (Garcia-Pichel and Castenholz 1991), 100% acetonitrile (Matsui et al. 2012), 80% tetrahydrofuran (Ferroni et al. 2010) and methanol:ethyl acetate (1:1 v/v) (Pandey et al. 2020; Kumari et al. 2021). Scytonemin is conserved in deep-sea sediments because it is not degraded by transport and erosion (Fulton et al. 2012), so, it is regarded as one of the key biomarkers in both paleoclimatological reconstructions and terrestrial extreme environments (Lalić et al. 2020). It protects against UV radiation by the following mechanisms: Reduces reactive oxygen species and thymine dimer formation, absorbs UV radiation and contains antioxidative properties, dissipates heat from absorbed UV radiation and enhances the temperature of the soil surface and mitigates the photosynthetic inhibition (Dillon and Castenholz 2003; Rastogi et al. 2014; Couradeau et al. 2016). As a natural sunscreen compound, its application for skin protection has attracted great interest among dermatologists and in cosmetics (Siezen 2011).

Scytonemin contains anti-proliferative and anti-inflammatory activities. Human fibroblast and endothelial cell proliferation is inhibited by scytonemin through selectively inhibiting kinases (Stevenson et al. 2002). Inflammation of skin can be inhibited by blocking the expression of inflammatory mediators and down-regulating the activity of NF-κB (Kang et al. 2020). Scytonemin can be used in anti-cancerous drugs, as the human polo-like kinase 1 can act as a target site (Strebhardt and Ullrich 2006). Also, scytonemin provides high tolerance to cyanobacteria against desiccation. This may be due to its ability to stabilize the exopolysaccharide matrix (Gao 2017). Therefore,

DOI: 10.1201/9781003398387-10

scytonemin is a cyanobacterial secondary metabolite with a high potential market value and a promising small-molecule drug.

10.2 EQUIPMENT

1. A 295-nm UV cutoff filter (Ultraphan, Digefra, Munich, Germany) to provide the desired wavelength of photosynthetically active radiation (PAR, 400–700 nm), UV-A (315–400 nm) and UV-B (280–315 nm) radiation.
2. Measuring cylinder.
3. Glass Petri dishes (Corning, catalog number: 3160-102).
4. UV treatment chamber equipped with PAR (55.08 ± 9.18 μmol photon m^{-2} s^{-1}) (OSRAM L 36 W: 32 Lumilux deluxe warm white and Radium NL 36 W: 26 Universal white, Germany), UV-A (Philips Ultraviolet-A TL 40 W: 12, Philips Lighting, model: TL-K 40W/10-R UV-A) and UV-B (Philips Ultraviolet-B TL 40 W: 12, Philips Lighting, model: TL 40W/12 RS SLV/25) lamps.
5. Glass rod (Fisher Scientific, catalog number: 11-380A).
6. Ultrasonicator (SONICS Vibra cell VCX 750).
7. Refrigerated centrifuge (Remi Electrotechnik, model: CM-12 PLUS).
8. Reverse-phase HPLC (Waters 2998, Photodiode Array, pump L-7100, USA).
9. Microcentrifuge tubes.
10. Shaker.
11. Solvent filtration unit.
12. UV-VIS spectrophotometer (Hitachi High-Technologies, model: U-2600).
13. Quartz cuvette (3.5 mL) (Cole-Parmer, JENWAY, catalogue number: 035 028).

10.3 MATERIALS

1. Cyanobacterial cells (*Scytonema* HKAR-8).
   Note: Scytonema HKAR-8 is an autotrophic, filamentous and heterocystous cyanobacterium. It is a member of Nostocales, forming dark mats, and is characterized by false branching. It was grown under axenic conditions in BG-11 medium without nitrogen (Rippka et al. 1979) at 20 ± 2°C and provided continuous white light (12 ± 2 Wm^{-2}) and in the exponential growth phase had an OD_{750} (0.8–0.9) that was measured by spectrophotometer.
2. Aqueous Millipore filter (0.45 μm, catalog number: HATF04700).
3. Organic Millipore filter (0.5 μm, catalog number: FHUP04700).
4. Sterilized syringe Millipore filters (0.2 μm pore size) (AxivaSichem Biotech, New Delhi; catalog number: FGLP02500) (to filter the sample before performing HPLC).
5. Chromatography vials and closures (Borosil, catalog number: VC02C109PBS112).
6. Acetonitrile, HPLC grade (Merck, catalog number: 114291).
7. Methanol, HPLC grade (Merck, catalog number: 106018).
8. Tetrahydrofuran, HPLC grade (Merck, catalogue number: 108101).
9. Double distilled water.

10.4 METHODS

Steps used in the isolation and purification of scytonemin are shown in Figure 10.1.

1. Place 100 mL of cyanobacterial culture in sterile glass Petri dishes using a measuring cylinder. Cover the Petri dishes with a 295-nm UV cutoff filter and keep them in a UV light treatment chamber.

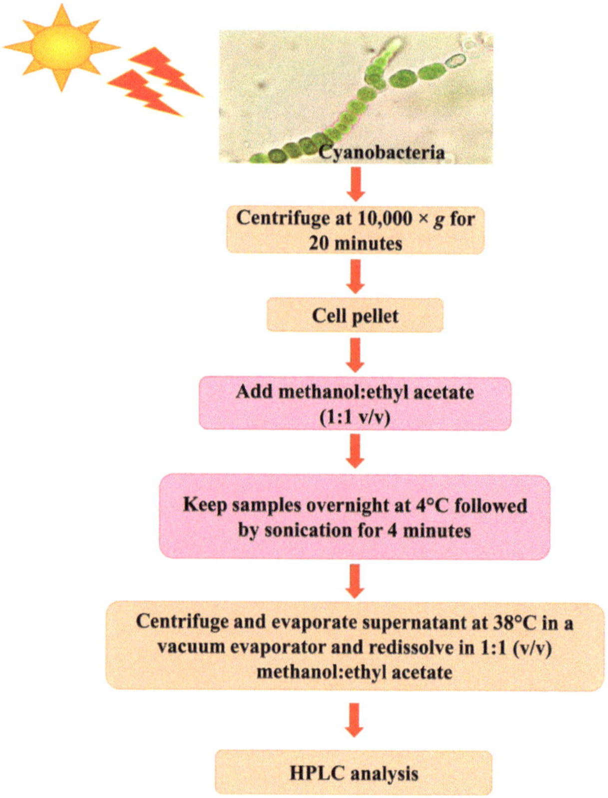

FIGURE 10.1 Steps used in the isolation and purification of scytonemin (color figure in e-book).

2. Provide radiation of PAR, UV-A and UV-B to the cyanobacteria and keep the temperature of the UV light treatment chamber at 25 ± 2°C to avert the heating effect. Mix the culture with the help of a glass rod at regular intervals to avoid self-shading of the cells.
3. Take the sample and centrifuge at 10,000 × *g* for 20 minutes at room temperature.
4. Dissolve the cell pellet in methanol/ethyl acetate (1:1 v/v) and keep the samples overnight at 4°C followed by sonication for 4 minutes.
5. Centrifuge the samples at 10,000 × *g* for 5 minutes, evaporate supernatants at 38°C in a vacuum evaporator and redissolve in 500 µL of 1:1 (v/v) methanol:ethyl acetate.
6. Filter the samples through sterilized syringe filters before the HPLC analysis.
7. Inject 20 µL of the samples into the HPLC column via an autoinjector (Waters 717 plus).
8. The flow rate of elution is 1.0 mL min^{-1} using the mobile phase consisting of solvent A (ultra-pure water) and solvent B (acetonitrile-methanol-tetrahydrofuran, 75:15:10 v/v), and it is fixed with a linear increase (0–15 minutes) from 15% solvent A to 100% solvent B and 15–30 minute at 100% solvent B.
9. The detection wavelength for scytonemin is 386 nm with a PDA scan wavelength ranging from 250 to 750 nm.

10.5 DATA ANALYSIS

An HPLC chromatogram of the sample reveals the presence of scytonemin, showing a prominent peak at a retention time of 5.388 minutes and strong absorbance in the UV region, having major peaks in the UV-A (384 ± 2 nm), UV-B (300 ± 2 nm) and UV-C regions (252 ± 2 nm) (Figure 10.2).

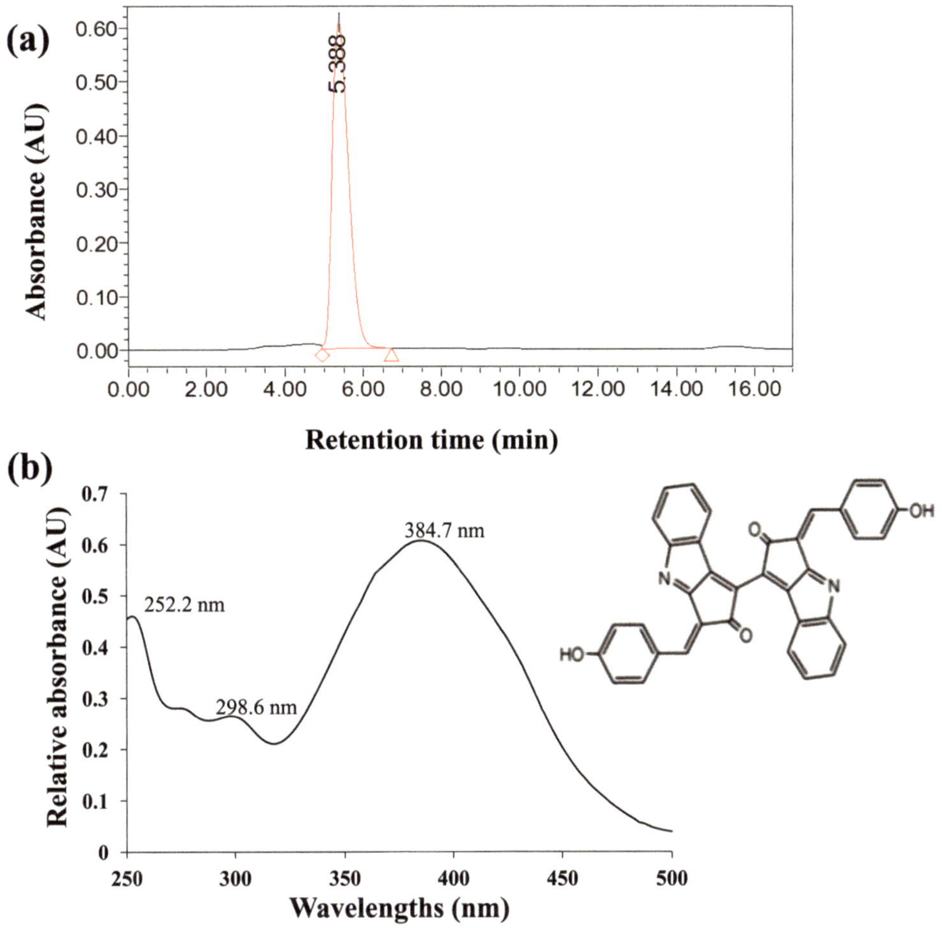

FIGURE 10.2 HPLC chromatogram of partially purified scytonemin (a) and its corresponding absorption spectrum (b) from a cyanobacterial sample.

10.6 NOTES/PRECAUTIONS

1. The samples are filtered through sterilized syringe filters (0.2 μm pore size) prior to the HPLC analysis.
2. The solvent should be filtered before performing HPLC.

ACKNOWLEDGMENTS

N. Kumari and P.R. Singh are thankful to CSIR, New Delhi, India, for providing financial support in the form of a senior research fellowship. S. Mishra and J. Jaiswal are thankful to the University Grants Commission (UGC), New Delhi, India, for the financial assistance in the form of a junior research fellowship.

REFERENCES

Bultel-Poncé V, Félix-Théodose F, Sarthou C, Ponge JF and Bodo B (2004) New pigments from the terrestrial cyanobacterium *Scytonema* sp. collected on the Mitaraka Inselberg, French Guyana. *Journal of Natural Products* 67: 678–681.

Couradeau E, Karaoz U, Lim HC, Da Rocha UN, Northen T, Brodie E and Garcia-Pichel F (2016) Bacteria increase arid-land soil surface temperature through the production of sunscreens. *Nature Communications* 7: 1–7.
Dillon JG and Castenholz RW (2003) The synthesis of the UV-screening pigment, scytonemin, and photosynthetic performance in isolates from closely related natural populations of cyanobacteria (*Calothrix* sp.). *Environmental Microbiology* 5: 484–491.
Ferroni L, Klisch M, Pancaldi S and Häder D-P (2010) Complementary UV-absorption of mycosporine-like amino acids and scytonemin is responsible for the UV-insensitivity of photosynthesis in *Nostoc flagelliforme*. *Marine Drugs* 8: 106–121.
Fulton JM, Arthur MA and Freeman KH (2012) Subboreal aridity and scytonemin in the Holocene Black Sea. *Organic Geochemistry* 49: 47–55.
Gao Q and Garcia-Pichel F (2011) Microbial ultraviolet sunscreens. *Nature Reviews Microbiology* 9: 791–802.
Gao X (2017) Scytonemin plays a potential role in stabilizing the exopolysaccharidic matrix in terrestrial cyanobacteria. *Microbial Ecology* 73(2): 255–258.
Gao X, Jing X, Liu X and Lindblad P (2021) Biotechnological production of the sunscreen pigment scytonemin in cyanobacteria: Progress and strategy. *Marine Drugs* 19: 129.
Garcia-Pichel F and Castenholz RW (1991) Characterization and biological implications of scytonemin, a cyanobacterial sheath pigment. *Journal of Phycology* 27: 395–409.
Kang MR, Jo SA, Lee H, Yoon YD, Kwon JH, Yang JW, Choi BJ, Park KH, Lee MY, Lee CW and Lee KR (2020) Inhibition of skin inflammation by scytonemin, an ultraviolet sunscreen pigment. *Marine Drugs* 18: 300.
Kumari N, Pathak J, Dwivedy AK and Sinha RP (2021) Bioprospection of UV-screening compounds from lichens inhabiting the Indian state of Sikkim. *Plant Archives* 21: 1168–1177.
Lalić D, Meriluoto J, Zorić M, Dulić T, Mirosavljević M, Župunski M and Svirćev Z (2020) Potential of cyanobacterial secondary metabolites as biomarkers for paleoclimate reconstruction. *Catena* 185: 104283.
Matsui K, Nazifi E, Hirai Y, Wada N, Matsugo S and Sakamoto T (2012) The cyanobacterial UV-absorbing pigment scytonemin displays radical-scavenging activity. *The Journal of General and Applied Microbiology* 58: 137–144.
Pandey A, Pathak J, Singh DK, Ahmed H, Singh V, Kumar D and Sinha RP (2020) Photoprotective role of UV-screening pigment scytonemin against UV-B-induced damages in the heterocyst-forming cyanobacterium *Nostoc* sp. strain HKAR-2. *Brazilian Journal of Botany* 3: 1–4.
Rastogi RP, Sonani RR and Madamwar D (2014) The high-energy radiation protectant extracellular sheath pigment scytonemin and its reduced counterpart in the cyanobacterium *Scytonema* sp. R77DM. *Bioresource Technology* 171: 396–400.
Rippka R, Deruelles J, Waterbury JB, Herdman M and Stainer RY (1979) Generic assignments, strain histories and properties of pure cultures of cyanobacteria. *The Journal of General Microbiology* 111: 1–6.
Siezen RJ (2011) Microbial sunscreens. *Microbial Biotechnology* 4: 1–7.
Stevenson CS, Capper EA, Roshak AK, Marquez B, Grace K, Gerwick WH, Jacobs RS and Marshall LA (2002) Scytonemin-a marine natural product inhibitor of kinases key in hyperproliferative inflammatory diseases. *Inflammation Research* 51: 112.
Strebhardt K and Ullrich A (2006) Targeting polo-like kinase 1 for cancer therapy. *Nature Reviews Cancer* 6: 321–330.

11 Isolation and Purification of Mycosporine-Like Amino Acids

Sonal Mishra, Neha Kumari, Jyoti Jaiswal, and Rajeshwar P. Sinha

11.1 INTRODUCTION

Over the last decades, an increase in ultraviolet-B (UV-B; 280–315 nm) and ultraviolet-A (UV-A; 315–400 nm) radiation at Earth's surface has raised widespread concern about its harmful effects on both aquatic and terrestrial life forms consisting of cyanobacterial and algal populations (Williamson et al. 2014; Häder et al. 2015; Mishra et al. 2019a). Cyanobacteria are considered substantial biomass producers that play an important role in maintaining the trophic energy dynamics of an ecosystem (Abed et al. 2009; Rastogi and Sinha 2009). These organisms have ensured defense mechanisms in response to harmful UV radiation that are mainly correlated with the biosynthesis of photoprotective compounds: Mycosporine-like amino acids (MAAs) (Garcia-Pichel and Castenholz 1993; Rastogi et al. 2010; Rastogi and Incharoensakdi 2014).

MAAs and their associated group of organic compounds have a low molecular weight of 400 Da or less, and they are colorless, water-soluble, multifunctional compounds. They are composed of either an aminocyclohexanone or an aminocyclohexinamine chromophore conjugated with a nitrogen substitution of an amino acid or its imino alcohol. There are around 40 MAAs, with their allied groups studied so far with absorption maxima ranging from 309 to 365 nm (Carreto et al. 1990; Wada et al. 2015). It has been observed that the MAA porphyra-334 can block harmful UV radiation and protect the organisms (Mishra et al. 2019b). MAAs appeared to have bioactive properties in both the pharmaceutical and cosmetic industries and emerged as a natural replacement for artificial sun blockers due to their UV-absorbing potential. To date, a broadly applicable method for the presence, isolation and purification of MAAs in cyanobacteria based on UV-VIS spectrophotometry and high-performance liquid chromatography (HPLC) systems has been reported (Matsui et al. 2011; Waditee-Sirisattha et al. 2014; Hu et al. 2015; Kageyama and Waditee-Sirisattha 2018; Mishra et al. 2019a,b).

11.2 EQUIPMENT

1. Ultraviolet cutoff filter (Ultraphan, Digefra, Munich, Germany) to select the desired wavelength of ultraviolet radiation (280–400 nm) and photosynthetically active radiation (PAR; 400–700 nm).
2. Ultraviolet treatment chamber equipped with UV-B (Philips TL 40 W: 12, Philips Lighting, model: TL 40W/12 RS SLV/25), UV-A (Philips TL 40 W: 12, Philips Lighting, model: TL-K 40W/10-R UV-A) and PAR (55.08 ± 9.18 μmol photon m^{-2} s^{-1}) (OSRAM L 36 W: 32 Lumilux deluxe warm white and Radium NL 36 W: 26 Universal white, Germany) lamps.
3. Light microscope (OLYMPUS, model number: CX21i-TR).
4. Glass Petri dishes (Corning, catalogue number: 3160-102).
5. Measuring cylinder.
6. Reverse-phase HPLC (Photodiode Array Detector, Waters 2998, 515 HPLC pump, Nova-Pak® C18 reverse-phase guard column (4.6 mm × 150 mm inside diameter).
7. Microcentrifuge tubes (Tarsons, Cliklok microcentrifuge tube).

DOI: 10.1201/9781003398387-11

8. Refrigerated centrifuge (Remi Electrotechnik, model: CM-12 PLUS).
9. Shaker.
10. Glass rod (Fisher Scientific, catalogue number: 11-380A).
11. UV-VIS spectrophotometer (Shimadzu, Japan, model: U-2600, Double beam spectrophotometer).
12. Cuvette (quartz, 3.5 mL) (Cole-Parmer, JENWAY, catalogue number: 035028).
13. Ultrasonicator (SONICS Vibra Cell, Model: VC X 750; power: 750 watts, frequency: 20 KHz).
14. 100 mL glass beaker (Borosil, model number 1000D16).

11.3 MATERIALS

15. HPLC-grade methanol (Merck, catalogue number: 106009).
16. Chloroform (Merck, catalogue number: 102445).
17. Acetic acid (Merck, catalogue number: 5438080250).
18. Double distilled water (DDW).
19. Microcentrifuge syringe-driven filter, filter type: 0.22 µm pore size, 25 mm (Catalogue number: FGLP02500).
20. Aqueous micropore filter, filter type: 0.45 µm pore size (Merck, catalogue number: HATF04700).
21. Cyanobacterial cells, e.g., *Scytonema* sp. strain HKAR-16.
    Note: Scytonema sp. strain HKAR-16, a photosynthetic, filamentous and nitrogen-fixer cyanobacterium, was grown under axenic conditions in BG-11$^-$ (nitrate negative) medium (Safferman and Morris 1964). The temperature was maintained at 20 ± 2°C under continuous white light (12 ± 2 W m^{-2}) until the OD_{750} of culture was reached between 0.7 and 0.8.
22. Exmire micro syringe (Code number: MS-N500).
23. 2 mL Borosil chromatography vials (Catalogue number: VC02C109PBS112).

11.4 RECIPES

11.4.1 For MAA Extraction

2 mL cell sample (cyanobacterial cell pellet) dissolved in methanol.

11.4.2 MAA Isolation Solvent

80% water + 20% chloroform.

11.4.3 MAA Purification Solvent

0.02% v/v acetic acid in 1000 mL DDW.

11.5 METHODS

A flowchart for the isolation, presence and purification of MAA is shown in Figure 11.1.

1. Transfer 50 mL of cyanobacterial culture to sterile glass Petri dishes using a measuring cylinder. These Petri dishes are covered with a UV cutoff filter of 295 nm and transferred to the UV light treatment chamber.

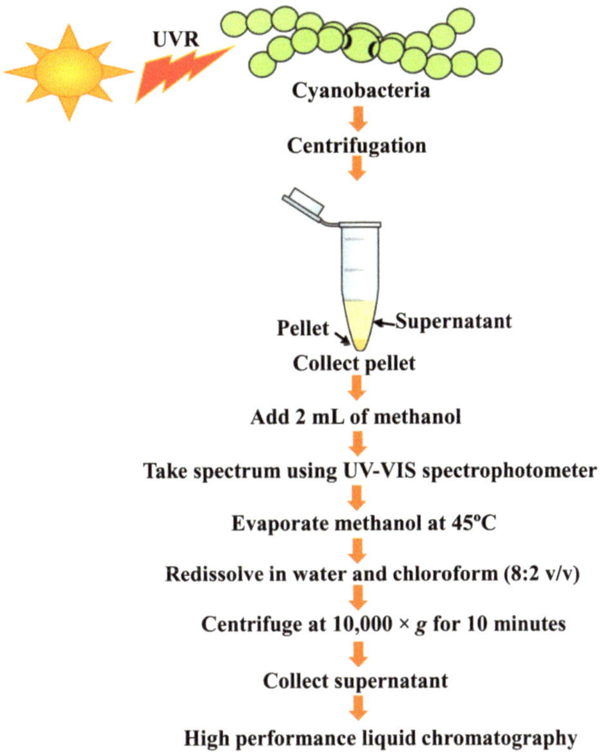

FIGURE 11.1 Flowchart showing the different steps involved in the protocol (color figure in e-book).

2. Cyanobacteria are irradiated with UV radiation (UV-B, UV-A and PAR), and the temperature of the chamber is maintained at a constant rate to avert a heating effect. Stir the culture with a glass rod at regular time intervals to avoid clump formation.
3. After the desired time (hereafter 24 hours and 36 hours) of exposure, take 2 mL of sample and collect cells by centrifugation at 10,000 × *g* for 10 minutes at room temperature.
4. Collect the cell pellet and extract MAAs overnight in 2 mL of 100% HPLC-grade methanol at 4°C.
5. Measure the spectrum with a UV-VIS spectrophotometer to confirm the presence of MAAs in methanolic extract after centrifugation at 10,000 × *g* for 10 minutes at room temperature.
6. For UV-VIS spectrophotometry, take a 3.5-mL quartz cuvette and add 2 mL of sample. Use methanol as a reference, and the spectrum is measured between 250 and 700 nm (Figure 11.2).
7. Evaporate the sample by placing it into a watch glass until complete dryness occurs at 45°C. One can also use a vacuum evaporator, if available.
8. After complete evaporation, dissolve the extract in 80% DDW + 20% chloroform to separate the water-soluble compounds from organic-soluble compounds.
9. After vigorous mixing, centrifuge the sample at 10,000 × *g* for 15 minutes.
10. Carefully collect the upper aqueous phase containing water-soluble MAAs and further purify by HPLC.
11. For HPLC analysis, filter the sample through a microcentrifuge syringe-driven filter (0.22 µm) while the solvent (see recipe) is filtered through the micropore filter (0.45 µm). Thereafter, inject the sample (20 µL) through the autosampler, and absorbance is observed at 334 nm (Figure 11.3). Use 0.02% v/v acetic acid as a mobile phase.

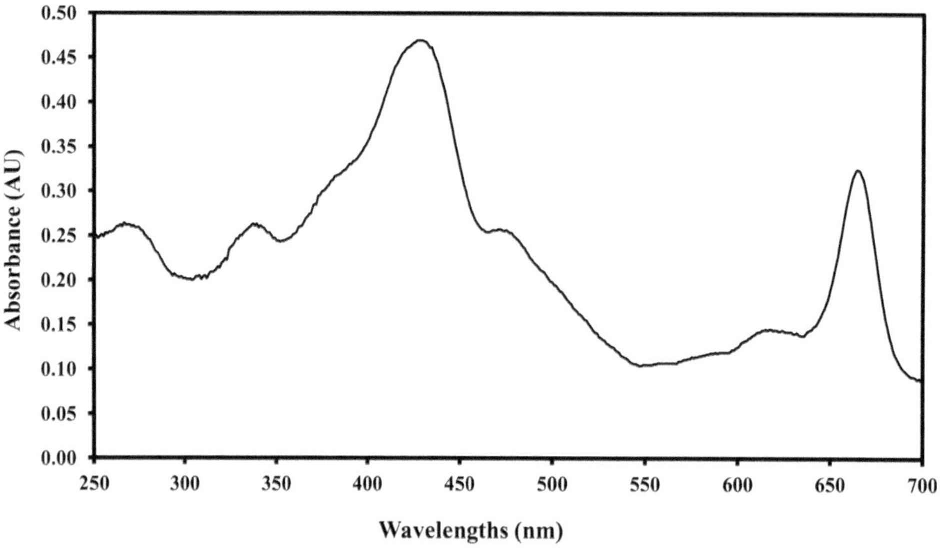

FIGURE 11.2 Absorption spectrum of methanolic extract of cyanobacterium *Scytonema* sp. strain HKAR-16.

11.6 DATA ANALYSIS

The UV-irradiated cells were extracted in methanol and analyzed with a UV-VIS spectrophotometer (250–700 nm). The absorption spectrum of the methanolic extract of the cyanobacterial sample showed a peak in the range of 309–365 nm, which suggested the presence of MAA in the sample. In addition to the peak for MAA, the methanolic extract also showed the peaks for chlorophyll *a* (420 and 660 nm), carotenoids (474 nm) and phycobiliproteins (620 nm) (Figure 11.2).

HPLC analysis was performed with the partially purified MAA. The HPLC chromatogram showed a peak with retention time (RT) corresponding to MAA (Figure 11.3a and 11.3b). Identification and purification of the photoprotective compound MAA porphyra-334 in the sample was done based on RT and absorption maxima (λ_{max} ~334 nm; RT 1.68 min).

11.7 NOTES/PRECAUTIONS

1. All solutions were filtered through a 0.2-μm pore size sterilized microcentrifuge syringe–driven filter before use.
2. The solvent prepared for HPLC was kept in an axenic environment to avoid bacterial contamination and could be stored for more than 1 month (freshly prepared solvent is preferred).

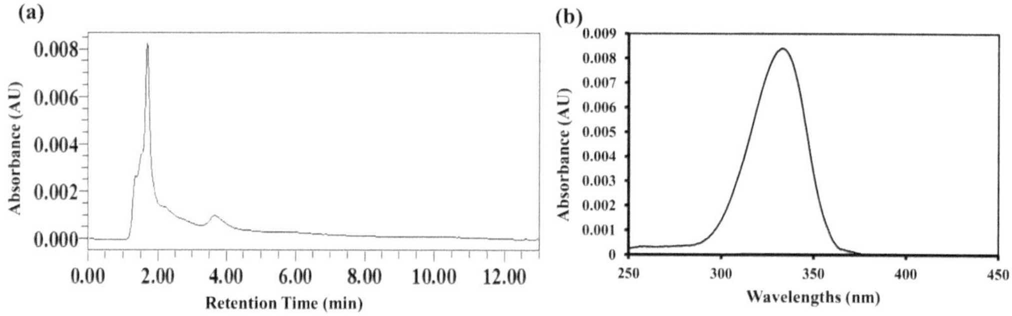

FIGURE 11.3 HPLC chromatogram and its corresponding absorption spectrum of the purified MAA porphyra-334 showing a peak at 334 nm. (Adapted from Mishra et al. 2019b.)

ACKNOWLEDGMENTS

Sonal Mishra and Jyoti Jaiswal are thankful to the University Grants Commission (UGC), New Delhi, India, for the financial assistance in the form of a junior research fellowship. Neha Kumari is grateful to CSIR, New Delhi, India, for providing financial support in the form of a senior research fellowship.

REFERENCES

Abed RM, Dobretsov S and Sudesh K (2009) Applications of cyanobacteria in biotechnology. *Journal of Applied Microbiology* 106: 1–12.

Carreto JI, Carignan MO, Daleo G and Marco SD (1990) Occurrence of mycosporine-like amino acids in the red-tide dinoflagellate *Alexandrium excavatum*: UV-photoprotective compounds? *Journal of Plankton Research* 12: 909–921.

Garcia-Pichel F and Castenholz RW (1993) Occurrence of UV-absorbing, mycosporine-like compounds among cyanobacterial isolates and an estimate of their screening capacity. *Applied and Environmental Microbiology* 59: 163–169.

Häder D-P, Williamson CE, Wängberg SÅ, Rautio M, Rose KC, Gao K, Helbling EW, Sinha RP and Worrest R (2015) Effects of UV radiation on aquatic ecosystems and interactions with other environmental factors. *Photochemical and Photobiological Sciences* 14: 108–126.

Hu C, Völler G, Süßmuth R, Dittmann E and Kehr JC (2015) Functional assessment of mycosporine-like amino acids in *Microcystis aeruginosa* strain PCC 7806. *Environmental Microbiology* 17: 1548–1559.

Kageyama H and Waditee-Sirisattha R (2018) Mycosporine-like amino acids as multifunctional secondary metabolites in cyanobacteria: From biochemical to application aspects. *Studies in Natural Products Chemistry* 59: 153–194.

Matsui K, Nazifi E, Kunita S, Wada N, Matsugo S and Sakamoto T (2011) Novel glycosylated mycosporine-like amino acids with radical scavenging activity from the cyanobacterium *Nostoc commune. Journal of Photochemistry and Photobiology B: Biology* 105: 81–89.

Mishra S, Pandey A, Ahmed H, Kumar D, Singh V and Sinha RP (2019a) Mycosporine-like amino acids (MAAs) profile of cyanobacteria from different historical kunds of Varanasi, India. *International Journal of Applied Sciences and Biotechnology* 7: 317–326.

Mishra S, Pandey A and Sinha RP (2019b) Characterization of photoprotective and pharmaceutically important compounds from cyanobacteria and algal assemblages from historical kunds of Varanasi, India. *Journal of Chemical and Pharmaceutical Research* 11: 24–36.

Rastogi RP and Incharoensakdi A (2014) Characterization of UV-screening compounds, mycosporine-like amino acids, and scytonemin in the cyanobacterium *Lyngbya* sp. CU2555. *FEMS Microbiology Ecology* 87: 244–256.

Rastogi RP and Sinha RP (2009) Biotechnological and industrial significance of cyanobacterial secondary metabolites. *Biotechnology Advances* 27: 521–539.

Rastogi RP, Sinha RP, Singh SP and Häder D-P (2010) Photoprotective compounds from marine organisms. *Journal of Industrial Microbiology and Biotechnology* 37: 537–558.

Safferman RS and Morris ME (1964) Growth characteristics of the blue-green algal virus LPP1. *Journal of Bacteriology* 88: 771–775.

Wada N, Sakamoto T and Matsugo S (2015) Mycosporine-like amino acids and their derivatives as natural antioxidants. *Antioxidants* 4: 603–646.

Waditee-Sirisattha R, Kageyama H, Sopun W, Tanaka Y and Takabe T (2014) Identification and upregulation of biosynthetic genes required for accumulation of mycosporine-2-glycine under salt stress conditions in the halotolerant cyanobacterium *Aphanothece halophytica. Applied and Environmental Microbiology* 80: 1763–1769.

Williamson CE, Zepp RG, Lucas RM, Madronich S, Austin AT, Ballaré CL, Norval M, Sulzberger B, Bais AF, McKenzie RL and Robinson SA (2014) Solar ultraviolet radiation in a changing climate. *Nature Climate Change* 4: 434–441.

12 Protein Estimation in Cyanobacterial Samples

Sonal Mishra, Jyoti Jaiswal, and Rajeshwar P. Sinha

12.1 INTRODUCTION

A larger portion of the biomass of actively growing microalgae and cyanobacteria consists of protein; despite this, they are commonly underestimated in comparison to minor products like omega fatty acids. The protein content of the cyanobacterial biomass is estimated either by methods involved in the measurement of elemental nitrogen concentration or by colorimetric methods (Lowry et al. 1951; Bradford 1976; Smith et al. 1985). Colorimetric analysis depends on the presence of a chromophore due to either involvement of the protein in a redox reaction or a dye may bind to protein. Estimation of protein by a colorimetric method has been reviewed by Sapan et al. (1999).

Colorimetric tests show sensitivity toward intervention (Peterson 1979), and their efficiency relies extremely on the procedures involved in the pretreatment of the sample (Guerlava et al. 1998). The Lowry method is one of the most accurate methods for the quantification of proteins (Peterson 1979). This method involves comparing the spectrophotometric absorbance of the unknown sample at a wavelength of 750 nm to a calibration curve prepared using standard solutions of a protein. Bovine serum albumin (BSA) is the most commonly used standard.

For the accurate measurement of microbial protein content, the pretreatment of cells is required to fully discharge the intracellular proteins. In pretreatments, mainly the cells are disrupted by physical or chemical methods (Guerlava et al. 1998; Barbarino and Lourenço 2005). Chemicals like sodium dodecyl sulfate (SDS) and hydrolytic enzymes can be used to lyse the cell walls (Meijer and Wijffels 1998), but cells are not equally susceptible to chemicals and enzymes (Chisti and Moo-Young 1986).

One of the most efficient methods to release intracellular proteins from even the hardiest of microbial cells is granulating cell slurry in the presence of glass beads or fine ceramic particles (Chisti and Moo-Young 1986; Garrido et al. 1994). Bead milling is mainly effective for microalgae (Chisti and Moo-Young 1986). In the laboratory, milling generally involves the use of pestle and mortar. In this chapter we present a method for estimating protein content in cyanobacterial samples.

12.2 EQUIPMENT

1. RNase and DNase free centrifuge tubes (Thermo Fisher Scientific, Nunc™, catalog number 339653).
2. Measuring cylinder (Borosil, catalog number 3021).
3. Aluminum foil.
4. Vortex shaker (Spinx™ Tarsons, model 3020).
5. Refrigerated centrifuge (Remi Electrotechnik, model CM-12 PLUS).
6. 3.5 mL quartz cuvette (Cole-Parmer, JENWAY, catalog number 035028).
7. UV-VIS spectrophotometer (Shimadzu High-Technologies, model UV-2600).

 DOI: 10.1201/9781003398387-12

12.3 MATERIALS

1. Cyanobacterial culture.
2. Sodium carbonate (Sigma-Aldrich, CAS number 497-19-8).
3. Sodium hydroxide (Sigma-Aldrich, CAS number 1310-73-2).
4. Copper (II) sulfate (Sigma-Aldrich, CAS number 7758-98-7).
5. Potassium sodium tartrate tetrahydrate (Sigma-Aldrich, CAS number 9381-59-5).
6. Folin and Ciocalteu's phenol reagent (Sigma-Aldrich, catalog number F9252).
7. BSA (Sigma-Aldrich, CAS number 9048-46-8).
8. Double distilled water (DDW).

12.4 RECIPES

12.4.1 Complex-Forming Reagent

Prepare immediately before use by mixing the following stock solutions A and B in the proportion of 50:1 (v/v), respectively.

Solution A: 2% (w/v) sodium carbonate in 0.1 N sodium hydroxide.
Solution B: 0.5% (w/v) copper sulfate in 1% sodium potassium tartrate.

12.4.2 Folin and Ciocalteu's Phenol Reagent

Folin and Ciocalteu's phenol reagent is prepared by mixing with distilled water in the ratio 1:1 (v/v).

12.5 METHODS

1. A flowchart of protein estimation from cyanobacterial culture is shown in Figure 12.1.
2. Take 5 mL of cyanobacterial culture in a sterile centrifuge tube with the help of a measuring cylinder.

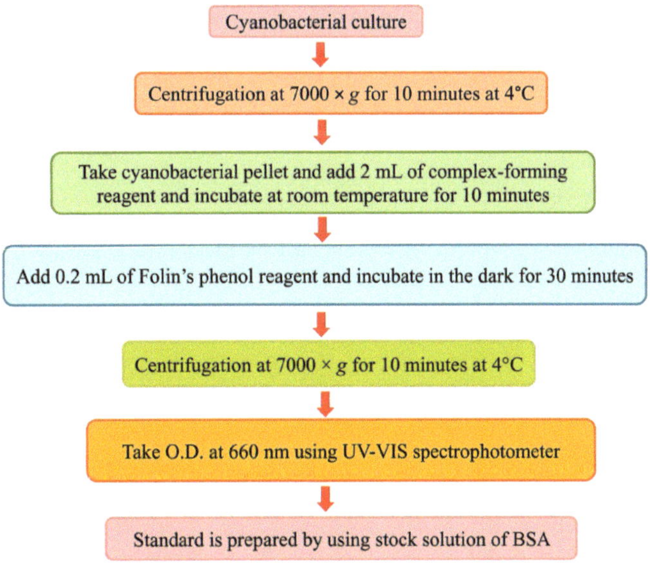

FIGURE 12.1 Steps involved in the estimation of protein content from cyanobacterial samples (color figure in e-book).

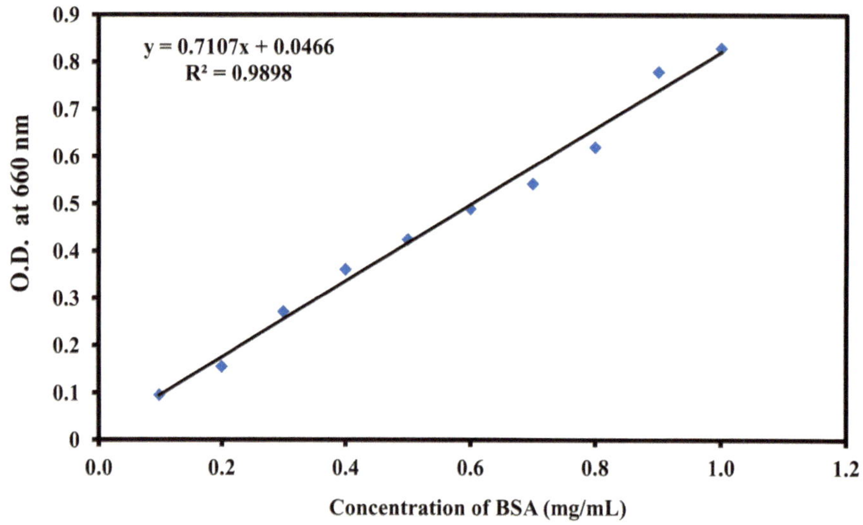

FIGURE 12.2 Standard plot of BSA protein for the estimation of protein content in cyanobacterial samples.

3. Cyanobacterial cells are harvested by centrifugation at 7000 × *g*, 4°C for 10 minutes, and the supernatant is discarded.
4. Add 2 mL of complex-forming reagent to the cyanobacterial pellet followed by vortex and incubation at room temperature for 10 minutes.
5. Add 0.2 mL of Folin and Ciocalteu's phenol reagent and incubate in the dark for 30 minutes.
6. The supernatant is collected after centrifugation at 7000 × *g*, 4°C for 10 minutes, and absorbance is measured at 660 nm using a UV-VIS spectrophotometer.
7. Stock solution of a standard protein is prepared using BSA containing 2 mg/mL BSA protein in distilled water.
8. Plot a standard curve of absorbance by diluting the stock solution of BSA with distilled water in the concentration range of 0.1–1.0 mg/mL of protein (Figure 12.2).
9. The concentration of unknown protein is calculated using a standard graph.

12.6 DATA ANALYSIS

Concentration of an unknown protein is estimated using the standard plot of BSA.

12.7 NOTES/PRECAUTIONS

1. The incubation time is very critical for a reproducible assay.
2. The reaction is also dependent on pH, and a working range of pH 9–10.5 is essential.
3. While performing the experiment, handle chemicals carefully.

ACKNOWLEDGMENTS

Sonal Mishra (Joint CSIR-UGC JRF-2019/NTA Ref. No.: 191620046790) and Jyoti Jaiswal (926/CSIR-UGC-JRF DEC, 2018) are thankful to the University Grants Commission, New Delhi, India, for the financial assistance in the form of junior research fellowships.

REFERENCES

Barbarino E and Lourenço SO (2005) An evaluation of methods for extraction and quantification of protein from marine macro- and microalgae. *Journal of Applied Phycology* 17: 447–460.

Bradford MM (1976) A rapid and sensitive method for the quantitation of microgram quantities of protein utilizing the principle of protein dye binding. *Analytical Biochemistry* 72: 248–254.

Chisti Y and Moo-Young M (1986) Disruption of microbial cells for intracellular products. *Enzyme and Microbial Technology* 8: 194–204.

Garrido F, Banerjee UC, Chisti Y and Moo-Young M (1994) Disruption of a recombinant yeast for the release of galactosidase. *Bioseparation* 4: 319–328.

Guerlava P, Izac V and Tholozan J (1998) Comparison of different methods of cell lysis and protein measurements in *Clostridium perfringens*: Application to the cell volume determination. *Current Microbiology* 36: 131–135.

Lowry OH, Rosebrough NJ, Farr AL and Randall RJ (1951) Protein measurement with the Folin phenol reagent. *Journal of Biological Chemistry* 193: 265–275.

Meijer EA and Wijffels RH (1998) Development of a fast, reproducible and effective method for the extraction and quantification of proteins of micro-algae. *Biotechnology Techniques* 12: 353–358.

Peterson GL (1979) Review of the Folin phenol protein quantitation method of Lowry, Rosebrough, Farr and Randall. *Analytical Biochemistry* 100: 201–220.

Sapan CV, Lundblad RL and Price NC (1999) Colorimetric protein assay techniques. *Biotechnology and Applied Biochemistry* 29: 99–108.

Smith PK, Krohn RI and Hermanson GT (1985) Measurement of protein using bicinchoninic acid. *Analytical Biochemistry* 150: 76–85.

13 Proteomic Analysis of Cyanobacteria Employing Two-Dimensional Gel Electrophoresis

Garvita Singh, Preeti Agarwal, Pritam Kaur, Akanksha Madan, and Rajeshwar P. Sinha

13.1 INTRODUCTION

Photosynthetic Gram-negative cyanobacteria have been developing on Earth for 2.8–3.5 billion years and initiated the existence of aerobic life because of their ability to fix atmospheric CO_2 and evolve O_2. Cyanobacteria are adapted to survive in extremely diverse habitats such as aquatic and terrestrial; from hot springs to polar regions; at extreme salt, pH, and light conditions. Therefore, multiple forms (unicellular and filamentous; free-living and symbiotic; freshwater and marine) can be observed in cyanobacteria (Sánchez-Baracaldo and Cardona 2020).

An additional inherent capability to fix atmospheric N_2 (due to the presence of the prokaryotic enzyme dinitrogenase) makes cyanobacteria ecologically and economically important biofertilizers, especially in nitrogen-deficit soils. In nutrient-deficient or stress-posing environments, cyanobacteria increase their nutrient-scavenging efficiency, initiate metabolism of internal nutrient reserves and limit metabolic activities to a basal level essential for survival, in addition to certain responses specific to the limiting factor (*viz.* elevated synthesis of transport systems and enzymes to increase the cellular availability of the limiting nutrient). Such properties of cyanobacteria make them excellent models to study various stress responses (abiotic/biotic), which can be correlated with higher plant systems (Pade and Hagemann 2015).

Despite ample information on physiological and biochemical responses, there is a scarcity of sensitive tools to capture the impact of minute nutrient fluctuations on cyanobacterial cellular physiology (Xiong et al. 2016). Proteomics is one such breakthrough field involving large-scale screening of proteins of a cell, organism or biological fluid (Figure 13.1). This involves a process that requires stringently controlled steps of sample preparation, two-dimensional electrophoresis (2-DE), image detection and analysis, spot identification and database searches. At present, there is no other technique that is capable of simultaneously resolving thousands of proteins in one separation procedure. The core technique of proteomics is 2-DE, which is unique in its ability to detect post- and co-translational modifications that cannot be predicted from the genome sequence. Applications of 2-D gel electrophoresis include proteome analysis, cell differentiation and detection of disease markers, monitoring therapies, drug discovery, cancer research, purity checks and microscale protein purification.

Two-dimensional polyacrylamide gel electrophoresis (2-D-PAGE) (O'Farrell 1975), where proteins are separated according to the isoelectric point (pI) by isoelectric focusing (IEF) in the first dimension and according to size (MW) by sodium dodecyl sulfate polyacrylamide gel electrophoresis (SDS-PAGE) in the second dimension, has the unique capacity to resolve complex mixtures of proteins, permitting the simultaneous analysis of hundreds or even thousands of gene products.

 DOI: 10.1201/9781003398387-13

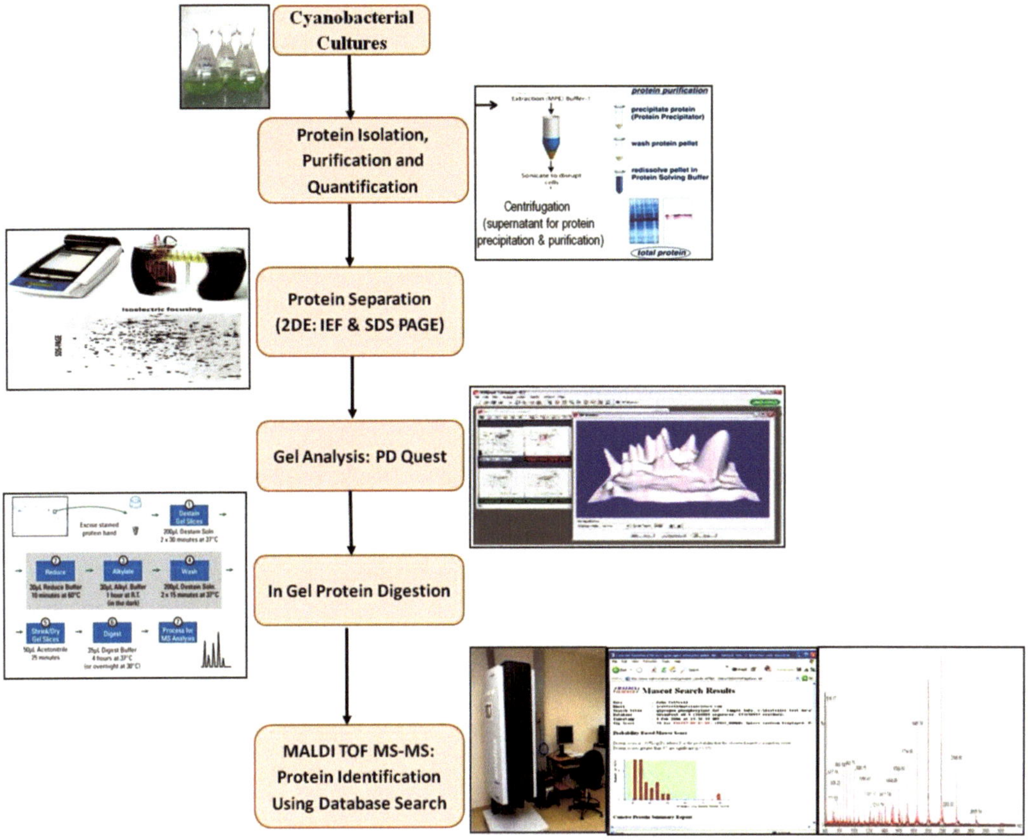

FIGURE 13.1 Flowchart representing the procedure for proteomic analysis of cyanobacteria using two-dimensional gel electrophoresis (2-DE) (color figure in e-book).

The study of proteome dynamics of these microbes can be analyzed through 2-D gel electrophoresis, matrix-assisted laser desorption/ionization time-of-flight mass spectrometry (MALDI-TOF/MS) of trypsin-digested protein fragments and N-terminal amino acid sequencing. Such results and observations can be used for cell differentiation, detection of important metabolic pathways up-regulated and/or down-regulated under specific stress conditions, understanding novel protein expression under stress conditions, monitoring their emergence and genes responsible, drug discovery for exploring novel compounds in medical and agricultural research, with purity checks, and microscale protein purification (Babele et al. 2017; Khalifa et al. 2021).

13.2 EQUIPMENT AND SOFTWARE

1. Sorvall RC-5B refrigerated superspeed centrifuge (DuPont Instruments, USA).
2. Rehydration tray.
3. Ettan IPGphor 3 (GE Healthcare UK Limited).
4. SE 600 ruby multiple gel-electrophoresis unit (Amersham Biosciences, GE Healthcare).
5. AlphaImager gel documentation (Alpha Innotech, USA).

6. MALDI TOF/TOF MS (Autoflex Speed, Bruker Daltonics, Bremen, Germany).
7. Vacuum centrifuge.
8. Ultraviolet (UV) treatment chamber fitted with UV-B (Philips Ultraviolet-B TL 40 W: 12, Philips Lighting, model: TL 40W/12 RS SLV/25 Cat No. 3-4408, Fotodyne, Inc., USA), UV-A (Philips Ultraviolet-A TL 40 W: 12, Philips Lighting, model: TL-K 40W/10-R UV-A) and photosynthetically active radiation (PAR; 55.08 ± 9.18 µmol m^{-2} s^{-1}) (OSRAM L 36 W: 32 Lumilux deluxe warm white and Radium NL 36 W: 26 Universal white, Germany) lamps.
9. PDQuest 2-D analysis software version 8.0.1 (BioRad Laboratories, USA).
10. FlexControl software version 3.3 (Bruker Daltonics, Bremen, Germany).
11. Biotools software version 3.2 (Bruker Daltonics, Germany).

13.3 MATERIALS AND REAGENTS

1. Immobiline DryStrip pH 4–7, 13 cm (GE Healthcare Bio-Sciences AB, Sweden).
2. IPG strip.
3. Focused strips.
4. MALDI target plate (MTP 384 ground steel, Bruker Daltonics, Germany).
5. Sorvall tubes.
6. Glass beads (0.17–0.18 mm).
7. 10 mM HEPES.
8. NaOH pellets for NaOH buffer (pH 7.2).
9. Phenylmethylsulfonyl fluoride (PMSF; stock 100 mM; working 1 mM).
10. Sodium dodecyl sulfate (SDS).
11. Acrylamide (stock 30%; working 1%).
12. Coomassie blue G-250 (stock 5%; working 1%).
13. Bromophenol blue.
14. Urea.
15. Thiourea.
16. 3-([3-cholamidopropyl]dimethylammonio)-1-propanesulfonate (CHAPS) (SigmaAldrich) (2% w/v).
17. α-Cyano-4-hydroxycinnamic acid (CCA) (stock 10%; working 1%).
18. IPG buffer (0.5% v/v).
19. 0.3% (w/v) dithiothreitol (DTT).
20. Trichloroacetic acid (TCA) (stock 100%; working 10–15%).
21. Acetone.
22. 2.5% (w/v) idoacetamide.
23. β-mercaptoethanol.
24. Bradford reagent.
25. Acetonitrile (ACN).
26. Ammonium bicarbonate.
27. Ammonium sulfate.
28. Methanol (stock 100%; working 40–100%).
29. Glycine.
30. Tris base.
31. Tris-HCl.
32. Glycerol.
33. Trifluoroacetic acid (TFA).
34. Peptide calibrant standard (Bruker Daltonics).

13.4 ENZYMES AND KITS

1. Trypsin 25 ng μl^{-1}.
2. 2-D Clean-Up Kit (GE Healthcare, UK).

13.5 RECIPES

1. Colloidal Coomassie blue G-250 dye—Ammonium sulfate (8% w/v), phosphoric acid (0.8% v/v), CBB G-250 (0.08% w/v) and methanol (20% v/v).
2. Bradford's solution—CBB G-250 (0.05% w/v), methanol (25% v/v) and phosphoric acid (42.5% v/v).
3. SDS-PAGE running buffer—Glycine (1.92 M), tris base (250 mM) and SDS (1% w/v).
4. SDS reducing buffer—Tris-HCl (100 mM, pH 6.8), SDS (4% w/v), bromophenol blue (0.2% w/v), glycerol (20% v/v) and DTT/β-mercaptoethanol (200 mM). Store the SDS gel-loading buffer without DTT/β-mercaptoethanol at room temperature (RT). Add DTT/β-mercaptoethanol from a 1 M stock just before the buffer is used.
5. HEPES (10 mM)-NaOH buffer (pH 7.2) supplemented with 1 mM PMSF.
6. 2-D rehydration buffer—Urea (7 M), thiourea (2 M), CHAPS (2% w/v), DTT (0.3% w/v) and IPG buffer (0.5% v/v).
7. Equilibration buffer [50 mM tris/HCl, pH 8.8; 6 M urea; 30% (v/v) glycerol; 2% (w/v) SDS] containing (1) 1% (w/v) DTT in the first part and (2) 2.5% (w/v) iodoacetamide in the second part.
8. In-gel trypsin digestion.
   a. Washing solution—Ammonium bicarbonate (NH_4HCO_3) 100 mM/acetonitrile (1:1 v/v).
   b. Reduction solution—10 mM DTT in 100 mM ammonium bicarbonate.
   c. Alkylation solution—55 mM iodoacetamide solution in 100 mM ammonium bicarbonate.
   d. Extraction solution—Acetonitrile (100%)/TFA (1%) in water (1:1 v/v).
   e. Digestion solution—20 ng/μL trypsin in 25 mM ammonium bicarbonate.

13.6 METHODS

13.6.1 Growth of Test Organisms and UV Treatment

1. *Anabaena* sp., a filamentous, heterocystous cyanobacterium strain L31, was cultured diazotrophically in BG-11 medium in axenic conditions (Rippka et al. 1979).
2. The cyanobacterium was grown in the culture room at 27 ± 2°C and illuminated with white light (Sylvania 40 WT 12 cool fluorescent tubes) at an intensity of 14.4 W m^{-2} for 14 hours of light and 10 hours of dark. All experiments were done with the initial dry weight of culture as 0.1 mg mL^{-1}.
3. The cultures were irradiated with UV radiation at maximum output 312.67 nm (intensity measured by Black-Ray J-221, Long Wave Ultraviolet Intensity Meter [UVP, Inc., San Gabriel, CA]) using UV-B lamps. The survival percentage of cultures with UV-B and white light or UV-B alone was monitored as previously described by Singh et al. 2013.
4. For further analysis, the homogenous culture suspension (200 mL) was divided equally into two sets in two glass Petri dishes—120 mm.
5. Set 1 (treated as an experimental culture) was irradiated for 6 hours, 9 hours and 12 hours with UV-B radiation (1 W m^{-2}) and white light (14.4 W m^{-2}) in a specially fabricated UV chamber. The culture was stirred magnetically to facilitate uniform exposure to light.
6. Set 2 (treated as a control) was exposed to white light (14.4 W m^{-2}) in culture room growth conditions. After irradiation, 30 mL of culture suspension was removed from both sets followed by immediate processing for protein extraction.

13.6.2 Two-Dimensional Polyacrylamide Gel Electrophoresis

13.6.2.1 Preparation of Protein Sample

1. To analyze the proteome profile, the whole-cell protein was extracted by the modified method of Kurian et al. 2006.
2. Cells were harvested from UV-B treated and untreated or control cultures by centrifugation at 10,000 × *g* for 10 minutes at 4°C.
3. The pellet obtained was suspended in 10 mM HEPES-NaOH buffer (pH 7.2) supplemented with 1 mM PMSF, and cells were broken by vortexing using glass beads (0.17–0.18 mm) at 4°C.
4. Glass beads and unbroken cells were then removed via centrifugation at 10,000 × *g* for 20 minutes.
5. The supernatant thus obtained contained cytoplasmic proteins. To this supernatant chilled 10% TCA in acetone was added.
6. The samples were kept at –20°C for 1 hours for protein precipitation, followed by centrifugation at 15,000 × *g* for 30 minutes at 4°C.
7. The pellet obtained was washed thrice with chilled acetone (95% v/v) and then the sample was air-dried and kept at –20°C for further use.
8. Salts and insoluble impurities, if present, were removed using the 2-D Clean-Up Kit.

13.6.2.2 Protein Separation by Sodium Dodecyl Sulfate-Polyacrylamide Gel Electrophoresis

1. First-dimension IEF was performed using an Immobiline DryStrip of 13 cm at pH 4–7.
2. Protein samples (500 µg) dissolved in rehydration buffer [7 M urea, 2 M thiourea, 2% (w/v) CHAPS, 0.3% (w/v) DTT, and 0.5% (v/v) IPG buffer] were loaded onto each IPG strip that was rehydrated at 4°C in the dark overnight (Babele et al. 2015).
3. IEF was completed with Ettan IPGphor 3 after rehydration, and the focusing time was adjusted to 60,000 Vh using the following IEF program: 100 V, 2 hours (gradient); 200 V, 2 hours (gradient); 500 V, 2 hours (gradient); 500 V, 2 hours (linear); 1500 V, 2 hours (gradient); 3000 V; 2 hours (gradient); 6000 V, 2 hours (gradient); 6000 V and 8000 V up to 60,000 Vh (linear).
4. After completion of IEF, IPG strips were equilibrated in buffer [50 mM tris/HCl, pH 8.8; 6 M urea; 30% (v/v) glycerol; 2% (w/v) SDS] containing (1) 1% (w/v) DTT in the first part and (2) 2.5% (w/v) iodoacetamide in the second part (Nouwens et al. 2002).
5. Focused strips were placed for 20 minutes initially in the first part of the equilibration buffer containing DTT followed by 20 minutes in the second part containing iodoacetamide.
6. After the equilibration step, strips were positioned on top of an SDS-polyacrylamide gel (12%) and electrophoresis was performed using an SE 600 Ruby multiple gel-electrophoresis unit at 20 mA per gel (Babele et al. 2015).
7. Three gels were run for each protein sample (UV-B treated and untreated control).
8. Gels were stained using colloidal Coomassie blue G-250 stain (Candiano et al. 2004) and images were photographed by the AlphaImager Gel Documentation unit.

13.6.3 Analysis of 2-D Gels

1. The captured gel images were then further analyzed via PDQuest 2-D analysis software for visualizing the protein spots, quantification, background subtraction and spot matching among gels.
2. Spot quantification in the control (untreated) and UV-B exposed culture (treated) gels was done through spot volumes (intensity × mm^2) as described by Agrawal et al. 2014.

3. The relative spot volume with respect to the UV-B treated and untreated samples at desired time intervals were compared using Student's t-test at statistically significant P values less than 0.05 (Babele et al. 2015).

13.6.4 Protein Spot Digestion and Mass Spectroscopic Analysis

1. The selected spots were excised from the gel via manual picking using a sterile one-touch spot picker into separate microcentrifuge tubes.
2. Prior to MS analysis, CBB-stained gel pieces were distained using the protocol mentioned by Sumner et al. (2002).
3. All the excised protein spots were then digested with trypsin by means of a modified protocol of Shevchenko et al. (1996).
4. The distained gel pieces were dehydrated by incubating them in 100% acetonitrile (ACN) for 15 minutes.
5. Protein spots were reduced by incubating in reduction solution (10 mM DTT in 100 mM ammonium bicarbonate; 50 μL) for 1 hour at 55°C. After cooling to RT, samples were alkylated in 55 mM iodoacetamide solution in 100 mM ammonium bicarbonate for 30 minutes in the dark.
6. Next the gel pieces were washed with 100 μL water followed by dehydration in 100% ACN (100 μL).
7. Dehydrated gel pieces were dried in a vacuum centrifuge to remove traces of ACN and were rehydrated in digestion solution (20 ng/μL trypsin in 25 mM ammonium bicarbonate) by incubating overnight at 37°C.
8. Peptides were extracted from the gel bands/spots as described by Sprenger and Horrevoets (2007).
9. The peptides thus extracted were subjected to a MALDI TOF/TOF MS analysis. MALDI-TOF MS analysis of protein spots was done commercially from the Centre for Genomic Application (CGA), New Delhi.
10. For mass spectra, 1 μL of alpha cyano-cinnamic acid (CCA) matrix (10 mg mL^{-1} in 50% ACN and 0.1% TFA) was mixed with 1 μL of digested protein samples and spotted on the MALDI target plate (MTP 384 ground steel) and was allowed to air dry so that crystals formed.
11. Data acquisition was done using FlexControl software. All MS survey scans were acquired over the mass range m/z 700-4000 Da in the reflection positive-ion mode and accumulated from 2500 laser shots.
12. The MS spectra were calibrated externally using Peptide Calibrant Standard mass ranging between 757.3991 and 3147.4710 Da and calibrated internally using trypsin autolytic products (m/z 842.509, 1045.564 and 2211.104).
13. The data were analyzed using Flex Analysis software.
14. Peptide precursor ions corresponding to contaminants, including keratin and the trypsin autolytic products, were excluded in a mass tolerance of 100 ppm.
15. The filtered precursor ions were searched using the Mascot (http://www.matrixscience.com) online web server for a peptide mass fingerprint (PMF) search and the matched peaks were subjected to further fragmentation.
16. Fragmentation of precursor ions was done using the LIFT positive mode. MS/MS spectra were accumulated from 4000 laser shots.
17. The resultant MS and MS/MS spectra were added using Biotools software, and again an MS-MS search was performed using Mascot. The parameters used for the Mascot search were as follows: Database used NCBInr; taxonomy other bacteria; peptide mass tolerance ±100 ppm; MS/MS tolerance ±1.0 Da; partial miscleavage 0; fixed modification carbamidomethyl of cysteine; variable modification oxidation of methionine. Proteins with a Mascot score >77 (at $p < 0.05$) were considered a significant hit.

13.7 DATA ANALYSIS

13.7.1 Database Search and Bioinformatics Analyses

1. The proteins were identified by comparing PMFs at the NCBInr database using the Mascot search engine (http://www.matrixscience.com). The search parameters allowed for oxidation of methionines, carbamidomethylation of cysteines, one miscleavage of trypsin and ± 30 ppm mass accuracy.
2. The proteins were identified based on the first ranking result and Mascot scores >74, which indicated that the hits were significant.
3. After obtaining protein sequences, a database of *Nostoc* sp. PCC 7120 was used as a reference to analyze the physicochemical properties using the Protparam tool (web.expasy.org/protparam/).
4. To identify the signal transduction pathway for hypothetical proteins, a KEGG BLAST search (http://www.genome.jp/tools/blast/, http://biocyc.org/blast.htmL, http://www.genome.jp/kegg/genes.htmL) was used.
5. To predict the putative signal peptides and their cleavage sites, SignalP 4.0 Server (www.cbs.dtu.dk/services/SignalP/) was used, and to predict transmembrane helices and topology of proteins, HMMTop (www.enzim.hu/hmmtop/) was used.
6. For the elucidation of the protein family, InterProScan (version 4.8) server (http://www.ebi.ac.uk/Tools/pfa/iprscan/) was used.
7. Similarity, a search of hypothetical proteins against different species from the National Centre for Biotechnology Information (NCBI) database was performed by Basic Local Alignment Search Tool (BLAST) server (http://blast.ncbi.nlm.nih.gov/Blast.cgi), and the tree was constructed by the Unweighted Pair Group Method with the Arithmetic Mean (UPGMA) method.
8. In addition, similar proteins for the hypothetical proteins were used as input for motif elucidation (highly conserved regions) by using the MEME tool (https://meme-suite.org/meme/).
9. To analyze the promoter/*cis*-acting elements of hypothetical proteins (all3797, all4050), upstream sequences were retrieved using genome blasts (http://genome.kazusa.or.jp/cyanobase). An untranslated upstream region was taken as an input sequence to detect the promoter/*cis*-acting elements using the BPROM tool from the Softberry server (http://www.softberry.com/).
10. Detection of post-translational modification sites was performed using the CBS prediction server (http://www.cbs.dtu.dk/services/NetPhosBac-1.0/).
11. Protein structure modeling and stereochemical properties of hypothetical proteins were performed by using CPHModels (http://www.cbs.dtu.dk/services/CPHmodels/) and PDBSum (www.ebi.ac.uk/pdbsum/).

13.8 NOTES/PRECAUTIONS

1. Always wear gloves when handling IPG strips, SDS polyacrylamide gels, gel buffer strips or any equipment that these items will contact. The use of gloves will reduce protein contamination that can produce spurious or false spots or bands in 2-D patterns.
2. Before starting the experiments, cleaning all assemblies involved in gel or sample preparation is recommended and should be rinsed well with distilled water. This is a very critical step when highly sensitive MS techniques are to be employed for spot identification and characterization, as it may affect the results.
3. Always use the highest-quality reagents and double distilled water that is free from impurities.

4. Several chemicals used in the procedures like acrylamide, N,N'-methylenebisacrylamide, TEMED, ammonium persulfate, SDS and acids are extremely hazardous and should be carefully handled, taking all necessary precautions mentioned in manuals provided with the reagents.

ACKNOWLEDGMENTS

The authors acknowledge the DST, Govt. of India, New Delhi (project SR/SO/PS-49/09) for financial support. We thank Prof. Ashok Kumar, School of Biotechnology, Banaras Hindu University, Varanasi, for his continuous support and guidance in providing instruments and lab facilities. The authors also thank Gargi College, University of Delhi, for support during manuscript preparation.

REFERENCES

Agrawal C, Sen S, Singh S, Rai S, Singh PK, Singh VK and Rai LC (2014) Comparative proteomics reveals association of early accumulated proteins in conferring butachlor tolerance in three N_2-fixing *Anabaena* spp. *Journal of Proteomics* 96: 271–90.

Babele P, Singh G, Kumar A and Tyagi M (2015) Induction and differential expression of certain novel proteins in *Anabaena* L31 under UV-B radiation stress. *Frontiers in Microbiology* 6: 133.

Babele P, Singh G, Singh A, Kumar A, Tyagi M and Sinha R (2017) UV-B radiation and temperature stress-induced alterations in metabolic events and defense mechanisms in a bloom-forming cyanobacterium *Microcystis aeruginosa*. *Acta Physiologiae Plantarum* 39: 1–11.

Candiano G, Bruschi M, Musante L, Santucci L, Ghiggeri GM, Carnemolla B, Orecchia P, Zardi L and Righetti PG (2004) Blue silver: A very sensitive colloidal Coomassie G-250 staining for proteome analysis. *Electrophoresis 25*: 1327–1333.

Khalifa SA, Shedid ES, Saied EM, Jassbi AR, Jamebozorgi FH, Rateb M and El-Seedi HR (2021) Cyanobacteria—From the oceans to the potential biotechnological and biomedical applications. *Marine Drugs* 19(5): 241.

Kurian D, Phadwal K and Maenpaa P (2006) Proteomic characterization of acid stress response in *Synechocystis* sp. PCC 6803. *Proteomics* 6: 3614–3624.

Nouwens AS, Willcox MDP, Walsh BJ and Cordwell SJ (2002) Proteomic comparison of membrane and extracellular proteins from invasive (PAO1) and cytotoxic (6206) strains of *Pseudomonas aeruginosa*. *Proteomics* 2: 1325–1346.

O'Farrell PH (1975) High-resolution two-dimensional electrophoresis of proteins. *Journal of Biological Chemistry* 250(10): 4007–4021.

Pade N and Hagemann M (2015) Salt acclimation of cyanobacteria and their application in biotechnology. *Life 5*(1): 25–49.

Rippka R, Deruelles J, Waterbury JB, Herdman M and Stainer RY (1979) Generic assignments, strain histories and properties of pure cultures of cyanobacteria. *The Journal of General Microbiology* 111: 1–6.

Sánchez-Baracaldo P and Cardona T (2020) On the origin of oxygenic photosynthesis and cyanobacteria. *New Phytologist* 225(4): 1440–1446.

Shevchenko A, Wilm M, Vorm O and Mann M (1996) Mass spectrometric sequencing of proteins silver-stained polyacrylamide gels. *Analytical Chemistry* 68(5): 850–858.

Singh G, Babele PK, Sinha RP, Tyagi MB and Kumar A (2013) Enzymatic and non-enzymatic defense mechanisms against ultraviolet-B radiation in two *Anabaena* species. *Process Biochemistry* 48(5–6): 796–802.

Sprenger RR and Horrevoets AJG (2007) Proteomic study of caveolae and rafts isolated from human endothelial cells. *Methods in Molecular Biology* 357: 199–213.

Sumner LW, Wolf-Summer B, White SP and Asirvatham VS (2002) Silver stain removal using H_2O_2 for enhanced peptide mass mapping by matrix assisted laser desorption/ionization time-of-flight mass spectrometry. *Rapid Communication in Mass Spectrometry* 16: 160–168.

Xiong Q, Chen Z and Ge F (2016) Proteomic analysis of post-translational modifications in cyanobacteria. *Journal of Proteomics* 134: 57–64.

14 LC-MS/MS–Based Proteomics Study of Cyanobacteria

Soumila Mondal, Subodh Kumar, Phulen Sarma, Bikash Medhi, Anjali Gupta, Sapna Tiwari, Pankaj K. Maurya, and Shailendra P. Singh

14.1 INTRODUCTION

Cyanobacteria are a monophyletic group of Gram-negative, oxygen-producing photosynthetic microorganisms that are well known for their significant contribution to global CO_2 and N_2 fixation (Zahra et al. 2020). Several cyanobacteria can be easily genetically modified and have simple growth requirements as compared to algae and plants (Berla et al. 2013; Singh and Sinha 2020). A cyanobacterial biomass of a wild type as well as genetically engineered strains have commercial and industrial importance for the production of various secondary metabolites such as long-chain alcohols, alkanes, fatty acids, ethylene, 2,3-butanediol, ethanol, polyhydroxybutyrate and hydrogen (Knoot et al. 2018). Due to the obligatory requirement of light for their growth, cyanobacteria are exposed to a dynamic light environment, both diurnally and seasonally in their natural habitat, as well as in large-scale cultivation systems (Ooms et al. 2016). In addition to light, several other abiotic factors such as salinity, drought, temperature, pH and nutrient availability affect their overall fitness, growth and development. Also, installation of novel metabolic pathways and associated metabolite production could alter the overall regulatory network of the organism and its response to environmental factors (Ungerer et al. 2018; Gupta et al. 2020). Therefore, transcriptomics and proteomics studies have become important tools for the identification and quantification of transcriptomes and proteomes of native as well as genetically engineered strains to understand their response to different environmental conditions. However, information obtained from transcriptomic study may not necessarily translate to a protein level. Therefore, proteomic studies have emerged as a powerful tool to understand the total protein complement of any genome and its role as a part of a complex network system (Aebersold and Mann 2003; Karpievitch et al. 2010).

Proteomics studies can help to understand the regulation of cellular processes such as proliferation, apoptosis, motility and development and a cellular or organism's response to various stimuli or stressors (Han et al. 2008). The study of the proteome using mass spectrometry (MS) has become a popular high-throughput platform to analyze and quantify accumulation of proteins in different samples (Cox and Mann 2011). Although traditional protein quantification methods like antibodies, gel electrophoresis and labeling with radioactive molecules are still used in biological research, these methods lack molecular specificity and the range of proteome coverage that is far better in the case of MS-based proteomics studies (Hawkridge 2014).

Among different types of MS-based methods such as surface-enhanced laser desorption ionization (SELDI), matrix-assisted laser desorption ionization (MALDI) coupled with time-of-flight (TOF) and liquid chromatography coupled to tandem mass spectrometry (LC-MS/MS) to study

DOI: 10.1201/9781003398387-14

the proteome, LC-MS/MS is most often utilized for online separation of complex protein samples in high-throughput proteomics (Han et al. 2008; Karpievitch et al. 2010). LC-MS/MS–based proteomics studies are widely used to examine post-translational modifications (PTMs), identification and quantification of proteins, identification of different components of protein complexes and intracellular localization of proteins (Han et al. 2008). This technique also allows comparing thousands of proteins and can be used to characterize different biochemical processes to understand systems biology (Cox and Mann 2011).

A schematic diagram of a typical LC-MS/MS instrument is shown in Figure 14.1. The LC-MS/MS technique is a combination of high-performance liquid chromatography (HPLC) and MS. LC is required for the purification, pre-fractionation and separation of proteins prior to their MS-based detection, which involves the conversion of molecules to an ionized state followed by their analysis based on their mass to charge ratio (m/z) (Karpievitch et al. 2010). The HPLC system contains a pump, a sample injector, column, detectors and a recorder. The purpose of the pump is to maintain a constant flow of the mobile phase through the HPLC system. The HPLC column is a stationary phase that consists of silica with carbon chains, the length of which varies between 50 and 300 nm (Figure 14.1). There are different types of detectors used in an HPLC system for the detection of compounds separated by HPLC. Generally, reverse-phase (RP) HPLC is used as an 'LC' component in LC-MS/MS because it provides a

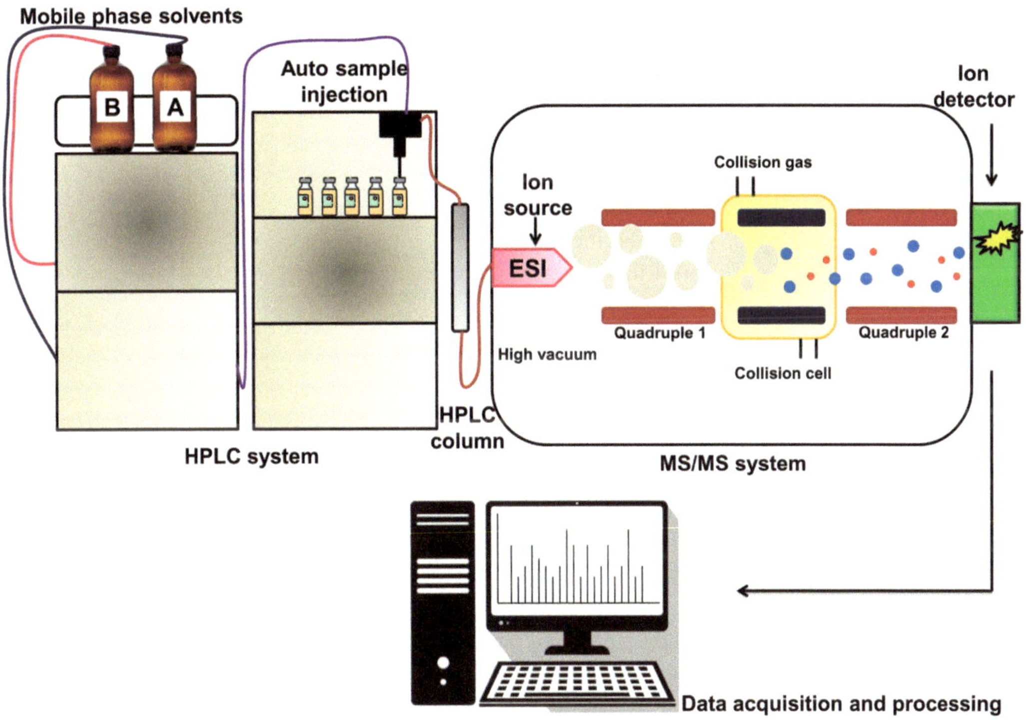

FIGURE 14.1 Schematic representation of a liquid chromatography coupled to a tandem mass spectrometry (LC-MS/MS) system used in proteomics studies. The system includes high-performance liquid chromatography (HPLC) and a tandem mass spectrophotometer (MS/MS). Separation of peptides is achieved by HPLC. After separation, peptides are fragmented into ions and passed through the MS based on their mass to charge (m/z) ratio and finally detected by an ion detector. After completion of the procedure, peptides are assembled and analyzed using data acquisition and processing software (color figure in e-book).

better separation of peptides digested with trypsin. Digested peptides are further transferred into the ion source of an MS (Figure 14.1) where samples can be ionized using different ionization sources such as atmospheric pressure ionization (API), electrospray ionization (ESI), atmospheric-pressure ionization (APCI), thermospray and plasma spray ionization (TSPI) and atmospheric pressure photo ionization (APPI). Generally, ESI is widely used for ionization in proteomics study, which was invented in 1984 (Yamashita and Fenn 1984). Later, nano-electrospray (nESI) was introduced, which enabled the characterization of a low concentration of tryptic peptides by enhancing the sensitivity of ESI (Wilm and Mann 1996). After ionization, ions are further transferred into a mass analyzer, where separation of ions is done based on their mass to charge, i.e., m/z ratio (Figure 14.1).

In proteomics studies, a quadruple analyzer is generally used for separation of ions generated in the MS. Analysis of peptides using a single MS provides information about only a mass of peptides, while peptides are further fragmented into daughter ions in tandem MS (MS/MS), and therefore, provides further information about their amino acid composition. The information obtained from tandem MS also helps in localizing the post-translational or chemical modification of amino acids. In this chapter, we present a method that covers sample preparation, digestion, sample cleanup, data analysis and representation for proteomics studies in cyanobacteria by LC-MS/MS. The method can also be used for proteomics studies in organisms other than cyanobacteria.

14.2 EQUIPMENT

1. Orbi-trap Fusion Mass Spectrophotometer (Thermofisher Scientific, USA).
2. Sonicator (Samarth Electronics, India).
3. Water bath (Borosil, India).
4. Cooling centrifuge (Remi, India).
5. Incubator (Shivam Lab Instruments, India).
6. Nano-Drop (Thermofisher Scientific, USA).
7. Rotary evaporator (Thermofisher Scientific, USA).

14.3 MATERIALS

1. 1× TBS buffer (pH 7.5) (see recipe).
2. 1 mM protease inhibitor (see recipe).
3. Cell lysis buffer (see recipe).
4. 100 mM DTT stock solution (see recipe).
5. 500 mM IAA stock solution (see recipe).
6. MS-grade water (Thermofisher Scientific, catalog number: 85189).
7. pH paper (Thermofisher Scientific, catalog number: 38141).
8. Trypsin stock solution (0.5 $\mu g\ \mu L^{-1}$) (see recipe).
9. 10% TFA stock solution (see recipe).
10. C18 Spin Column (Thermofisher Scientific, catalog number: 89870).
11. Solution A (see recipe).
12. Solution B (see recipe).
13. Solution C (see recipe).
14. 5% ACN, MS grade (see recipe).
15. 50% Methanol, MS grade (see recipe).
16. 1.5 mL microcentrifuge tubes (Abdos, catalog number: P10202)
17. 2.0 mL microcentrifuge tubes (Abdos, catalog number: P10203).
18. 1.5 mL amber microcentrifuge tubes (Abdos, catalog number: P10204A).

19. Micropipettes (Eppendorf, Germany).
20. 1.0 mL pipette tips (Abdos, catalog number: P10109).
21. 200 μL pipette tips (Abdos, catalog number: P10140).
22. 10 μL pipette tips (Abdos, catalog number: P10116).
23. 0.2 μm size Millipore membrane filter (Merck, catalog number: GSWP04700).

14.4 RECIPES

14.4.1 1× Tris Buffer Saline (TBS; pH 7.5)

To prepare 1× TBS, dissolve 6.05 g Tris and 8.76 g sodium chloride (NaCl) in 800 mL of milli-Q water. Adjust the pH to 7.5 with 1 M hydrochloric acid (HCl) and make the final volume to 1 L by adding milli-Q water. TBS can be stored at 4°C for 3 months.

14.4.2 Cell Lysis Buffer (6 M Guanidine Hydrochloride/0.1 M Tris; pH 8.5)

To prepare the cell lysis buffer, dissolve 28.659 g guanidine hydrochloride and 0.6057 g Tris in 50 mL of milli-Q water and adjust the pH to 8.5 with 1 M HCl or 1 M NaOH. The cell lysis buffer can be stored at 4°C for 6 months.

14.4.3 1 mM Protease Inhibitor Solution

To prepare 1 mM protease inhibitor solution, weigh 0.17419 mg phenylmethylsulfonyl fluoride (PMSF) and dissolve it in 1 mL of cell lysis buffer. Dissolve completely by vortexing. Always prepare fresh protease inhibitor solution.

14.4.4 100 mM DTT Stock Solution

Prepare DTT stock solution by dissolving 15.4253 mg of DTT in 1 mL of milli-Q water by vortex and sterilize the stock solution by passing it through a 0.22-μm syringe filter. Make aliquots of 100 μL in 1.5-mL amber microcentrifuge tubes and store at –20°C. The stock solution of DTT can be stored for 1 year.

14.4.5 500 mM IAA Stock Solution

Dissolve 92.48 mg IAA in 1 mL of milli-Q water in 1.5-mL amber microcentrifuge tubes and store at –20°C. Always prepare fresh IAA solution before use.

14.4.6 10% TFA Solution

Add 100 μL of TFA in 900 μL of milli-Q water and mix properly. Keep in the dark at room temperature (RT; 25°C).

14.4.7 Trypsin Stock Solution

Briefly centrifuge the lyophilized recombinant trypsin vial to collect the solid powder at the bottom before removing the cap. Prepare the trypsin solution (0.5 μg/mL) by dissolving the solid (0.1 μg) powder in 200 μL of 1 M HCl (pH 3.0). The prepared trypsin stock solution is stable for 2 years at –20°C.

14.4.8 50% Methanol Solution

Add 25 mL MS-grade methanol in 25 mL of MS-grade water and store at 4°C.

14.4.9 Solution A

Mix 16 mL of MS-grade acetonitrile and 20 μL of MS-grade formic acid in 4 mL of MS-grade water and store at 4°C.

14.4.10 Solution B

Mix 400 μL of MS-grade acetonitrile and 20 μL of formic acid in 19.6 mL of MS-grade water and store at 4°C.

14.4.11 Solution C

Mix 5 mL of MS-grade acetonitrile and 10 μL of MS-grade formic acid in 5 mL of MS-grade water and store at 4°C.

14.4.12 5% Acetonitrile Solution

Mix 250 μL of MS-grade acetonitrile and 10 μL of MS-grade formic acid in 4.75 mL of MS-grade water and store at 4°C.

14.5 METHODS

14.5.1 Sample Preparation

1. Take 100 mL of exponentially growing cyanobacterial cell culture. Centrifuge at 8000 × *g* at 4°C for 20 minutes and remove the supernatant.
2. Wash the cell pellet twice with 1× TBS and centrifuge at 8000 × *g* at 4°C for 20 minutes to obtain a cell pellet after removing the supernatant.
3. Add 100 μL of cell lysis buffer containing 1 mM protease inhibitor to the cell pellet and mix properly by vortexing.
4. Incubate samples immediately at 90°C in a water bath for 20 minutes.
5. Sonicate samples for 1 minutes (2 × 50 kHz pulse for 30 seconds).
6. After sonication, incubate samples at 90°C in a water bath for 5 minutes.
7. Centrifuge samples at 15,000 × *g* at 4°C for 30 minutes.
8. Carefully collect the supernatant without touching the pellet.
9. Estimate the concentration of protein in the collected supernatant using the Bradford or Lowry method. Use lysis buffer as a blank.
10. Store protein samples at –80°C until further processing.

14.5.2 Trypsin Digestion

1. Take 20–25 μg of protein from step 10 in a 1.5-mL amber microcentrifuge tube.
2. Add 2 μL of DTT from stock solution (see recipe). Incubate for 45 minutes in the dark at RT.
3. Add 2 μL of IAA from stock solution (see recipe). Incubate for another 45 minutes in the dark at RT.
4. Dilute the sample 10-fold using MS-grade water.
5. Check the pH of the sample with a pH paper. It should be around 8.0–8.5.
6. Add 2 μL of trypsin from stock solution (see recipe) and incubate samples at 37°C for 20 hours.
7. Add 2–3 μL of 10% TFA. Check the pH with a pH paper. It should be around 2.0. If not, add a few more drops of TFA and maintain the pH at 2.0.

14.5.3 Sample Clean-Up for LC-MS/MS

1. Wash the C18 spin column with 200 µL of 50% methanol (see recipe). Centrifuge at 1000 × *g* for 1 minute. Repeat this step three times.
2. Wash the spin column with 200 µL of solution A (see recipe). Centrifuge at 1000 × *g* for 1 minutes. Repeat this step three times.
3. Equilibrate the spin column with 200 µL of solution B (see recipe). Centrifuge at 1000 × *g* for 1 minutes. Repeat this step three times.
4. Load the digested protein samples into the spin column. Centrifuge at 800 × *g* for 1 minutes. Repeat this step eight times using flow-through. Store flow-through at –20°C until further analysis.
5. Wash the spin column with 200 µL of solution B. Centrifuge at 1000 × *g* for 1 minute. Repeat this step five times.
6. Elute the bound proteins using 150 µL of solution C (see recipe). Centrifuge at 800 × *g* for 1 minute.
7. Elute the bound proteins two times using 150 µL and 100 µL of solution A, respectively. Centrifuge at 800 × *g* for 1 minute.
8. Dry the samples completely using a rotary evaporator for 1–2 hours.
9. Add 30–50 µL of 5% ACN in the dried sample tube. Mix well.
10. Sonicate the samples using a bath sonicator for 1 minute (two cycles of 30 seconds each).
11. Estimate the protein concentration using the Bradford or Lowry method setting in a Nano-Drop.
12. Inject an equal concentration of sample in LC-MS/MS.
13. Wait until the process is completed (Figure 14.2). Identify the obtained peptide sequences with the help of a public database such as Uniprot Proteome or Cyanobase database.

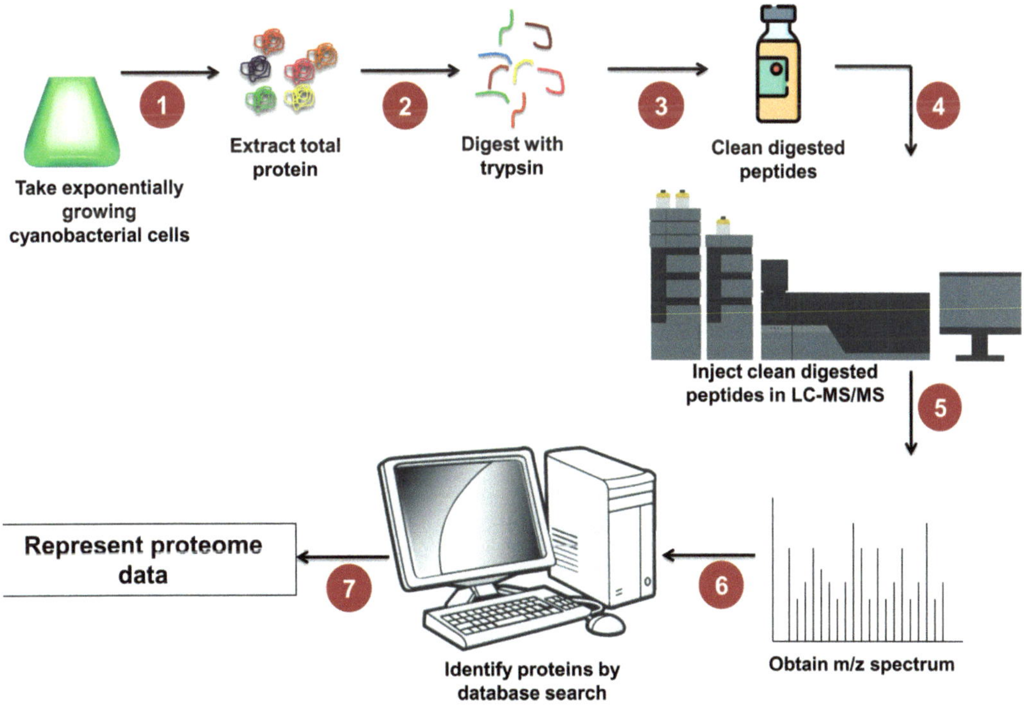

FIGURE 14.2 Schematic diagram of LC–MS/MS–based proteomics study in cyanobacteria. The brief procedure involves sample preparation, protein identification and data analysis (color figure in e-book).

14. After the database search, arrange the identified proteins in an Excel file on the basis of their accumulation in terms of the number of peptides identified for a particular protein. Use the developed file for gene ontology and KEGG pathway analysis using the STRING database.

14.6 DATA ANALYSIS

Proteomics studies using LC-MS/MS provide information about the accumulation of different proteins in any organism, including cyanobacteria. The data obtained from the LC-MS/MS–based experiment can be represented as a list of proteins accumulating under a particular growth condition. Similarly, proteomics data can also be represented as a heat-map picture depicting a differential accumulation of distinct proteins in cyanobacteria grown under different growth conditions. The obtained information can also be used for gene ontology and KEGG pathway analysis, which can further identify biological processes that are affected under different environmental conditions in native or genetically engineered strains of cyanobacteria.

14.7 NOTES/PRECAUTIONS

LC-MS/MS is a highly sensitive analytical machine that can be affected by the presence of impurities in the chemicals that are used for sample preparation. These impurities can also disturb the accuracy and reproducibility of the data. Therefore, only MS-grade chemicals and reagents should be used in LC-MS/MS–based experiments.

ACKNOWLEDGMENTS

This work was supported by funding from the Institute of Eminence Incentive Grant, Banaras Hindu University (R/Dev/D/IOE/Incentive/2021-2022/32399) and Science and Engineering Research Board (SERB), New Delhi (SCP/2022/000201) to Shailendra P. Singh. Pankaj K. Maurya, Soumila Mondal, Anjali Gupta and Sapna Tiwari are thankful to the UGC, New Delhi, India (3616/NET-DEC2014), UGC and BHU, ICMR, New Delhi, India (3/1/3/JRF-2019/HRD-LS) and UGC, New Delhi, India (NTA Ref. No.: 201610202339), respectively.

REFERENCES

Aebersold R and Mann M (2003) Mass spectrometry-based proteomics. *Nature* 422(6928): 198–207.

Berla BM, Saha R, Immethun CM, Maranas CD, Moon TS and Pakrasi HB (2013) Synthetic biology of cyanobacteria: Unique challenges and opportunities. *Frontiers in Microbiology* 4: 246.

Cox J and Mann M (2011) Quantitative, high-resolution proteomics for data-driven systems biology. *Annual Review of Biochemistry* 80: 273–299.

Gupta JK, Rai P, Jain KK and Srivastava S (2020) Overexpression of bicarbonate transporters in the marine cyanobacterium *Synechococcus* sp. PCC 7002 increases growth rate and glycogen accumulation. *Biotechnology for Biofuels* 13(1): 1–12.

Han X, Aslanian A and Yates JR III (2008) Mass spectrometry for proteomics. *Current Opinion in Chemical Biology* 12(5): 483–490.

Hawkridge AM (2014) Practical considerations and current limitations in quantitative mass spectrometry-based proteomics. In: New Developments in Mass Spectrometry: Quantitative Proteomics (Eds. C.E. Eyers and S. Gaskell). The Royal Society of Chemistry, Cambridge, pp. 1–25. ISBN: 978-1-84973-808-8

Karpievitch YV, Polpitiya AD, Anderson GA, Smith RD and Dabney AR (2010) Liquid chromatography mass spectrometry-based proteomics: Biological and technological aspects. *The Annals of Applied Statistics* 4(4): 1797–1823.

Knoot CJ, Ungerer J, Wangikar PP and Pakrasi HB (2018) Cyanobacteria: Promising biocatalysts for sustainable chemical production. *The Journal of Biological Chemistry* 293(14): 5044–5052.

Ooms MD, Dinh CT, Sargent EH and Sinton D (2016) Photon management for augmented photosynthesis. *Nature Communications* 7: 1–13.

Singh SP and Sinha RP (2020) Marine photosynthetic microorganisms: Valuable compound production and bioenergy applications. In: Encyclopedia of Marine Biotechnology (Eds. S.K. Kim). John Wiley & Sons, Inc., Hoboken, pp. 2229–2245. ISBN: 9781119143772

Ungerer J, Wendt KE, Hendry JI, Maranas CD and Pakrasi HB (2018) Comparative genomics reveals the molecular determinants of rapid growth of the cyanobacterium *Synechococcus elongatus* UTEX 2973. *Proceedings of the National Academy of Sciences, USA* 115(50): E11761–E11770.

Wilm M and Mann M (1996) Analytical properties of the nanoelectrospray ion source. *Analytical Chemistry* 68(1): 1–8.

Yamashita M and Fenn JB (1984) Electrospray ion source. Another variation on the free-jet theme. *The Journal of Physical Chemistry* 88(20): 4451–4459.

Zahra Z, Choo DH, Lee H and Parveen A (2020) Cyanobacteria: Review of current potentials and applications. *Environments* 7(2): 13.

15 Monitoring Photosynthetic Performance in Cyanobacteria by a Modulated Fluorometer

Sapna Tiwari, Pankaj K. Maurya, Soumila Mondal, Anjali Gupta, Donat-P. Häder, and Shailendra P. Singh

15.1 INTRODUCTION

Photoautotrophs are the only biological organisms that are capable of reducing atmospheric carbon dioxide (CO_2) using sunlight as a source of energy during photosynthesis. Being the primary producers in the ecosystems, photoautotrophs are a suitable source of energy for higher trophic levels (Bryant 1994). In addition to higher plants, algae and cyanobacteria are the major contributors to the production of oxygen on the Earth. Also, these photosynthetic microorganisms are emerging as a model system for the sequestration of CO_2 and production of biofuels and valuable chemicals in a carbon-neutral manner (Field et al. 1998a; Barber 2009; Maurya et al. 2021). Cyanobacteria, formerly known as blue-green algae, are oxygen-producing, photosynthetic, Gram-negative bacteria that use proteinaceous macrocomplexes called phycobilisomes (PBSs) for harvesting solar radiation to drive photosynthesis. PBSs are composed of chromophore-binding phycobiliproteins such as allophycocyanin (APC; λ_{max} ~650 nm), phycocyanin (PC; λ_{max} ~620 nm) and phycoerythrin (PE; λ_{max} ~565 nm) (Sanfilippo et al. 2019).

Hemidiscoidal-shaped PBSs have core and rod parts, where the core always consists of the APC while the rod part is composed of PC and/or PE. The core part of PBSs is attached to the thylakoid membrane at the site of the photosystem (PS) via anchor polypeptides, while the rod part radiates from the core part (Anderson and Toole 1998; Samsonoff and MacColl 2001). All the PBSs do the job of 'light harvesting', acting as antenna. Similar to algae and plants, cyanobacteria possess PSII and PSI but use PBSs as a major light-harvesting complex (LHC) for the excitation of PSII, while chlorophyll *a* and carotenoids-based LHCs are mainly used for the excitation of PSI (Stirbet et al. 2019; Calzadilla and Kirilovsky 2020). In contrast to cyanobacteria, algae and plants use chlorophylls and carotenoids containing LHCII and LHCI for the excitation of PSII and PSI, respectively (Stirbet et al. 2019; Calzadilla and Kirilovsky 2020). However, under fluctuating light environments, PBSs can move between PSII and PSI (state transitions) for balanced excitation of the two photosystems (Watanabe et al. 2014; Stirbet et al. 2019; Calzadilla and Kirilovsky 2020). The photochemistry in both PSII and PSI brings the electrons to the linear electron transport chain (Z-scheme) and results in the production of NADPH and adenosine triphosphate (ATP) (Blankenship and Prince 1985; Govindjee and Shevela 2011; Govindjee et al. 2017). In this chapter, we describe the application of pulse amplitude modulation (PAM) fluorometry in cyanobacteria to measure non-invasively the *in vivo* chlorophyll *a* fluorescence from which photosynthetic parameters can be calculated.

15.1.1 THE PULSE AMPLITUDE MODULATION (PAM) FLUOROMETER

Due to the low intensity of Chl fluorescence, the PAM is considered the best alternative to detect weak signals of Chl fluorescence. The PAM-2500 (Walz, Germany) instrument consists of two basic

 DOI: 10.1201/9781003398387-15

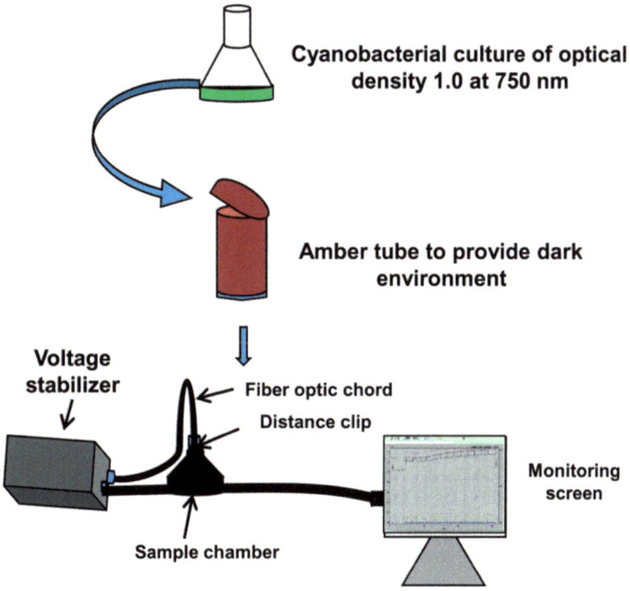

FIGURE 15.1 Schematic diagram of steps involved in measuring photosynthetic parameters of cyanobacterial cells using a pulse amplitude modulated (PAM) fluorometer. The fluorometer used is PAM-2500 (Walz, Germany). The basic components of this setup are an external battery (12 V, 2.1 Ah) and the control unit, which provides the socket for the different components of the setup. The fiber optic chord is connected to the front side of the PAM-2500 control unit with the help of a special plug that resembles an electrical connector. There are three 'fiber pins' with different optical cross-sections which fit into the corresponding holes at the front side of the PAM-2500 housing, where they interface the various light sources and the photodiode detector. The suspension cuvette has the carrying capacity of 500 μL of the sample. The monitoring screen digitally monitors the fluorescent changes in the graphical view (color figure in e-book).

parts, i.e., distance clip and fiber optic cord (Figure 15.1). The distance clip is kept over the sample in such a way that it is at an equal distance and angle away from the sample and supports the fiber optic cord, which functions as a transmitter. This whole setup makes use of a light beam in carrying information from one point to another through a fiber optic cord.

The PAM fluorometer uses three discrete light sources (light-emitting diodes [LEDs]): (1) a non-actinic measuring red light of low intensity <0.1 μmol photons m^{-2} s^{-1}, (2) a red actinic light of up to 4000 μmol photons m^{-2} s^{-1} or blue actinic light of up to 800 μmol photons m^{-2} s^{-1} and (3) a saturation pulse of red light of high intensity, i.e., 25,000 μmol photons m^{-2} s^{-1} for 0.1–0.8 seconds in duration. The organization and physiology of the cyanobacterial thylakoid membrane are typically different from that of the chloroplasts in higher plants, as components of the electron transport chain of both respiration and photosynthesis are located in the same thylakoid membrane of the cyanobacteria (Vermaas et al. 1994). The respiratory enzyme complexes [succinate dehydrogenase and NAD(P)H dehydrogenase] can reduce the photochemical quenching (PQ) pool, which increases the minimum fluorescence (F_0) (Mullineaux and Allen 1986; Schreiber et al. 1995). In addition, the state 1 to state 2 transition, due to reduction of the PQ pool, results in the association of PBS with PSI, which otherwise has a high content of Chl in comparison to that in PSII. Thus, the presence of PBSs and state transitions results in a higher value of F_0 in cyanobacteria (Schreiber et al. 1995; Jallet et al. 2012).

To overcome this problem, far-red light (exciting mainly PSI) for 5 seconds is given first to the dark-adapted cyanobacterial cells followed by a weak-intensity measuring beam of red light to measure F_0. This results in the measurement of the true minimum yield of fluorescence, as cells are photosynthetically 'quiet' while the PSII reaction centers are functionally open to driving photochemistry (PQ reduction) (Stirbet 2011). Further, applying a short flash (0.1–0.8 seconds) of high-light intensity of red light, i.e., a saturation pulse, to dichlorophenyl dimethylurea (DCMU)–treated cells results in the complete reduction of the first stable electron acceptor of PSII, i.e., Q_A. Consequently, all PSII reaction centers are closed and cannot drive photochemistry, while at the same time, the weak intensity of the measuring beam of red light in the background gives the maximum fluorescence F_m (Campbell et al. 1998; Campbell and Oquist 1996; Papageorgiou 1996; Papageorgiou et al. 2007; Govindjee and Shevela 2011; Ogawa and Sonoike 2016; Stirbet et al. 2019).

Table 15.1 shows different photosynthetic parameters that can be monitored by PAM. The difference in the maximum and minimum fluorescence, i.e., $(F_m - F_0)$, is the variable fluorescence (F_v), and the ratio of $F_m - F_0/F_m$, i.e., (F_v/F_m) gives the maximum photosynthetic efficiency of PSII of dark-adapted samples (Baker 2008; Brooks and Niyogi 2011; Stirbet et al. 2019). Thereafter, the continuous actinic light of a certain intensity gives $F_{(t)}$, which is considered F_s when at steady state, i.e., there is no change in the intensity of fluorescence with time (Figure 15.2). After illumination with light for a few minutes, a steady state of photosynthesis is reached where the whole photosynthetic apparatus is in equilibrium, and electron movement from any of the components of the electron transport chain gives an overall photosynthetic rate. The fluorescent signal of chlorophyll at this point indicates the overall performance of photosynthesis (Kalaji et al. 2017). To learn whether the quenching of F_m under steady state is due to PQ or non-photochemical quenching (NPQ), a saturation light pulse is given to close all PSII reaction centers. This causes maximum reduction of the PQ pool, and the maximal fluorescence is indicated as F_m' (Brooks and Niyogi 2011). The difference between the fluorescence yield at F_m and F_m' is due to the NPQ,

TABLE 15.1
Different Parameters Monitored during PAM Analysis

F_0	The minimum fluorescence measured using a measuring beam in the dark-adapted sample (the open state of PSII).
F_m	The maximum fluorescence measured after the saturation pulse (the closed state of PSII).
$F_v = (F_m - F_0)$	The variable fluorescence is the measurement of the fluorescence that changes during the saturation pulse. It signifies the ability of PSII to perform photochemistry.
F_v/F_m	The maximum photosynthetic efficiency of PSII.
F_0'	The minimum fluorescence measured in the light-adapted sample when reaction centers are open.
F_m'	The maximum fluorescence measured in the light-adapted sample when all reaction centers are closed.
$F_v' = F_m' - F_0'$	The variable fluorescence in the light-adapted sample. It signifies the ability of PSII to perform photochemistry in the light-adapted state.
$[F_v'/F_m']$	The maximum photosynthetic efficiency of PSII in a light-adapted sample.
ΦPSII or Y(II) $F_m' - F_s/F_m'$	The operating efficiency of PSII (the efficiency of PSII to utilize the absorbed light for photochemistry).
F_s, F, $F_{(t)}$	F_s is the fluorescence measured at steady state, F is the steady-state fluorescence measured during the light curve and $F_{(t)}$ is the fluorescence at time t.
Y(NPQ)	$(F_{(t)}/F_m') - (F_{(t)}/F_m)$. The collective measure of non-photochemical quenching of Chl *a* fluorescence due to several factors (state transition, thermal dissipation and photoinhibition).
qN	$(F_m - F_0) - (F_m' - F_0')/F_m - F_0$ The coefficient for thermal dissipation.
qP	$F_m' - F_t/F_m' - F_0'$ The coefficient for photochemical quenching.

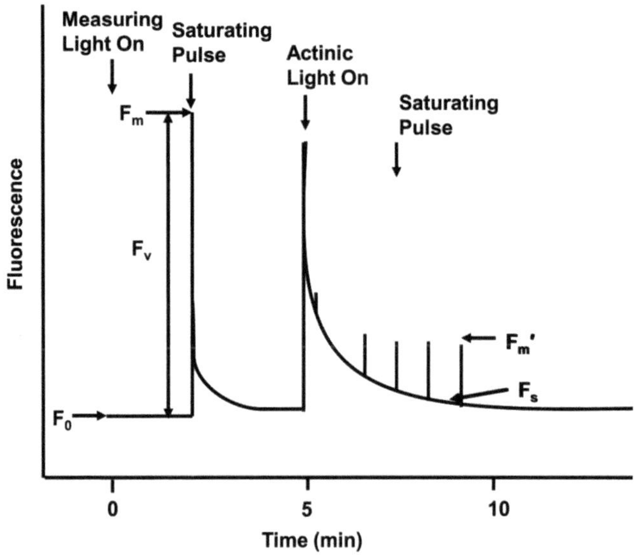

FIGURE 15.2 Graph showing slow kinetics of *Arabidopsis thanliana*, where F_0 is the minimum and F_m the maximum fluorescence for the dark-adapted sample. F_v is the variable fluorescence for the dark-adapted sample, which is determined by the difference between F_m and F_0 ($F_m - F_0$). F_m' is the maximum fluorescence for the light-adapted sample, and F_s is the steady-state fluorescence. (From Brooks and Niyogi 2011.)

while the difference between F_s and F_m' is due to PQ (Figure 15.2). The $F_m' - F_s/F_m'$ gives Y(II), which signifies the operating efficiency of PSII (Cailly 1996; Brooks and Niyogi 2011). Figure 15.2 shows that the maximal fluorescence under actinic light (F_m') is lower than the maximal fluorescence of the dark-adapted state (F_m), which signifies the presence of NPQ processes like thermal dissipation, state transitions or photoinhibition (Baker 2008; Brooks and Niyogi 2011; Stirbet et al. 2019).

The actinic light is the source of energy for the cyanobacterial cells to perform light reactions yielding high-energy compounds ATP and NADPH. The response of cyanobacterial photosynthesis to a range of light intensities is detected by the light response curve, which includes ETR (electron transport rate) and fluorescence ratio parameters. The ETR is the measure of the rate of number of electrons passing through PSII, and it is a product of Y(II) and photosynthetically active radiation (PAR) (Kolber and Falkowski 1993; Schreiber et al. 2012). However, it is difficult to measure the partitioning of actinic light between PSII and PSI, and therefore, a factor of 0.5 is used for the calculation of ETR, which accounts for the almost equal partitioning of energy between PSII and PSI (Kolber and Falkowski 1993; Schreiber et al. 2012).

Any further increase in actinic light increases the rate of electron transport, as light is a limiting factor at low intensities; however, ETR attains saturation at higher intensities of light (Blackman 1905). In cyanobacteria, CO_2 is generally not a limiting factor due to the presence of a CO_2 concentrating mechanism (CCM), and therefore, ETR is considered a function of Y(II) of PSII (Schreiber et al. 2012; Price et al. 2013). However, at higher light intensities, a marked increase in the wasteful processes, such as photoinhibition of PSII and thermal dissipation due to dissociated PBSs, ultimately result in decreased Y(II) (Gwizdala et al. 2016). We note that the absorptance of the cell suspension is not calculated in the case of cyanobacteria, and the light absorption fraction, i.e., 0.84, generally used for leaves, is not included in the ETR calculations (Singh et al. 2013). Instead, the term relative ETR (rETR) is used in the place of ETR to define the rate of the number of electrons passing through the PSII.

15.2 EQUIPMENT AND SOFTWARE

1. PAM-2500 (Walz, Germany).
2. Benchtop centrifuge (Remi, NEYA 16R, India).
3. UV-visible spectrophotometer (Shimadzu, Japan).
4. PAM Win_3 software.

15.3 MATERIALS

1. 2 mL cyanobacterial cell culture adjusted to an optical density of 1.0 at 750 nm (OD_{750}).
2. 2 mL amber Eppendorf tube (Tarsons, catalogue number: 500013D) (for dark environment).
3. 1 mL micropipette (Accupipet Tarsons, India).
4. 1 mL microtip (Tarsons, catalogue number: 521020).

15.4 METHODS

1. Take 2 mL of cyanobacterial cell culture of OD_{750} 1.0. For low-density samples, concentrate the sample to OD_{750} 1.0 by centrifugation at 9000 × *g* for 2 minutes at room temperature. Dilute the sample to OD_{750} 1.0 in case of the high-density samples.
2. Transfer the sample to a 2-mL amber tube and keep it in the dark for 10 minutes for dark adaptation.
3. Open the PAMWin_3 software on the desktop connected to the PAM-2500 chlorophyll fluorometer. Go to 'ADVANCE'. Click on the 'windows' to select the slow kinetics.
4. Put a magnet stirrer in the suspension cuvette. The mixing of the cell culture by the magnet stirrer ensures homogenous exposure of all cells to light. Transfer 500 μL of the dark-adapted sample in the suspension cuvette.
5. Then, expose the cyanobacterial cells to low-pulse frequency of blue or far red light for 5 seconds that leads to a state transition, allowing PBS to attach to the PSII reaction center complex.
6. Press 'start' and 'stop' simultaneously to get the values of F_0, F_m, and F_v/F_m or start the 'slow kinetics' monitoring graph.
7. Take another 500 μL of the sample to measure the light response curve (LRC).
8. The LRC includes measurements of 'rETR' and 'Fluorescence' parameters.
9. To report the data in their original format, disconnect the PAM-2500 from the desktop and go to PAMWin_3 and select the PAM-2500.
10. Export the data as a comma-separated value (CSV) file.

15.5 DATA ANALYSIS

15.5.1 Slow Kinetics Graph

The slow kinetics data of *Synechococcus elongatus* PCC 7942 and *Synechocystis* sp. PCC 6803 are shown in Figures 15.3 and 15.4, respectively. It is important to mention that it is difficult to get true values of F_0 and F_m in cyanobacteria due to excess fluorescence from PSI (Stirbet et al. 2019; Schreiber and Klughammer 2021 and references therein). Therefore, in order to get a true value of F_0 in cyanobacteria, the dark-adapted samples are first illuminated with far-red light for 5 seconds, which is absorbed by PSI and causes oxidation of the PQ pool (Stirbet et al. 2019). This results in the open state of the PSII reaction center and gives the F_0 value for PSII Chl *a* fluorescence. Thereafter, use of a short-saturation flash of <1 second fully reduces the PQ pool and closes the PSII reaction center, which gives the maximum fluorescence (F_m) for dark-adapted samples (Figures 15.3 and 15.4). Further, applying the actinic blue light of 30 μmol photons m^{-2} s^{-1} activates PSII, which

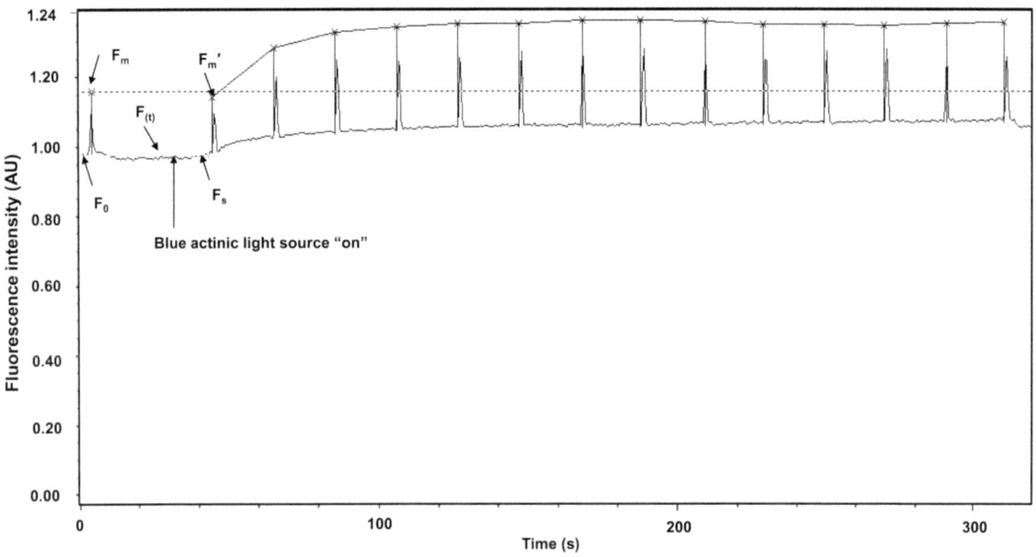

FIGURE 15.3 Time course of Chl *a* fluorescence representing the slow kinetics of the *Synechococcus elongatus* PCC 7942, where blue light (30 μmol photons m^{-2} s^{-1}) was used as the actinic source of light at regular intervals.

induces photochemistry and carotenoid-based NPQ in plants, which results in the decrease of F_m to F_m' (Figure 15.2). However, in contrast to plants, there is a gradual increase in light-adapted maximal fluorescence F_m' in comparison to F_m in *S. elongatus* PCC 7942 at a regular time interval (Figure 15.3). This is due to the fact that the blue actinic light preferentially excites PSI in cyanobacteria, which favor a state II to state I transition, and results in the increase of F_m' at regular intervals due to decoupling of PBSs (Calzadilla and Kirilovsky 2020). *S. elongatus* PCC 7942 lacks orange carotenoid protein (OCP)–based NPQ, and therefore, blue light causes photodamage. The damaged PSIIs dissociate PBSs from the reaction center, which results in higher fluorescence

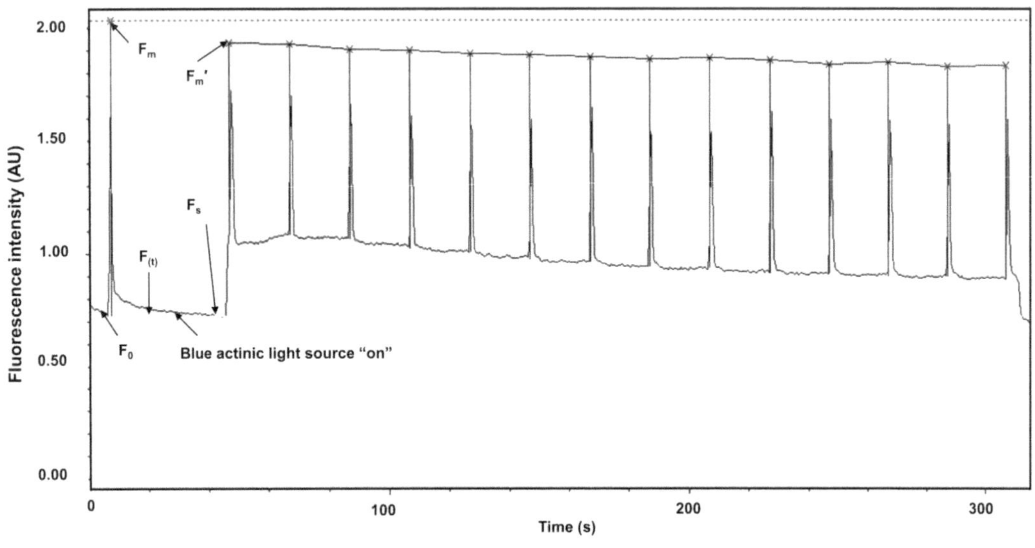

FIGURE 15.4 Time course of Chl *a* fluorescence showing the slow kinetics of the *Synechocystis* sp. PCC 6803, where blue light (30 μmol photons m^{-2} s^{-1}) was used as the actinic source of light at regular intervals.

(Gwizdala et al. 2016; Calzadilla and Kirilovsky 2020). This is also supported by the recent observation where growth under blue light resulted in appressed thylakoid membranes in *S. elongatus* PCC 7942 due to decoupling of PBSs (Maurya et al. 2023) Also, the reduction of the PQ pool from the respiratory electron transport chain and photochemistry gives higher steady-state fluorescence F_s compared to minimum fluorescence F_0 (Vernotte et al. 1979; Field et al. 1998b). However, in the case of *Synechocystis* sp. PCC 6803 there is a decrease in the F_m' at regular intervals, similar to plants (Figure 15.4). In contrast to *S. elongatus* PCC 7942, *Synechocystis* sp. PCC 6803 possesses an OCP-based NPQ, and blue light activates the OCP to dissipate the high energy in the form of heat (Stirbet et al. 2019; Polívka et al. 2005).

15.5.2 Light Response Curve

15.5.2.1 Fluorescence Ratio Parameters

The light response curves (see Figures 15.5 and 15.6) are the graphical representations of the fluorescence-based parameters, which collectively help to estimate the efficiency of the overall photosynthetic process. Figure 15.5 shows the fluorescence ratio parameters of *Synechocystis* sp. PCC 6803, and Figure 15.6 shows the same for *S. elongatus* PCC 7942. It is obvious that *Synechocystis* sp. PCC 6803 shows a better photosynthetic performance than *S. elongatus* PCC 7942 at higher fluxes of red light (Figures 15.5 and 15.6). It is evident from the slow kinetics data (Figure 15.4) that the rate of fluorescence decrease or decrease in the yield of PSII after 47 seconds onwards is slower for *Synechocystis* sp., which suggests its ability to survive at higher fluxes of light than *S. elongatus* PCC 7942 (Figures 15.3 and 15.4). The increase of the F_t or F (from 0.923 arbitrary units onwards) and the parallel decrease of F_m' (from 2.105 923 arbitrary units onwards) with increasing intensities of PAR (Figure 15.5) signify that the yield of PSII slowly decreases at successive fluxes. The reason

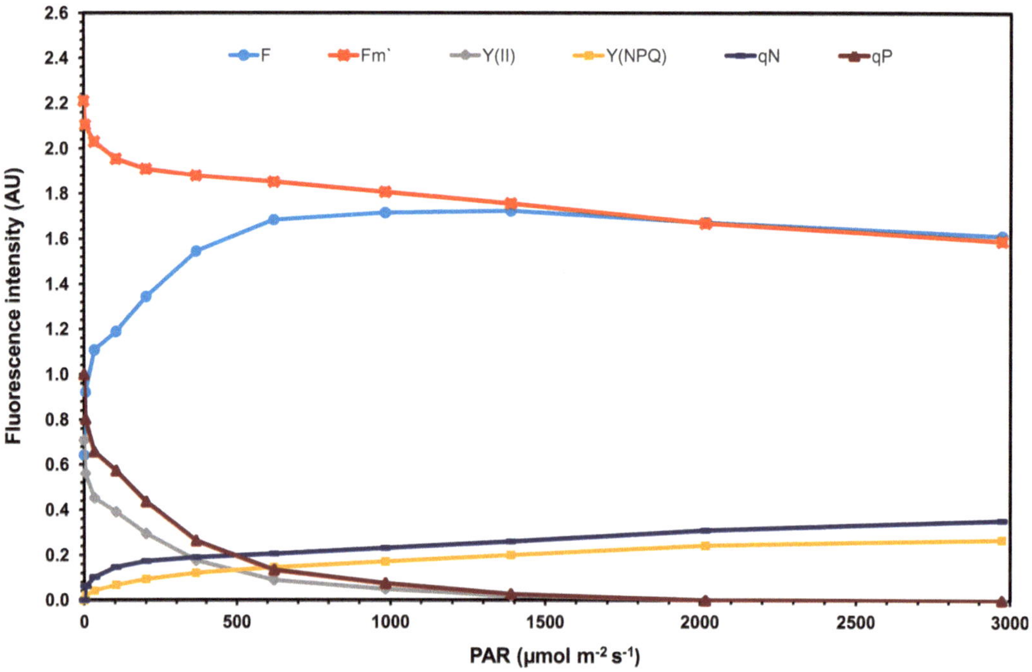

FIGURE 15.5 Light intensity dependence of variations in the fluorescence parameters (shown in the top part of the graph). The abscissa is photosynthetically active radiation (PAR) in micromole photons m^{-2} s^{-1}. The upper graph shows changes in different fluorescence parameters of *Synechocystis* sp. PCC 6803 with increasing PAR – see text for details (color figure in e-book).

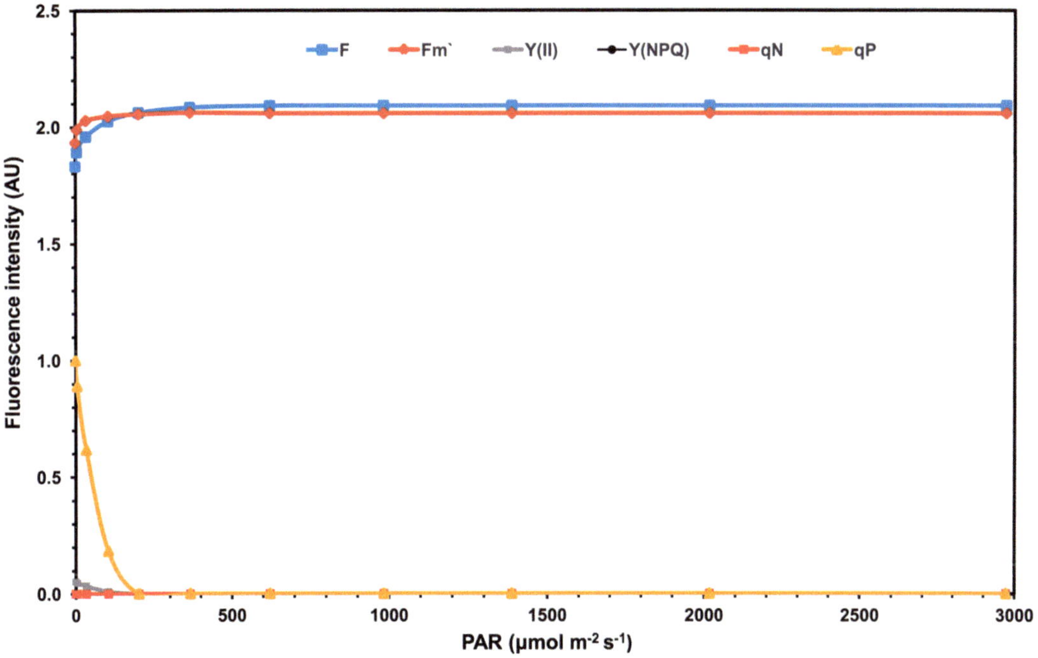

FIGURE 15.6 Light intensity dependence of variations in the fluorescence parameters (shown in the top part of the graph). The abscissa is photosynthetically active radiation (PAR) in micromole photons m^{-2} s^{-1}. The upper graph shows changes in different fluorescence parameters of *Synechococcus elongatus* PCC 7942 with increasing PAR – see text for details (color figure in e-book).

behind the increase of F_t is the faster rate of PQ pool reduction at higher fluxes of light, which consequently results in the increase of the quantum yield of Chl *a* fluorescence. However, the decrease of F_m' is due to both PQ and NPQ quenching. The increase in the coefficient of non-photochemical (qN) and decrease in the coefficient of photochemical quenching (qP) after 32 μmol photons m^{-2} s^{-1} suggest a decrease in Y(II) at higher light intensities (Figure 15.5). This decrease in Y(II) at higher light intensities is accompanied by non-photochemical quenching (indicated by the merging of F_t and F_m' at 2018 μmol photons m^{-2} s^{-1} in Figure 15.5) of Chl *a* fluorescence, which includes state transition, thermal dissipation and photoinhibition (Muller et al. 2001).

This phenomenon, shown in *Synechocystis* sp. PCC 6803, is not applicable to *S. elongatus* PCC 7942 as it lacks OCP. As shown in Figure 15.6, there is a slight decrease in the F_m' after exposure to ~362 μmol photons m^{-2} s^{-1}, but it was never less than F_m. However, the increase of F_m' more than F_m at a higher photosynthetic photon flux density (PPFD) shows that *S. elongatus* PCC 7942 cells do not perform non-photochemical–based quenching. But it has been found that if one applies high-intensity light for a longer period, there is a decrease in the synthesis of PBS in *S. elongatus* PCC 7942 (Gwizdala et al. 2016). Therefore, the progressive decrease in the Y(II) and qP (Figure 15.6) could be correlated with the thermal energy dissipation in a non-photochemical way by the dissociated PBSs (Calzadilla and Kirilovsky 2020). At an intensity of ~200 μmol photons m^{-2} s^{-1}, the decreased values of Y(II) and of qP suggest photoinhibition (Figure 15.6) (Boulay et al. 2008).

15.5.2.2 Relative Electron Transport Rate

The rETR is calculated using the formula:

$$\text{rETR} = \text{PAR}\left(\mu\text{mol photons m}^{-2}\text{s}^{-1}\right) \times \text{Y(II)} \times 0.5 \tag{15.1}$$

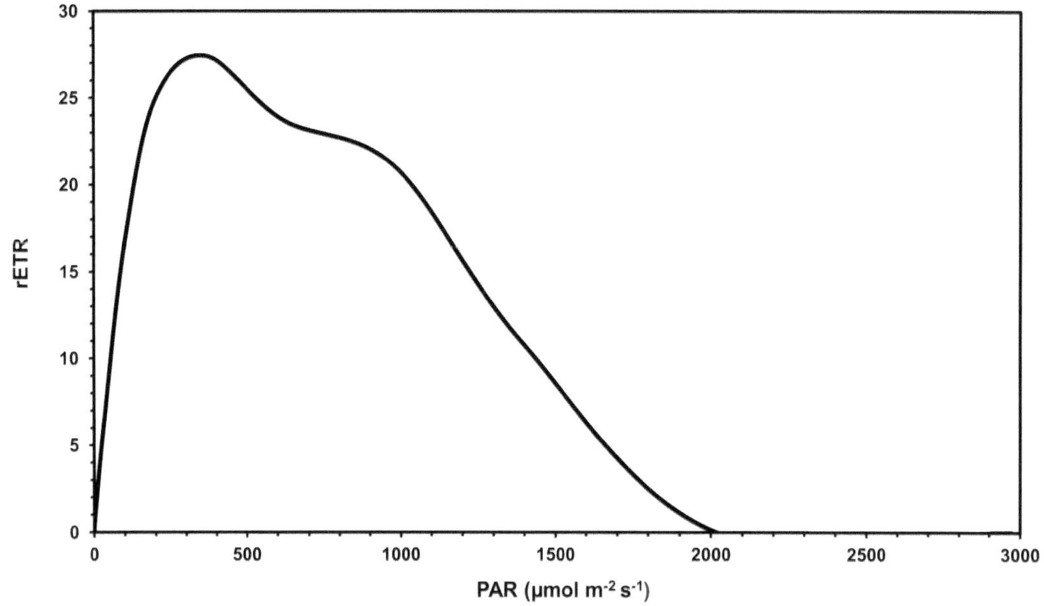

FIGURE 15.7 Light intensity dependence of saturation of the relative electron transport rate (rETR) showing the working efficiency of PSII. The abscissa is photosynthetically active radiation (PAR) in micromole photons m^{-2} s^{-1}. The saturation of the rETR shows the changes in the fluorescence parameters of *Synechocystis* sp. PCC 6803 with increasing PAR (see text for details).

where PAR is the photosynthetic active radiation, Y(II) is the effective quantum yield and 0.5 is a factor that accounts for the partitioning of energy between PSII and PSI. The graph of rETR versus photon flux density is shown in Figures 15.7 and 15.8.

Figure 15.7 shows the graph for rETR of *Synechocystis* sp. PCC 6803, rETR being the working efficiency of the PSII. As described earlier (Baker 2008), the saturation of rETR is the reflection of the change in the fluorescence parameters. The maximum value for rETR, i.e., 27.4, was achieved at 366 μmol photons m^{-2} s^{-1} (Figure 15.7). The value of rETR declined since a significant decrease in the Y(II) as a function of PAR was observed. However, the rate of decrease of Y(II) was slow due to the presence of OCP-dependent, non-photochemical quenching in *Synechocystis* sp. PCC 6803 at higher light intensities (Figure 15.7). The sudden decrease in rETR was not found in *Synechocystis* sp. PCC 6803 in contrast with that in *S. elongatus* PCC 7942, where OCP is absent (Figure 15.8). We observed that the maximum value for rETR, i.e., 0.6, was achieved at 60–80 μmol photons m^{-2} s^{-1} of light in *S. elongatus* PCC 7942 (Figure 15.8). This observation suggests that 60–80 μmol photons m^{-2} s^{-1} are sufficient to saturate PSII, while a further increase in light intensity inhibits the overall photochemistry due to photoinhibition associated with damage to PSII. In addition, the decoupling of PBSs from the reaction center causes a higher decrease in F_m' than in $F_{(t)}$ (Figure 15.6) due to heat dissipation at a high intensity of the saturation pulse (Calzadilla and Kirilovsky 2020), which results in low Y(II) and a drastic decrease of rETR (Figure 15.8).

15.6 NOTES/PRECAUTIONS

1. Cultures should be centrifuged in the amber tube to avoid any change in the physiology of the cells.
2. It is difficult to get the true value of F_0 in cyanobacteria due to the fluorescence contribution of PBSs and PSI.
3. The respiratory electron transport chain, present on the thylakoid membranes of cyanobacteria, causes non-photochemical reduction of the plastoquinone pool during dark

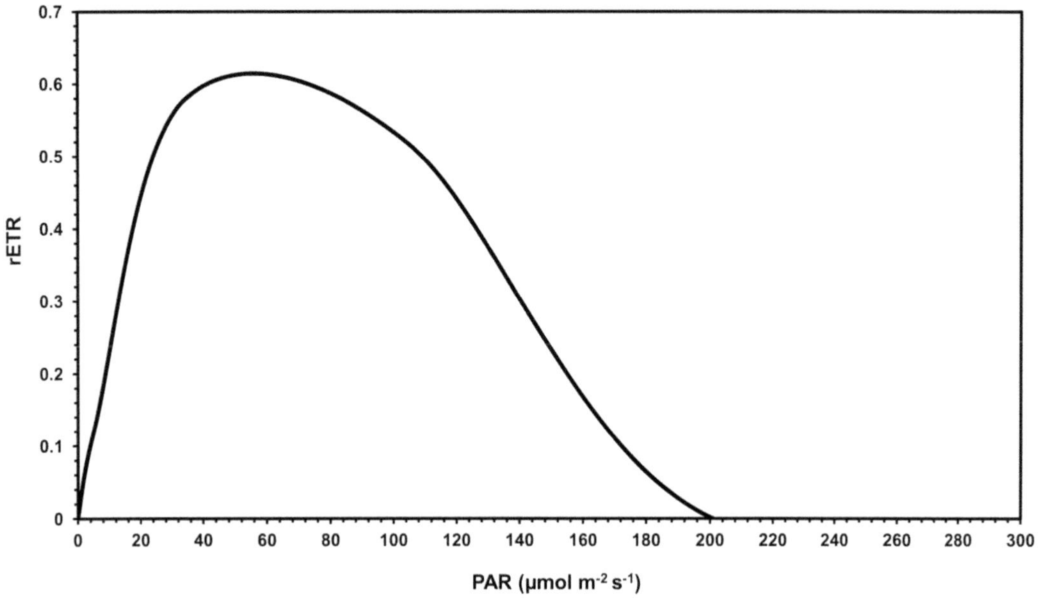

FIGURE 15.8 Light intensity dependence of the saturation of the relative electron transport rate (rETR) showing the working efficiency of PSII. The abscissa is photosynthetically active radiation (PAR) in micromole photons m^{-2} s^{-1}. The saturation of the rETR shows changes in the fluorescence parameters of *Synechococcus elongatus* PCC 7942 with increasing PAR (see text for details).

adaptation and causes state transitions which result in the association of PBS antenna molecules to PSI instead of PSII. This is the major cause of the higher value of F_0.

4. To minimize this effect, a low intensity of blue light or far-red light must be used, which causes oxidation of the PQ pool and state I transition.
5. The number of cells should be constant in all the experiments so that the calculated fluorescence parameters can be compared.
6. The cell density should be optimized for the measurement; otherwise, the fluorescence intensity will be difficult to measure with samples of lower density.
7. Once the measurement is done, remove the sample and magnet from the sample chamber and clean it properly.

ACKNOWLEDGMENTS

This work was supported by funding from the Institute of Eminence incentive grant, Banaras Hindu University (R/Dev/D/IOE/Incentive/2021-2022/32399). The financial support from the Science and Engineering Research Board (SERB), New Delhi (SCP/2022/000201) is greatly acknowledged. ST (NTA Ref. No.-201610202339) is thankful to the UGC, New Delhi, India for the junior research fellowship. AG thanks ICMR, New Delhi, for the senior research fellowship (3/1/3/JRF-2019/HRD-LS). We thank Govindjee of the University of Illinois at Urbana-Champaign < https://www.life.illinois.edu/govindjee/> for reading this manuscript and for making valuable suggestions during the preparation of this chapter.

REFERENCES

Anderson LK and Toole CM (1998) A model for early events in the assembly pathway of cyanobacterial phycobilisomes. *Molecular Microbiology* 30(3): 467–474.

Baker NR (2008) Chlorophyll fluorescence: A probe of photosynthesis in vivo. *Annual Review of Plant Biology* 59(89).

Barber J (2009) Photosynthetic energy conversion: Natural and artificial. *Chemical Society Reviews* 38(1): 185–196.

Blackman FF (1905) Optima and limiting factors. *Annals of Botany* 19(74): 281–295.

Blankenship RE and Prince RC (1985) Excited-state redox potentials and the Z-scheme of photosynthesis. *Trends in Biochemical Sciences* 10: 382–383.

Boulay C, Abasova L, Six C, Vass I and Kirilovsky D (2008) Occurrence and function of the orange carotenoid protein in photoprotective mechanisms in various cyanobacteria. *Biochimica et Biophysica Acta – Bioenergetics* 1777(10): 1344–1354.

Brooks MD and Niyogi KK (2011) Use of a pulse-amplitude modulated chlorophyll fluorometer to study the efficiency of photosynthesis in *Arabidopsis* plants. In: Chloroplast Research in *Arabidopsis* (Ed. R.P. Jarvis). Humana Press, Totowa, NJ 07512-1165 US. pp. 299–310. ISBN: 978-1-61779-236-6

Bryant DA (1994) The Molecular Biology of Cyanobacteria, First Ed. Springer science and Business media, Heidelberg.

Cailly AL (1996) Fate of excitation at PS II in leaves: The non-photochemical side. *Plant Physiology and Biochemistry* 31: 86.

Calzadilla PI and Kirilovsky D (2020) Revisiting cyanobacterial state transitions. *Photochemical and Photobiological Sciences* 19(5): 585–603.

Campbell D, Hurry V, Clarke AK, Gustafsson P and Oquist G (1998) Chlorophyll fluorescence analysis of cyanobacterial photosynthesis and acclimation. *Microbiology and Molecular Biology Reviews* 62(3): 667–683.

Campbell D and Oquist G (1996) Predicting light acclimation in cyanobacteria from nonphotochemical quenching of photosystem II fluorescence, which reflects state transitions in these organisms. *Plant Physiology* 111(4): 1293–1298.

Field CB, Behrenfeld MJ, Randerson JT and Falkowski P (1998a) Primary production of the biosphere: Integrating terrestrial and oceanic components. *Science* 281(5374): 237–240.

Field TS, Nedbal L and Ort DR (1998b) Nonphotochemical reduction of the plastoquinone pool in sunflower leaves originates from chlororespiration. *Plant Physiology* 116(4): 1209–1218.

Govindjee G and Shevela D (2011) Adventures with cyanobacteria: A personal perspective. *Frontiers in Plant Science* 2: 28.

Govindjee G, Shevela D and Bjorn LO (2017) Evolution of the z-scheme of photosynthesis: A perspective. *Photosynthesis Research* 133: 5–15.

Gwizdala M, Berera R, Kirilovsky D and Van Grondelle R and Krger TP (2016) Controlling light harvesting with light. *Journal of the American Chemical Society* 138(36): 11616–11622.

Jallet D, Gwizdala M and Kirilovsky D (2012) ApcD, ApcF and ApcE are not required for the Orange Carotenoid Protein related phycobilisome fluorescence quenching in the cyanobacterium *Synechocystis* PCC 6803. *Biochimica et Biophysica Acta – Bioenergetics* 1817(8): 1418–427.

Kalaji HM, Schansker G, Brestic M, Bussotti F, Calatayud A, Ferroni L and Bąba W (2017) Frequently asked questions about chlorophyll fluorescence, the sequel. *Photosynthesis Research* 132: 13–66.

Kolber Z and Falkowski PG (1993) Use of active fluorescence to estimate phytoplankton photosynthesis *in situ*. *Limnology and Oceanography* 38(8): 1646–1665.

Maurya PK, Mondal S, Kumar V and Singh SP (2021) Roadmap to sustainable carbon-neutral energy and environment: Can we cross the barrier of biomass productivity? *Environmental Science and Pollution Research* 28(36): 49327–49342.

Maurya PK, Mondal S, Kumar V and Singh SP (2023) Green and blue light-dependent morphogenesis, decoupling of phycobilisomes and higher accumulation of reactive oxygen species and lipid contents in *Synechococcus elongatus* PCC 7942. *Environmental and Experimental Botany* 205: 105105.

Muller P, Li XP and Niyogi KK (2001) Non-photochemical quenching. A response to excess light energy. *Plant Physiology* 125(4): 1558–1566.

Mullineaux CW and Allen JF (1986) The state 2 transition in the cyanobacterium *Synechococcus* 6301 can be driven by respiratory electron flow into the plastoquinone pool. *Federation of European Biochemical Societies Letters* 205(1): 155–160.

Ogawa T and Sonoike K (2016) Effects of bleaching by nitrogen deficiency on the quantum yield of photosystem II in *Synechocystis* sp. PCC 6803 revealed by Chl fluorescence measurements. *Plant and Cell Physiology* 57(3): 558–567.

Papageorgiou GC (1996) The photosynthesis of cyanobacteria (blue bacteria) from the perspective of signal analysis of chlorophyll *a* fluorescence. *Journal of Scientific and Industrial Research (C)* 55: 596–617.

Papageorgiou GC, Tsimilli-Michael M and Stamatakis K (2007) The fast and slow kinetics of chlorophyll *a* fluorescence induction in plants, algae and cyanobacteria: A viewpoint. *Photosynthesis Research* 94(2): 275–290.

Polívka T, Kerfeld CA, Pascher T and Sundström V (2005) Spectroscopic properties of the carotenoid 3′-hydroxyechinenone in the orange carotenoid protein from the cyanobacterium *Arthrospira maxima*. *Biochemistry* 44(10): 3994–4003.

Price GD, Pengelly JJ, Forster B, Du J, Whitney SM, von Caemmerer S and Evans JR (2013) The cyanobacterial CCM as a source of genes for improving photosynthetic CO_2 fixation in crop species. *Journal of Experimental Botany* 64(3): 753–768.

Samsonoff WA and MacColl R (2001) Biliproteins and phycobilisomes from cyanobacteria and red algae at the extremes of habitat. *Archives of Microbiology* 176(6): 400–405.

Sanfilippo JE, Garczarek L, Partensky F and Kehoe DM (2019) Chromatic acclimation in cyanobacteria: A diverse and widespread process for optimizing photosynthesis. *Annual Review of Microbiology* 73: 407–433.

Schreiber U and Klughammer C (2021) Evidence for variable chlorophyll fluorescence of photosystem I *in vivo*. *Photosynthesis Research* 149: 213–231.

Schreiber U, Klughammer C and Kolbowski J (2012) Assessment of wavelength-dependent parameters of photosynthetic electron transport with a new type of multi-color PAM chlorophyll fluorometer. *Photosynthesis Research* 113(1): 127–144.

Schreiber UBWN, Bilger W and Neubauer C (1995) Chlorophyll fluorescence as a nonintrusive indicator for rapid assessment of *in vivo* photosynthesis. In: Ecophysiology of Photosynthesis (Eds. E.-D. Schulze and M.M. Caldwell). Springer, Inc, Berlin, Heidelberg. pp. 49–70. ISBN: 978-3-642-79354-7.

Singh SP, Rastogi RP, Sinha RP and Häder D-P (2013) Photosynthetic performance of *Anabaena variabilis* PCC 7937 under simulated solar radiation. *Photosynthetica* 51(2): 259–266.

Stirbet A (2011) On the relation between the Kautsky effect (chlorophyll *a* fluorescence induction) and photosystem II: Basics and applications of the OJIP fluorescence transient. *Journal of Photochemistry and Photobiology* 104(1–2): 236–257.

Stirbet A, Lazar D, Papageorgiou G and Govindjee G (2019) Chlorophyll *a* fluorescence in cyanobacteria: Relation to photosynthesis. In: Cyanobacteria: From Basic Science to Applications (Eds. A.K. Mishra, D.N. Tiwari and A.N. Rai). Elsevier Publishers Academic Press, Inc, Poole UK. pp. 79–130. ISBN: 978-0-12-814667-5

Vermaas WF, Shen G and Styling S (1994) Electrons generated by photosystem II are utilized by an oxidase in the absence of photosystem I in the cyanobacterium *Synechocystis* sp. PCC 6803. *Federation of European Biochemical Societies letters* 337(1): 103–108.

Vernotte C, Etienne AL and Briantais JM (1979) Quenching of the system II chlorophyll fluorescence by the plastoquinone pool. *Biochimica et Biophysica Acta – Bioenergetics* 545(3): 519–527.

Watanabe M, Semchonok DA, Webber-Birungi MT, Ehira S, Kondo K, Narikawa R and Ikeuchi M (2014) Attachment of phycobilisomes in an antenna–photosystem I supercomplex of cyanobacteria. *Proceedings of the National Academy of Sciences, USA* 111(7): 2512–2517.

16 Reactive Oxygen Species Measurement in Cyanobacteria Using 2′,7′-Dichlorodihydrofluorescein Diacetate (DCFH-DA) Dye and Fluorescence Microscope or Flow Cytometer

Soumila Mondal, Anjali Gupta, Sapna Tiwari, Pankaj K. Maurya, Priyul Pandey, and Shailendra P. Singh

16.1 INTRODUCTION

Cyanobacteria are prokaryotic photosynthetic organisms that produce oxygen (O_2) during photosynthesis similar to algae, bryophytes, pteridophytes, gymnosperms and angiosperms (Sánchez-Baracaldo and Cardona 2020). These Gram-negative bacteria have a long evolutionary history of survival under harsh environmental conditions that supported their presence in almost every ecological niche (Sinha and Häder 1996). Cyanobacteria possess different morphological forms like unicellular, unicellular colonial, differentiated and undifferentiated multicellular filamentous forms with true or false branching (Whitton and Potts 2012). Apart from O_2 production, cyanobacteria contribute significantly to the productivity of aquatic and terrestrial ecosystems by fixing carbon dioxide (CO_2) and nitrogen (N_2).

Cyanobacteria can also be used to produce various secondary metabolites of industrial importance such as long-chain alcohols, alkanes, fatty acids, ethylene, 2,3-butanediol, ethanol, polyhydroxybutyrate and hydrogen either naturally or by using a metabolic engineering approach (Lai and Lan 2015; Gao et al. 2016). Due to the availability of genome sequences, tools for their genetic modification and simple growth requirements, several cyanobacteria are potential candidates for the production of bioenergy and valuable chemicals (Vermaas 2004; Lai and Lan 2015; Al-Haj et al. 2016; Gao et al. 2016; Kitchener and Grunden 2018). However, several abiotic stressors such as dynamic light environment, pH, salinity, drought, temperature and nutrient limitation affect their overall fitness and survival by disturbing the redox homeostasis (Sudhir and Murthy 2004; Fatma 2009).

The disturbance in the oxidation-reduction processes occurring inside the cyanobacterial cell leads to the generation of reactive oxygen species (ROS). An increased level of ROS can damage different biological molecules such as membrane lipids, nucleic acids, proteins, photosynthetic pigments, etc. ROS also damages the D2 protein of the photosystem II, which leads to photoinhibition

 DOI: 10.1201/9781003398387-16

(Yamamoto et al. 2008). Thus, high levels of ROS production under different environmental conditions have a negative effect on the survival and growth of cyanobacteria. Therefore, a fast and accurate ROS detection method is required to assess the fitness and long-term survival of small- or large-scale cyanobacterial cultivation systems.

ROS can be measured directly by visualizing the cells under a fluorescence microscope after treatment with a fluorescent probe. 2′,7′-dichlorodihydrofluorescein diacetate (DCFH-DA) is a well-known fluorescent probe used for the detection of ROS in live cells (Hakkila et al. 2014; Liu et al. 2016). DCFH-DA is converted to DCFH by the activity of the esterase enzyme in a live cell. DCFH is mainly non-fluorescent, but it emits green fluorescence after oxidation to DCF by intracellular ROS such as peroxides, superoxide anion radicals, hydroxy radicals, hydrogen peroxide, hypochlorous acid, peroxyl, alkoxyl, hydroperoxyl, singlet oxygen and peroxynitrite (Rastogi et al. 2010; Hakkila et al. 2014; Liu et al. 2016). In addition to DCFH-DA, CellROX and MitoRox are also used for the detection of ROS. However, the presence of photosynthetic pigments in cyanobacteria restricts the use of MitoRox Red or CellROX deep red dye, as their fluorescence emission overlaps with the fluorescence of chlorophyll *a* and phycobiliproteins. The fluorescence emission of CellROX green dye is between 500 and 550 nm, but it can only detect a limited number of oxidizing agents such as hydroxyl radicals and superoxides. In contrast, DCFH-DA is a perfect fluorescent probe to detect a broader range of ROS in cyanobacteria and other organisms (Choi et al. 2015).

A flow cytometer (FCM) is widely used to perform fluorescence-based experiments, as it can analyze millions of cells within a short period. FCM can measure the optical and fluorescence properties of a single cell present in a fluid system when it passes through a laser beam (Macey 2010). The principle of an FCM is associated with the scattering of light and emission of fluorescence that occurs when the laser beam (excitation source) strikes the cells. Scattering of light is directly related to the structural and morphological properties of the cell, whereas fluorescence emission is proportional to the amount of fluorescent probe bound to its target in a cell (Macey 2010). A standard FCM is equipped with four fluorescence channels and two optical detectors. The two optical detectors are forward scatter (FSC) and side scatter (SSC), which can detect the relative cell size and cellular complexity, respectively. The different fluorescent channels (FL) can detect the fluorescence emissions of different wavelengths.

Here, we present a detailed method to detect and measure the levels of ROS in cyanobacteria using a fluorescence microscope and FCM. The method includes sample preparation, acquisition, visualization and data analysis.

16.2 EQUIPMENT

1. Centrifuge (Remi, India).
2. Nikon 90i Eclipse fluorescence microscope (Nikon, Japan).
3. UV-VIS spectrophotometer 1800 (Shimadzu, Japan).
4. BD FACSCalibur (Becton Dickinson, USA).
5. CellQuest Pro software (Becton Dickinson, USA).
6. NIS elements AR software 4.0 (Nikon, Japan).

16.3 MATERIALS

1. 0.1 M PBS pH 7.4 (100 mL) (see Recipes).
2. 0.2 μm size Millipore membrane filter (Merck, catalog number: GSWP04700).
3. 1.5 mL amber microcentrifuge tubes (Abdos, catalog number: P10204A).
4. 10 μL pipette tips (Abdos, catalog number: P10116).

5. 1 cm quartz cuvette (Shimadzu, catalog number: 226-85010-92).
6. 1.0 mL pipette tips (Abdos, catalog number: P10109).
7. 200 µL pipette tips (Abdos, catalog number: P10140).
8. 20 mL 1.0 M HEPES pH 8.0 (Himedia, catalog number: RM380-500G).
9. 2.0 mM (w/v) DCFH-DA stock solution (1 mL) (see Recipes).
10. 3.0 mL round bottom polystyrene tubes (Becton Dickinson, catalog number: 156758).
11. BD FACS clean solution (Becton Dickinson, catalog number: 340345).
12. Coverslips (Borosil, catalog number: 9115S01).
13. DCFH-DA (Sigma-Aldrich, catalog number: D6883).
14. Disodium hydrogen phosphate (Na_2HPO_4) (Merck, catalog number: 1.93622.0521).
15. Ethanol (Merck, catalog number: 1.00983.0511).
16. Exponentially growing cyanobacterial cell culture.
17. Glass slides (Borosil, catalog number: BT409100P02).
18. Hydrogen peroxide (Merck, catalog number: 1.93408.0521).
19. Autoclaved Milli-Q or double distilled water.
20. Sodium chloride (NaCl) (SRL, catalog number: 41721).
21. Sodium dihydrogen phosphate (NaH_2PO_4) (Merck, catalog number: 1.93624.0521).

16.4 RECIPES

16.4.1 2.0 mM (w/v) DCFH-DA Stock Solution (1 mL)

Dissolve 0.974 mg of DCFH-DA in 1.0 mL of 100% molecular-grade ethanol. Vortex thoroughly until DCFH-DA dissolves completely. Keep the stock solution in the dark at −20°C. The stock can be stored for 3 months.

16.4.2 0.1 M PBS pH 7.4 (100 mL)

Prepare 0.2 M disodium hydrogen phosphate stock solution by dissolving 2.8392 g Na_2HPO_4 in 100 mL Milli-Q water. Prepare 0.2 M sodium dihydrogen phosphate (NaH_2PO_4) stock solution by dissolving 2.3996 g NaH_2PO_4 in 100 mL Milli-Q water. Mix 40 mL of 0.2 M Na_2HPO_4 stock with 10 mL of NaH_2PO_4 stock. Add 0.9 g NaCl and dissolve it completely by stirring. Adjust the volume to 100 mL with Milli-Q water and maintain the pH at 7.4. Filter the solution through a 0.2-µm size Millipore membrane filter and store at 4°C. The buffer can be stored for 3 months.

16.5 METHODS

16.5.1 ROS Estimation by Flow Cytometry (Mondal AND Singh 2022)

Step 1: Optimization of different parameters of FCM

1. Take 10 mL of exponentially growing cyanobacterial culture. Centrifuge at 8000 × *g* for 10 minutes at room temperature (RT; 25°C). Discard the supernatant.
2. Wash the harvested cells twice with 0.1 M PBS and obtain a cell pellet by centrifugation at 8000 × *g* for 10 minutes at RT.
3. Suspend the cell pellet in 10 mL 0.1 M PBS. Record the optical density at 750 nm.
4. Make 1.0-mL aliquots of cells having 0.1, 0.2, 0.3, 0.4, 0.5, 0.6, 0.7 and 0.8 optical density at 750 nm.
5. Wash the BD FACSCalibur machine twice with BD FACS clean solution. Fill the sheath tank with sheath fluid and empty the waste tank. Pour 100 mL BD FACS clean solution into the waste tank.

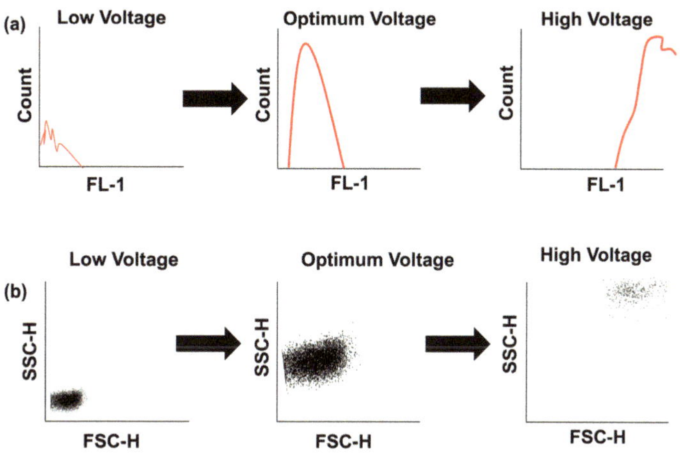

FIGURE 16.1 Optimization of PMT voltages of different channels of a flow cytometer (FCM). FCM histogram showing optimization of the PMT voltage of fluorescence channel (a). The FCM dot plot shows the optimization of FSC and SCC voltages (b) (color figure in e-book).

6. Open BD CellQuest Pro software. Draw two histogram plots of FL-1 (green fluorescence, emission range 515–545 nm) and FL-3 (red fluorescence, emission range >670 nm) and two dot plots of FSC vs. SSC and FSC vs. FL-3.
7. Take the sample in a polystyrene tube. Acquire the samples with a low (12 μL min^{-1}) flow rate.
8. Adjust the PMT voltages of FSC, SSC, FL-1 and FL-3 channels as shown in Figure 16.1.
9. Acquire 50,000–100,000 events.
10. Note the concentration of cells having a sample acquisition rate of 1000 events/second.
11. Save the data in FCS format.

Step 2: Optimization of DCFH-DA concentration

1. Take 10 mL of exponentially growing cyanobacterial culture. Record the optical density at 750 nm. Adjust the optical density which was previously determined in **Step 1**.
2. Treat cells with 100 mM H_2O_2 for 1 hour.
3. Centrifuge cells at 8000 × *g* for 10 minutes at RT. Discard the supernatant.
4. Wash the cells with 0.1 M PBS twice by centrifugation at 8000 × *g* for 10 minutes at RT.
5. Suspend cells in 10 mL of 0.1 M PBS and divide into aliquots of 1.0 mL.
6. Treat aliquots of cells with different concentrations of DCFH-DA, i.e., 0 μM, 5 μM, 10 μM, 15 μM, 20 μM, 25 μM, 30 μM, 35 μM and 40 μM. Vortex briefly to mix.
7. Incubate samples in the dark with shaking at ~150 rpm for 1 hour at RT.
8. After incubation, acquire the samples immediately with the help of a flow cytometer.
9. Set different flow cytometer parameters and draw different plots as described in Step 1.
10. Acquire samples at a low flow rate. Save the files in FCS format.
11. Prepare a titration curve by plotting FL-1 mean fluorescence intensity (MFI) values against different concentrations of dye in an Excel file to determine the minimum concentration of DCFH-DA required for *in vivo* detection of ROS.

Step 3: Detection of ROS using a flow cytometer

1. Collect cyanobacterial samples from different experimental conditions. Centrifuge at 8000 × *g* for 10 minutes at RT and discard the supernatant to obtain a cell pellet.
2. Wash the cell pellet twice with 0.1 M PBS by centrifugation at 8000 × *g* for 10 minutes at RT. After washing, resuspend the cell pellet in 0.1 M PBS buffer.

3. Adjust the optical density at 750 nm that was determined in Step 1.
4. Treat cells with a predetermined concentration of DCFH-DA from Step 2. Incubate the cells in the dark for 1 hour with shaking (~150 rpm) at RT.
5. After incubation, acquire the samples with a flow cytometer. Set all the parameters as described in Step 1.
6. Acquire at least 50,000–100,000 events.
7. Save the data in FCS format.

16.5.2 ROS Estimation by Fluorescence Microscope (Rastogi et al. 2010; Mondal et al. 2020; Mondal and Singh 2022)

Step 1: Optimization of DCFH-DA dye concentration

1. Take 10 mL of cyanobacterial cell culture and centrifuge at 8000 × *g* for 10 minutes at RT. Discard the supernatant.
2. Wash the cell pellet twice with 0.1 M PBS by centrifugation at 8000 × *g* for 10 minutes at RT.
3. Resuspend the cell pellet in 0.1 M PBS to obtain an optical density of 0.2 at 750 nm (OD_{750}) and divide the sample into aliquots of 1.0 mL.
4. Treat aliquots with different concentrations of DCFH-DA (0 μM, 5 μM, 10 μM, 15 μM, 20 μM, 25 μM, 30 μM, 35 μM and 40 μM) using a stock solution of DCFH-DA. Briefly vortex and incubate the samples in the dark at RT for 1 hour with shaking at ~150 rpm.
5. Set the exposure time of the FITC channel and visualize the cells under the FITC channel of a fluorescence microscope.
6. Measure the intensity of the DCF fluorescence coming from the cell. Prepare a titration curve by plotting DCF fluorescence values against different concentrations of DCFH-DA dye to determine the minimum concentration of dye required for *in vivo* ROS detection.

Step 2: Detection of ROS by fluorescence microscope

1. Collect cyanobacterial samples from different experimental conditions. After centrifugation at 8000 × *g* for 10 minutes at RT, discard the growth medium and wash the cells twice with 0.1 M PBS.
2. Resuspend the cell pellet in 0.1 M PBS to get an optical density of 0.2 at 750 nm (OD_{750}). Treat the samples with an optimized concentration of DCFH-DA (refer to Step 1). Incubate the samples for 1 hour in the dark at RT with ~150 rpm shaking.
3. Visualize the cells under FITC (Excitation 465–495 nm; Emission 515–550 nm) and TRITC (Excitation 540/25 nm; Emission 605/55 nm) channels of a fluorescence microscope. Preset the exposure time for each channel as mentioned in Step 1.
4. Take images and measure the intensity of fluorescence coming from the cells under both channels.
5. Export the data in Excel format.

16.6 DATA ANALYSIS

16.6.1 Measurement of ROS Using DCFH-DA Dye and Flow Cytometry (Mondal and Singh 2022)

The amount of ROS present in the cells is directly proportional to the DCF fluorescence. Hence, the amount of ROS can be represented by the histogram plot of DCF fluorescence (FL-1). First, gate the cyanobacterial population from the dot plots of FSC vs. FL-3 using BD CellQuest Pro software (Figure 16.2). Thereafter, document the DCF MFI of the gated cyanobacterial population from the histogram diagram of FL-1. Prepare a histogram plot to represent the ROS level of the

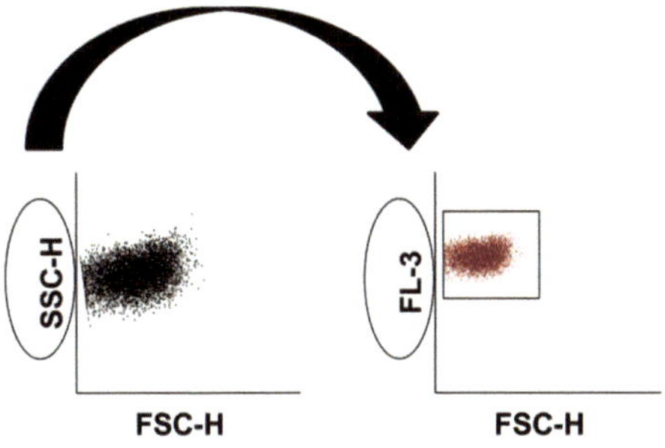

FIGURE 16.2 Dot plot image of gating cyanobacterial population for measurement of reactive oxygen species (ROS) levels using 2′,7′-dichlorodihydrofluorescein-diacetate (DCFH-DA) and flow cytometer. The dot plot diagram of FSC-H vs. SSC-H shows the relative size and cellular complexity, respectively, of the selected cyanobacterial population. Selection of a population is done with the help of chlorophyll *a* fluorescence (FL-3) and relative size (FSC-H) (color figure in e-book).

cyanobacterial population (Figure 16.3a). The MFI of the FL-1 channel histogram can be also used to develop a bar diagram to show levels of ROS in different samples.

16.6.2 Measurement of ROS Using DCFH-DA Dye and Fluorescence Microscope (Rastogi et al. 2010; Mondal et al. 2020; Mondal and Singh 2022)

The level of ROS measured by fluorescence microscope can be represented as a green/red (G/R) ratio together with image-based qualitative data acquired during ROS detection by fluorescence microscope (Figure 16.3b). The green fluorescence of DCF is represented by 'G', while 'R' is a quantitative representation of autofluorescence coming from cyanobacterial photosynthetic pigments, i.e., chlorophyll *a* and phycobiliproteins. A high G/R ratio represents increased levels of ROS. For fluorescence microscope–based ROS measurement, analyze at least 50 cells from images acquired for each replicate.

16.7 NOTES/PRECAUTIONS

1. A flow cytometer has limitations in terms of cell/filament size. A flow cytometer can analyze cells/filaments of approximately 1 μm to 50 μm in length. Therefore, it is important to measure the cell/filament length in a light microscope before conducting any flow cytometer–based experiment. Due to the size limitation, it is difficult to study filamentous cyanobacteria using flow cytometry because of their long filaments. To reduce the length of filaments, a shearing force like sonication or ultrasonic sound waves can be used. However, the impact of applied shearing force on cell survival and ROS levels should be analyzed before conducting an FCM-based experiment.
2. Samples to be analyzed by flow cytometer should be in either single-cell suspension or the filament length should be less than 50 μm. The clumping of cells/filaments can clog the nozzle of the flow cytometer.
3. The fluorescence probe DCFH-DA is sensitive to light. It is advisable to avoid direct exposure of the dye to light while performing the experiment.
4. Due to the small surface area, the cells having a small spherical morphology might not get an adequate amount of excitation energy for the excitation of the fluorescence molecules while analyzing by fluorescence microscope. However, FCM can detect fluorescence signals from any morphological form of cyanobacteria. Therefore, FCM is recommended while working with cyanobacteria having a small spherical morphology.

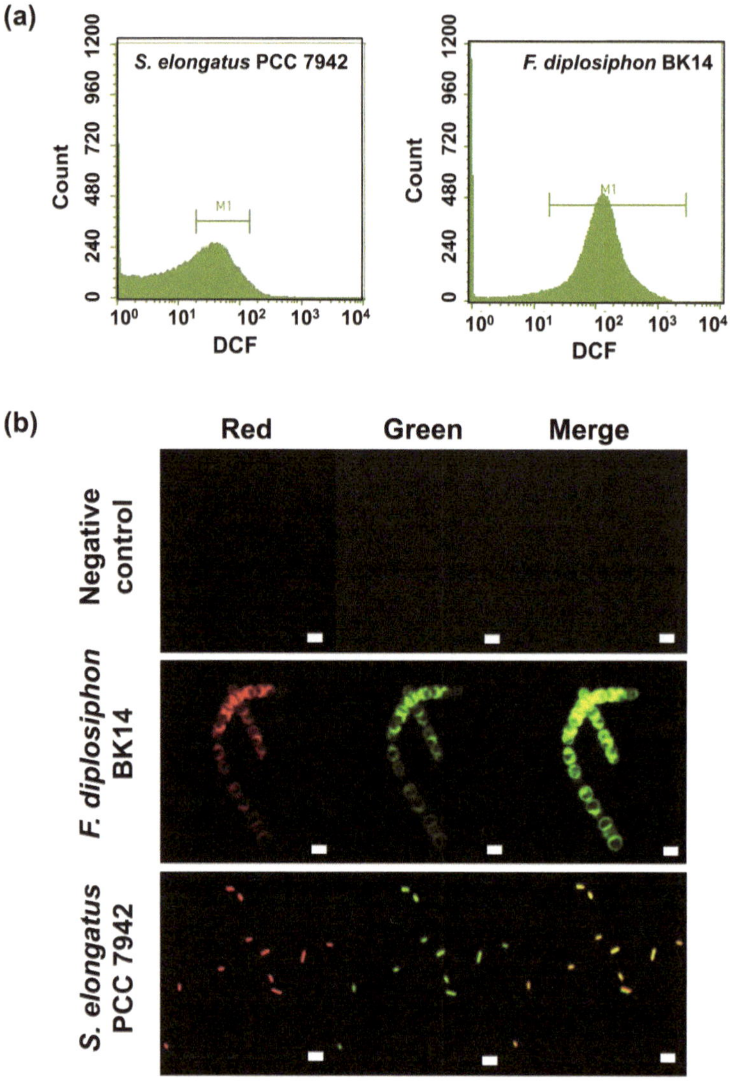

FIGURE 16.3 Measurement of reactive oxygen species (ROS) in *Synechococcus elongatus* PCC 7942 and *Fremyella diplosiphon* BK14 using 2′,7′-dichlorodihydrofluorescein-diacetate (DCFH-DA) and flow cytometry (FCM) or fluorescence microscopy. FCM histograms presenting fluorescence of dichlorofluorescein (DCF), i.e., ROS levels, in *S. elongatus* PCC 7942 and *F. diplosiphon* BK14 (a). DCF fluorescence showing ROS levels (green) and autofluorescence of photosynthetic pigments (red) in *F. diplosiphon* BK14 and *S. elongatus* PCC 7942 (b). Fluorescence from DCFH-DA only sample without cells is considered the negative control. Images were acquired using a ×40 objective. Scale bars, 5 μm (color figure in e-book).

5. This protocol has been tested with unicellular (*Synechococcus elongatus* PCC 7942 and *Synechocystis* sp. PCC 6803) and short filamentous (*Fremyella diplosiphon* BK 14) cyanobacteria. The same protocol can be used for other cyanobacteria after optimization of dye concentration and other parameters if required.

ACKNOWLEDGMENTS

This work was supported by funding from the Institute of Eminence Incentive Grant, Banaras Hindu University (R/Dev/D/IOE/Incentive/2021-2022/32399) and SERB, New Delhi, India, in the form of a Special Core Research Grant (SCP/2022/000201) to Shailendra P. Singh. Pankaj K. Maurya,

Soumila Mondal, Anjali Gupta, Sapna Tiwari and Priyul Pandey are thankful to the UGC, New Delhi, India (3616/NET-DEC2014), UGC and BHU, ICMR, New Delhi, India (3/1/3/JRF -2019/HRD-LS), UGC, New Delhi, India (NTA Ref. No.: 201610202339) and CSIR, New Delhi (File no: 09/0013(15476)/2022-EMR-I), respectively.

REFERENCES

Al-Haj L, Lui YT, Abed RM, Gomaa MA and Purton S (2016) Cyanobacteria as chassis for industrial biotechnology: Progress and prospects. *Life* 6: 42.

Choi H, Yang Z and Weisshaar JC (2015) Single-cell, real-time detection of oxidative stress induced in *Escherichia coli* by the antimicrobial peptide CM15. *Proceedings of the National Academy of Sciences USA* 112: E303–E310.

Fatma T (2009) Screening of cyanobacteria for phycobiliproteins and effect of different environmental stress on its yield. *Bulletin of Environmental Contamination and Toxicology* 83: 509–515.

Gao X, Sun T, Pei G, Chen L and Zhang W (2016) Cyanobacterial chassis engineering for enhancing production of biofuels and chemicals. *Applied Microbiology and Biotechnology* 100: 3401–3413.

Hakkila K, Antal T, Rehman AU, Kurkela J, Wada H, Vass I and Tyystjärvi T (2014) Oxidative stress and photoinhibition can be separated in the cyanobacterium *Synechocystis* sp. PCC 6803. *Biochimica et Biophysica Acta (BBA)-Bioenergetics* 1837: 217–225.

Kitchener RL and Grunden AM (2018) Methods for enhancing cyanobacterial stress tolerance to enable improved production of biofuels and industrially relevant chemicals. *Applied Microbiology and Biotechnology* 102: 1617–1628.

Lai MC and Lan EI (2015) Advances in metabolic engineering of cyanobacteria for photosynthetic biochemical production. *Metabolites* 5: 636–658.

Liu F, Du J, Song D, Xu M and Sun G (2016) A sensitive fluorescent sensor for the detection of endogenous hydroxyl radicals in living cells and bacteria and direct imaging with respect to its ecotoxicity in living zebra fish. *Chemical Communications* 52: 4636–4639.

Macey MG (2010) Principles of flow cytometry. In: Flow Cytometry: Principles and Applications (Ed. M.G. Macey). Humana Press, Totowa NJ. pp. 1–15. ISBN: 978-1-59745-415-3.

Mondal S, Kumar V and Singh SP (2020) Oxidative stress measurement in different morphological forms of wild-type and mutant cyanobacterial strains: Overcoming the limitation of fluorescence microscope-based method. *Ecotoxicology and Environmental Safety* 200: 110730.

Mondal S and Singh SP (2022) Flow cytometry-based measurement of reactive oxygen species in cyanobacteria. *Bio-Protocol* 12: e4417.

Rastogi RP, Singh SP, Häder D-P and Sinha RP (2010) Detection of reactive oxygen species (ROS) by the oxidant-sensing probe 2′,7′-dichlorodihydrofluorescein diacetate in the cyanobacterium *Anabaena variabilis* PCC 7937. *Biochemical and Biophysical Research Communications* 397: 603–607.

Sánchez-Baracaldo P and Cardona T (2020) On the origin of oxygenic photosynthesis and cyanobacteria. *New Phytologist* 225: 1440–1446.

Sinha RP and Häder D-P (1996) Photobiology and ecophysiology of rice field cyanobacteria. *Photochemistry and Photobiology* 64: 887–896.

Sudhir P and Murthy SDS (2004) Effects of salt stress on basic processes of photosynthesis. *Photosynthetica* 42: 481–486.

Vermaas WF (2004) Targeted genetic modification of cyanobacteria: New biotechnological applications. In: Handbook of Microalgal Culture: Biotechnology and Applied Phycology (Ed. A. Richmond). Blackwell Publishing, Iowa, USA. p. 457. ISBN: 0-632-05953-2.

Whitton BA and Potts M (2012) Introduction to the cyanobacteria. In *Ecology of Cyanobacteria II* (Ed. B. Whitton). Springer, Dordrecht. pp. 1–13. ISBN: 978-94-007-3854-6.

Yamamoto Y, Aminaka R, Yoshioka M, Khatoon M, Komayama K, Takenaka D and Yamamoto Y (2008) Quality control of photosystem II: Impact of light and heat stresses. *Photosynthesis Research* 98: 589–608.

17 Flow Cytometry–Based Methods for Estimating DNA Content and Live-Dead Cells in Cyanobacteria

Soumila Mondal, Anjali Gupta, Sapna Tiwari, Rinkesh Gupta, and Shailendra P. Singh

17.1 INTRODUCTION

Cyanobacteria are Gram-negative, photosynthetic, oxygen-producing bacteria that thrive in a variety of harsh environmental conditions (Whitton and Potts 2012). In addition to oxygen production, several cyanobacteria can fix atmospheric nitrogen due to the presence of nitrogenase enzymes. Therefore, cyanobacteria play an important role in sustaining life on Earth. Cyanobacteria are found in different morphological forms such as filamentous, unicellular and colonial forms. Filamentous forms generally show cellular differentiation during their development and may possess different cell types such as heterocysts for nitrogen fixation, akinetes as a resting stage cell and hormogonia as a reproductive short motile filament (Claessen et al. 2014; Herrero et al. 2016). Cyanobacteria are known for the production of industrially important metabolites and biofuels (Jones and Mayfield 2012; Neilan et al. 2013; Wijffels et al. 2013; Oliver and Atsumi 2014).

Generally, cyanobacteria contain a circular chromosome and one or more plasmids in each cell (Scanlan et al. 2009). However, the copy number of cyanobacterial genomic DNA may vary depending on growth conditions (Kaneko et al. 1995). For example, the copy number of genomic DNA in a cell can increase under a nutrient limitation or in a high-light environment to maintain genomic integrity (Meeks et al. 2001). Thus, an increased copy number of genomic DNA could help cyanobacteria to sustain lethal mutations caused by extreme environmental conditions such as ultraviolet radiation, nutrient limitation and high temperature (Singh et al. 2010; Cassier-Chauvat et al. 2021). Therefore, estimating DNA content in wild-type and mutant strains of cyanobacteria can provide valuable insights into their genome size, ploidy level and genetic characteristics. Cyanobacterial plasmids could harbor protein-encoding genes that are involved in various physiological and biochemical activities such as cysteine biosynthesis (Nicholson et al. 1995), heat tolerance (Kimura et al. 2002), metal stress resistance (Nagarajan et al. 2012) and DNA repair (Cassier-Chauvat et al. 2016).

Several cyanobacteria can form harmful algal blooms (HABs) and result in ecological imbalances due to the production of harmful toxins (Huisman et al. 2018). The quantification of live cells could help in the monitoring of cyanobacterial abundance and the detection of potential bloom events (Huisman et al. 2018). The quantification of live cyanobacterial cells can also provide valuable insights into the nutrient cycling and eutrophication processes occurring in aquatic systems. Also, the quantification of live cells can be used to study population dynamics under different environmental conditions (Huisman et al. 2018). Flow cytometry is a powerful technique for rapid and automated analysis of individual cells in a population. Cyanobacterial cells are stained

DOI: 10.1201/9781003398387-17

with a fluorescent dye, and cells are analyzed using a flow cytometer, which can measure the size, shape and fluorescence properties of a cell. Flow cytometry allows high-throughput analysis and provides detailed information on cell size, cell viability and several other cellular characteristics (Darzynkiewicz et al. 2017).

In this chapter, we describe methods for analyzing the genomic DNA content and live-dead cells using a flow cytometer and a nucleic acid binding dye, i.e., SyBr green I, in cyanobacteria.

17.2 EQUIPMENT

1. Centrifuge (Remi, India).
2. UV-VIS spectrophotometer 1800 (Shimadzu, Japan).
3. BD FACSCalibur (Becton Dickinson, USA).
4. CellQuest Pro software (Becton Dickinson, USA).

17.3 MATERIALS

1. 0.1 M PBS pH 7.4 (100 mL) (see Recipes).
2. 1.5 mL amber microcentrifuge tubes (Abdos, catalog number: P10204A).
3. Quartz cuvette (Shimadzu, catalog number: 226-85010-92).
4. 10 μL pipette tips (Abdos, catalog number: P10116).
5. 200 μL pipette tips (Abdos, catalog number: P10140).
6. 1 mL pipette tips (Abdos, catalog number: P10109).
7. 3.0 mL round bottom polystyrene tubes (Becton Dickinson, catalog number: 156758).
8. BD FACS clean solution (Becton Dickinson, catalog number: 340345).
9. Disodium hydrogen phosphate (Na_2HPO_4) (Merck, catalog number: 1.93622.0521).
10. Milli-Q (MQ) water.
11. Sodium dihydrogen phosphate (NaH_2PO_4) (Merck, catalog number: 1.93624.0521).
12. Cyanobacterial culture in the exponential growth phase.
13. 4% Paraformaldehyde (see Recipes).
14. RNase A solution (1 mg/mL).
15. SyBr Green I (Sigma-aldrich, catalog number: S9430).

17.4 RECIPES

17.4.1 0.1 M Phosphate Buffer (pH 7.4)

Solution A: Add 3.12 g sodium dihydrogen orthophosphate ($NaH_2PO_4.2H_2O$) to 100 mL of distilled water and vortex to dissolve properly.

Solution B: Add 2.84 g disodium hydrogen orthophosphate (Na_2HPO_4) to 100 mL distilled water and vortex to dissolve it completely.

Mix 19 mL of solution A and 81 mL of solution B to make 100 mL of 0.2 M phosphate buffer. To make a 0.1 M working solution, mix 50 mL of 0.2 M phosphate buffer with 50 mL MQ water and adjust the pH of the solution to 7.4 using HCl or NaOH. Phosphate buffer can be stored at 4°C for 3 months.

17.4.2 4% Paraformaldehyde

Add 4 g paraformaldehyde in 100 mL MQ water and dissolve completely at 60°C in a water bath. Add a few drops of 0.1 M NaOH solution while dissolving. Once paraformaldehyde is dissolved completely, cool it to room temperature and filter the solution using Whatman filter paper, grade 1. The solution can be kept at 4°C for 3 months.

17.4.3 SyBr Green I Stock Solution (1000×)

Dilute 10,000× concentrated SyBr Green I dye to 1000× using DMSO and store at –20°C in the dark.

17.4.4 RNase A Stock Solution (1 mg/mL)

Add 5 mg RNase A in 5 mL of 0.1 M phosphate buffer (recipe described earlier). After mixing by vortex, make aliquots of 500 μL. Store the aliquots at –20°C.

17.5 METHODS

17.5.1 Live-dead Cell Assay

Step 1: Flow cytometer parameter optimization

1. Take 10 mL exponentially growing cyanobacterial culture having an optical density 0.6 at 750 nm. Centrifuge the culture at 8000 × *g* for 10 minutes at room temperature (RT; 25°C) to harvest cells.
2. After discarding the supernatant, wash the cell pellet twice with 0.1 M phosphate buffer using centrifugation at 8000 × *g* for 10 minutes at RT.
3. After washing, suspend the cell pellet in 10 mL of 0.1 M phosphate buffer and measure the optical density of the cell suspension at 750 nm (OD_{750}).
4. Make 1-mL aliquots of cells having 0.2 OD_{750} (see Chapter 16).
5. Open BD CellQuest Pro software. Draw two histogram plots of FL-1 (green fluorescence, emission range 515–545 nm) and FL-3 (red fluorescence, emission range >670 nm) and two dot plots of FSC vs. SSC and FL-1 vs. FL-3.
6. Take a sample in a polystyrene tube. Acquire samples using flowcytometry (FCM) at a low flow rate (12 μL/min).
7. Adjust the photomultiplier tube (PMT) voltages of FSC, SSC, FL-1 and FL-3 channels as shown in Figure 17.1.
8. Acquire 50,000–100,000 events.
9. Save the data in FCS format.

Step 2: Optimization of dye concentration

1. Treat 1-mL aliquots (see Step 1) of cells with different volumes of SyBr Green I, i.e., 0 μL, 5 μL, 10 μL, 15 μL, 20 μL, 25 μL, 30 μL, 35 μL and 40 μL, and vortex briefly to mix.
2. Analyze the samples immediately using a flow cytometer.
3. Set different parameters of the flow cytometer and draw different plots as mentioned in Step 1.
4. Acquire samples at a low flow rate (<100 events/second).
5. Acquire 100,000 cells per sample. Save the files in FCS format.
6. Prepare a titration curve by plotting the mean fluorescence intensity (MFI) of the SyBr Green I channel (FL-1) against different concentrations of dye in an Excel file to determine the optimum concentration of SyBr Green I required for the detection of live-dead cells (Figure 17.1).

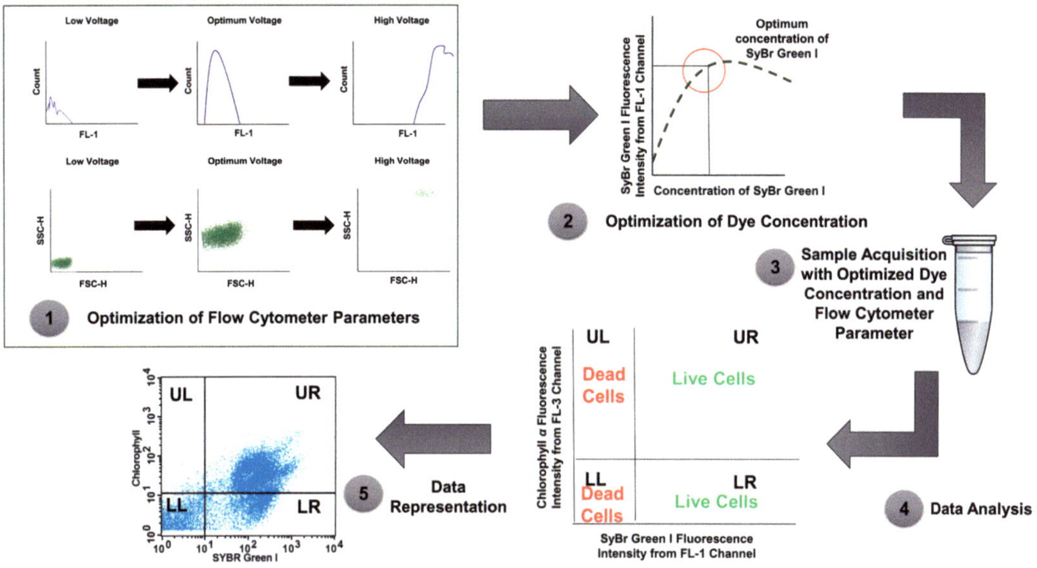

FIGURE 17.1 Schematic diagram showing different steps involved in the live-dead assay of cyanobacterial cells using SyBr Green I dye and flow cytometer. In step 1, the unstained cell suspension is used to adjust the voltages for different parameters of flow cytometer. The optimum concentration of SyBr Green I is then determined from the titration curve shown in step 2. Cyanobacterial samples from various experimental conditions are analyzed with a flow cytometer after optimization of flow cytometer parameters and dye concentration (step 3). The SyBr Green I–positive cells that appeared on the upper right (UR) and lower right (LR) sides of the dot plot are considered live cells. The dead cell population is shown at the lower left (LL) and upper left (UL) of the dot plot (steps 4 and 5) (color figure in e-book).

Step 3: Analysis of live cells in a cyanobacterial cell population

1. Collect cyanobacterial samples and wash the cells twice with 0.1 M phosphate buffer using centrifugation at 10,000 × *g* for 10 minutes at 4°C.
2. After washing, resuspend the cells in 0.1 M phosphate buffer.
3. Measure the optical density of the sample at 750 nm (OD_{750}). Adjust the OD_{750} of the sample to 0.2 using 0.1 M phosphate buffer.
4. Make 1-mL aliquots and stain samples with the optimized concentration of SyBr Green I (see Step 2).
5. Set different parameters of the flow cytometer and draw different plots as mentioned in Step 1.
6. Immediately analyze the samples using a flow cytometer.
7. Acquire samples at a low flow rate. Acquire 100,000 events.
8. Save the data in FCS file format for analysis.

17.5.2 Estimation of DNA Content

Step 1: Fixation of cyanobacterial cells

1. Harvest 10 mL of exponentially growing cyanobacterial cells (0.6 OD_{750}) by centrifugation at 10,000 × *g* for 20 minutes at 4°C and wash the cell pellet twice with 0.1 M phosphate buffer. After centrifugation, discard the supernatant.
2. Add 10 mL 4% paraformaldehyde to the cell pellet and vortex to resuspend the pellet.
3. Store the cell suspension overnight at 4°C.

4. After incubation, remove the fixative by centrifugation at 10,000 × *g* for 20 minutes at 4°C. Wash the cell pellet with 0.1 M phosphate buffer by centrifugation at 10,000 × *g* for 20 minutes at 4°C for complete removal of the fixative.
5. Resuspend the cell pellet in 10 mL of 0.1 M phosphate buffer and make 1-mL aliquots of cells having 0.2 OD_{750}.

Step 2: Optimization of dye concentration

1. Treat aliquots (Step 1) of cells with different volumes of SyBr Green I, i.e., 0 µL, 5 µL, 10 µL, 15 µL, 20 µL, 25 µL, 30 µL, 35 µL and 40 µL. Vortex briefly to mix.
2. Incubate samples in the dark with shaking (~150 rpm) for 1 hour at RT.
3. After incubation, analyze the samples immediately using a flow cytometer.
4. Open BD CellQuest Pro software. Draw two histogram plots of FL-1 (green fluorescence, emission range 515–545 nm) and FL-3 (red fluorescence, emission range >670 nm) and three dot plots of FSC vs. SSC, FSC vs. FL-3 and FL-1H vs. FL-1A.
5. Acquire samples at a low flow rate (<100 events/second).
6. Acquire 100,000 cells per sample. Save the files in FCS format.
7. Prepare a titration curve by plotting the MFI of the FL-1 channel value against different concentrations of dye in an Excel file to determine the minimum concentration of SyBr Green I required for *in vivo* detection of DNA (Figure 17.2).

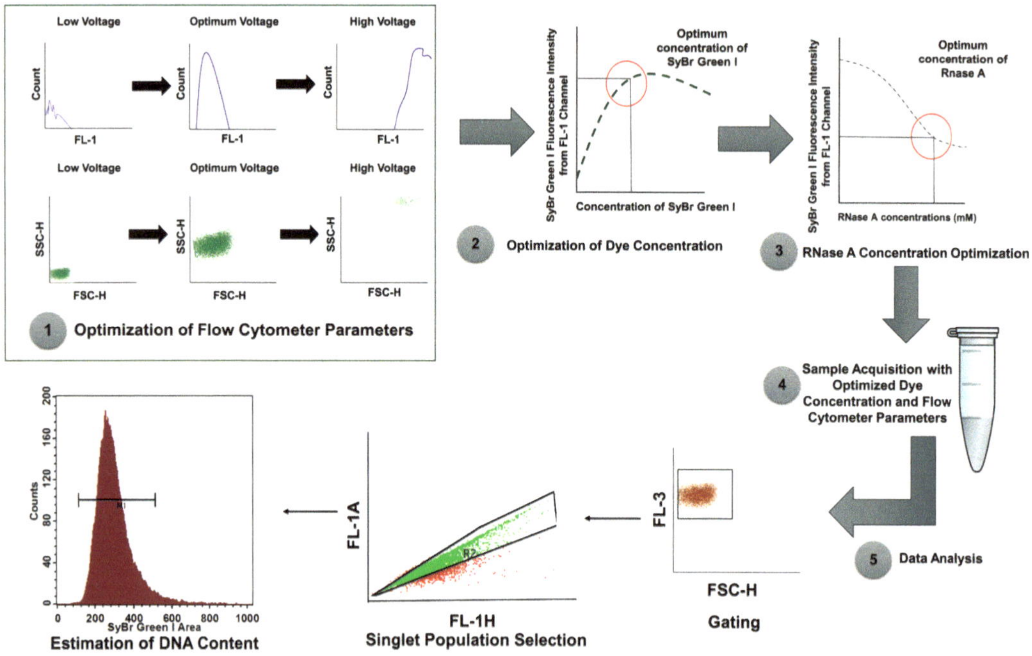

FIGURE 17.2 Schematic representations of different steps involved in the estimation of genomic DNA content in cyanobacterial cells using SyBr Green I and flow cytometer. In step 1, the unstained cell suspension is used to adjust the voltage for different parameters of the flow cytometer. The optimum concentration of SyBr Green I is then determined from the titration curve as shown in step 2. The required concentration of RNase A to digest RNA in cells is determined from the titration curve as shown in step 3. Cyanobacterial samples from various experimental conditions are analyzed with a flow cytometer after optimization of flow cytometer parameters, RNase A and SyBr Green I concentration (step 4). The cyanobacterial population is selected on the basis chlorophyll *a* (FL-3) fluorescence. The singlet cyanobacterial population is selected based on the FL-1A vs. FL-1H dot plot, and SyBr Green I fluorescence of selected singlet cells is calculated from the histogram (step 5) (color figure in e-book).

Step 3: Optimization of RNase A concentration

1. Treat 1-mL aliquots (Step 1) of cells with different concentrations of RNase A, i.e., 0 mM, 10 mM, 20 mM, 30 mM, 40 mM, 50 mM, 75 mM and 100 mM. Vortex briefly to mix.
2. Incubate samples in the dark with shaking (~150 rpm) for 1 hour at 37°C.
3. Wash the samples with 0.1 M phosphate buffer by centrifugation at 10,000 × *g* for 10 minutes at 4°C. Resuspend the cell pellet in 1 mL of 0.1 M phosphate buffer.
4. Stain samples with an optimized concentration of dye (see Step 2) and incubate for 1 hour with shaking (~150 rpm) at RT.
5. After incubation, analyze the samples using a flow cytometer.
6. Acquire the samples at a low flow rate (<100 events/second).
7. Acquire 100,000 cells per sample. Save the files in FCS format.
8. Prepare a titration curve by plotting the MFI of the FL-1 channel against different concentrations of RNase in an Excel file to determine the optimum concentration of RNase A required to digest the cellular RNA (Figure 17.2).

Step 4: Estimation of DNA concentration

1. Take an exponentially growing cyanobacterial cell suspension grown under different growth conditions.
2. Fix the cells as mentioned in Step 1.
3. Treat the cells with an optimized concentration of RNase A (see Step 3).
4. After treatment with RNase A, stain the cells with an optimum concentration of SyBr Green I (see Step 2).
5. After incubation, analyze the samples using a flow cytometer.
6. Acquire samples at a low flow rate (<100 events/second).
7. Acquire 100,000 cells per sample. Save the files in FCS format.

17.6 DATA ANALYSIS

The live cells from a population of cyanobacteria can be detected by the SyBr Green I fluorescence coming from the cells. SyBr Green I is a cell-permeable dye, and the cells with SyBr Green I fluorescence are considered live cells (Figure 17.1). The quantitative data of live cells can be obtained by applying quadrant stat to the FL-1 vs. FL-3 dot plot. The percentage of live cells can be calculated from the upper right (UR) and the lower right (LR) quadrant of the dot plot (Figure 17.1).

The amount of DNA present inside the cells is directly proportional to the SyBr Green I fluorescence. Hence, the amount of DNA can be represented by the histogram plot of SyBr Green I fluorescence coming from the cells. To obtain a histogram plot of SyBr green I fluorescence, first, gate the cyanobacterial population of interest in the dot plot of FSC vs. FL-3 using BD CellQuest Pro software (Figure 17.2). Thereafter, select the singlet population from the gated population using the dot plot of FL-1H vs. FL-1A to measure the SyBr Green I fluorescence of the selected singlet population. Prepare a histogram plot for SyBr Green I fluorescence, which represents the amount of genomic DNA content present in the cyanobacterial cells (Figure 17.2).

17.7 NOTES/PRECAUTIONS

1. A flow cytometer has a limitation in terms of the length of cell/filament to be analyzed. A flow cytometer can analyze cells/filaments of approximately 1 μm to 50 μm in length. Therefore, it is important to measure the cell/filament length in a light microscope before conducting any flow cytometer–based experiment. Due to the size limitation, it is difficult to study longer filaments of cyanobacteria using flow cytometry. The shearing force, like sonication or ultrasonic sound waves, can be used to break the longer filaments. However,

the impact of applied shearing forces on cell survival should be analyzed before conducting any flow cytometer–based experiment.

2. Samples to be analyzed using a flow cytometer should be in either single-cell suspension or the filament length should be less than 50 μm. The clumping of cells/filaments can clog the nozzle of the flow cytometer.
3. It is important to exclude doublets (two cells together) and select singlets (single-cell population) during DNA content estimation. A doublet is a single event that consists of two independent attached cells. The flow cytometer identifies these two cells as a single event because they pass together through the laser beam. The doublets can provide inaccurate information about DNA content, as two cells are considered single cells by the software.

ACKNOWLEDGMENTS

This work was supported by funding from an Institute of Eminence Incentive Grant, Banaras Hindu University (R/Dev/D/IOE/Incentive/2021-2022/32399) and SERB, New Delhi, India, in the form of a Special Core Grant (SCP/2022/000201) to Shailendra P. Singh. ST (NTA Ref. No.-201610202339) and RG (NTA Ref. No.-211610075296) are thankful to UGC, New Delhi, India for a junior research fellowship. AG thanks ICMR, New Delhi, for a senior research fellowship (3/1/3/JRF-2019/HRD-LS).

REFERENCES

Cassier-Chauvat C, Blanc-Garin V and Chauvat F (2021) Genetic, genomics, and responses to stresses in cyanobacteria: Biotechnological implications. *Genes* 12(4): 500.

Cassier-Chauvat C, Veaudor T and Chauvat F (2016) Comparative genomics of DNA recombination and repair in cyanobacteria: Biotechnological implications. *Frontiers in Microbiology* 7: 1809.

Claessen D, Rozen DE, Kuipers OP, Søgaard-Andersen L and Van Wezel GP (2014) Bacterial solutions to multicellularity: A tale of biofilms, filaments and fruiting bodies. *Nature Reviews Microbiology* 12(2): 115–124.

Darzynkiewicz Z, Huang X and Zhao H (2017) Analysis of cellular DNA content by flow cytometry. *Current Protocols in Immunology* 119(1): 5–7.

Herrero M, Henderson B, Havlík P, Thornton PK, Conant RT, Smith P, Wirsenius S, Hristov AN, Gerber P, Gill M, Butterbach-Bahl K, Valin H, Gamett T and Stehfest E (2016) Greenhouse gas mitigation potentials in the livestock sector. *Nature Climate Change* 6(5): 452–461.

Huisman J, Codd GA, Paerl HW, Ibelings BW, Verspagen JM and Visser PM (2018) Cyanobacterial blooms. *Nature Reviews Microbiology* 16(8): 471–483.

Jones CS and Mayfield SP (2012) Algae biofuels: Versatility for the future of bioenergy. *Current Opinion in Biotechnology* 23(3): 346–351.

Kaneko T, Tanaka A, Sato S, Kotani H, Sazuka T, Miyajima N, Hirosawa M, Sugiura M, Sasamoto S, Kimura T, Hosouchi T, Matsuno A, Muraki A, Nakazaki N, Naruo K, Okumura S, Shimpo S, Takeuchi C, Wada T, Watanabe A, Yamada M, Yasuda M and Tabata S (1995) Sequence analysis of the genome of the unicellular cyanobacterium *Synechocystis* sp. strain PCC 6803. I. Sequence features in the 1 Mb region from map positions 64% to 92% of the genome. *DNA Research* 2(4): 153–166.

Kimura A, Hamada T, Morita EH and Hayashi H (2002) A high temperature-sensitive mutant of *Synechococcus* sp. PCC 7002 with modifications in the endogenous plasmid, pAQ1. *Plant and Cell Physiology* 43(2): 217–223.

Meeks JC, Elhai J, Thiel T, Potts M, Larimer F, Lamerdin J, Predki P and Atlas R (2001) An overview of the genome of *Nostoc punctiforme*, a multicellular, symbiotic cyanobacterium. *Photosynthesis Research* 70: 85–106.

Nagarajan S, Sherman DM, Shaw I and Sherman LA (2012) Functions of the duplicated hik31 operons in central metabolism and responses to light, dark, and carbon sources in *Synechocystis* sp. strain PCC 6803. *Journal of Bacteriology* 194(2): 448–459.

Neilan BA, Pearson LA, Muenchhoff J, Moffitt MC and Dittmann E (2013) Environmental conditions that influence toxin biosynthesis in cyanobacteria. *Environmental Microbiology* 15(5): 1239–1253.

Nicholson ML, Gaasenbeek M and Laudenbach DE (1995) Two enzymes together capable of cysteine biosynthesis are encoded on a cyanobacterial plasmid. *Molecular and General Genetics MGG* 247: 623–632.
Oliver JW and Atsumi S (2014) Metabolic design for cyanobacterial chemical synthesis. *Photosynthesis Research* 120: 249–261.
Scanlan DJ, Ostrowski M, Mazard S, Dufresne A, Garczarek L, Hess WR, Hagemann M, Paulsen I and Partensky F (2009) Ecological genomics of marine picocyanobacteria. *Microbiology and Molecular Biology Reviews* 73(2): 249–299.
Singh SP, Häder D-P and Sinha RP (2010) Cyanobacteria and ultraviolet radiation (UVR) stress: Mitigation strategies. *Ageing Research Reviews* 9: 79–90.
Whitton BA and Potts M (2012). Introduction to the cyanobacteria. In: Ecology of Cyanobacteria II: Their Diversity in Space and Time (Ed. B.A. Whitton). Springer Dordrecht, pp. 1–13. ISBN: 978-94-007-3855-3
Wijffels RH, Kruse O and Hellingwerf KJ (2013) Potential of industrial biotechnology with cyanobacteria and eukaryotic microalgae. *Current Opinion in Biotechnology* 24(3): 405–413.

18 Genomic DNA and RNA Extraction from Cyanobacteria

Anjali Gupta, Soumila Mondal, Sapna Tiwari, Donat-P. Häder, Rajeshwar P. Sinha, and Shailendra P. Singh

18.1 INTRODUCTION

Cyanobacteria are an ancient group of Gram-negative, oxygenic photoautotrophs that contribute significantly to global carbon dioxide (CO_2) and nitrogen (N_2) fixation (Falkowski 1997; Whitehead et al. 2014). Phenotypic characterization combined with nucleic acid sequence–based molecular approaches is used to identify and classify a wide range of cyanobacteria occurring in diverse habitats (De Clerck et al. 2013; Sharma et al. 2014). The first and most important step in the molecular characterization of cyanobacteria, or any other organism, is to extract a sufficient amount of genomic DNA of high quality, which is further used as a template for the amplification of target genes for sequencing and analysis. Also, cyanobacteria are emerging model systems for the production of valuable chemicals by metabolic engineering that require alteration in the genetic makeup of the organism (Angermayr et al. 2009; Atsumi et al. 2009; Lindberg et al. 2010; Aikawa et al. 2014).

Several cyanobacterial strains produce high amounts of mucilaginous polysaccharides which impede cell lysis and therefore interfere with most of the nucleic acid extraction protocols (Porter 1988; Wilkins and Smart 1996; Margheri et al. 1999; Tillett and Neilan 2000). The most common problems associated with cyanobacterial DNA isolation are inefficient cell lysis and purification (Porter 1988; Billi et al. 1998; Fiore et al. 2000). Polyphenols, nucleases and photosynthetic pigments that are abundantly found in cyanobacteria are difficult to remove from nucleic acids. These impurities of nucleic acids inhibit enzymatic reactions such as reverse transcription, polymerase chain reaction (PCR) amplification, restriction digestion and manipulation of DNA (Cardellina et al. 1993; Lau et al. 1993; Cohen et al. 1994; Tillett and Neilan 2000). Therefore, several methods have been developed over time which involve various means of cell lysis and purification steps (Table 18.1). However, all nucleic acid extraction methods include cell lysis followed by deproteination, precipitation and elution. Different methods developed for the extraction of nucleic acids from cyanobacteria differ in terms of their extraction efficiency and time duration (Table 18.1).

Singh et al. (2011) developed a method for the extraction of genomic DNA from sheathed and non-sheathed cyanobacteria that does not require lysozyme treatment or sonication to lyse the cells (Figure 18.1). This method uses a lysis buffer to lyse the cells and does not require treatments to remove the outer layer of the cell wall or to break the clumps. The genomic DNA extracted by this method is of high quantity and quality and can be used for a variety of applications such as DNA sequencing, PCR amplification, genotyping, etc. This method is also applicable to a wide range of cyanobacterial samples (Singh et al. 2011). The first step of the extraction protocol uses a lysis buffer (see recipe) and excludes the requirement of enzymatic

DOI: 10.1201/9781003398387-18

TABLE 18.1
Different Methods Developed for the Extraction of Genomic DNA from Cyanobacteria

Methods	References
Boom's method or guanidium isothiocyanate/silica	Boom et al. 1990
Cesium-chloride and ultracentrifugation	de Lorimer et al. 1984; Reith et al. 1986; Golden et al. 1987; Joset 1988; Giovannoni et al. 1990; Leff et al. 1995
Enzyme digestion	de Lorimer et al. 1984; Golden et al. 1987; Joset 1988; Porter 1988; Smoker and Barnum 1988; Giovannoni et al. 1990; Cohen et al. 1994; Palenik 1994; Jackman and Mulligan 1995; Nishihara et al. 1997
Enzymatic and mechanical cell lysis	Morin et al. 2010
Freeze/Thaw	Motham et al. 2014
Glass beads and enzyme	Motham et al. 2014
Glass beads and liquid nitrogen	Motham et al. 2014
Glass beads and sonication	Singh et al. 2012; Motham et al. 2014
Glass beads in the presence of extraction buffer	Harrison and Thompson 2020
Glass beads in the presence of hot phenol	Billi et al. 1998
Dynabeads	Cler et al. 2006
Grinding under liquid nitrogen or dry ice	Van der Plas et al. 1989; Luo and Stevens 1997
High concentration of salt (NaCl) in the presence of isopropanol	Wu et al. 2000
Liquid nitrogen and polyvinylpolypyrrolidone	Yilmaz et al. 2009
Lysozyme	Pitcher et al. 1989; Ermakova-Gerdes and Vermaas 1999; Margheri et al. 1999; Wu et al. 2000; Saha et al. 2005; Morin et al. 2010
Mechanical cell breakage (homogenization, ultrasonication, French press)	Reith et al. 1986; Golden et al. 1987; Leff et al. 1995; Luo and Stevens 1997; West and Adams 1997; Sinha et al. 2001; Srivastava et al. 2007
Multiple organic solvent extractions	Giovannoni et al. 1990; Cohen et al. 1994; Ferris et al. 1996; Golden et al. 1987; Jackman and Mulligan 1995; Leff et al. 1995; Lotti et al. 1996; Mak and Ho 1991; Neilan 1995; Nishihara et al. 1997; Palenik 1994; Porter 1988; Smoker and Barnum 1988; West and Adams 1997; Zehr and McReynolds 1989
Salt (NaCl) and glass beads	Motham et al. 2014
Silica (glass beads)	Billi et al. 1998; Fiore et al. 2000; Guevara et al. 2002
Xanthogenate	Jhingan 1992
Xanthogenate-SDS nucleic acid isolation	Tillett and Neilan 2000
XS-3% PVPP-PEG	Yilmaz et al. 2009
Temporary heat shock with increased concentrations of detergent	Saba et al. 2016
CTAB	Murray and Thompson 1980
Potassium ethyl xanthogenate (XS buffer)	Moza and Postolache 2022
Chaotropic agents	Holm et al. 1986
Guanidine isothiocyanate	Lippke et al. 1987
Lysis buffer with vortex/Freeze/Thaw	Bhardwaj et al. 2019
Phenol/Chloroform	Gaget et al. 2017

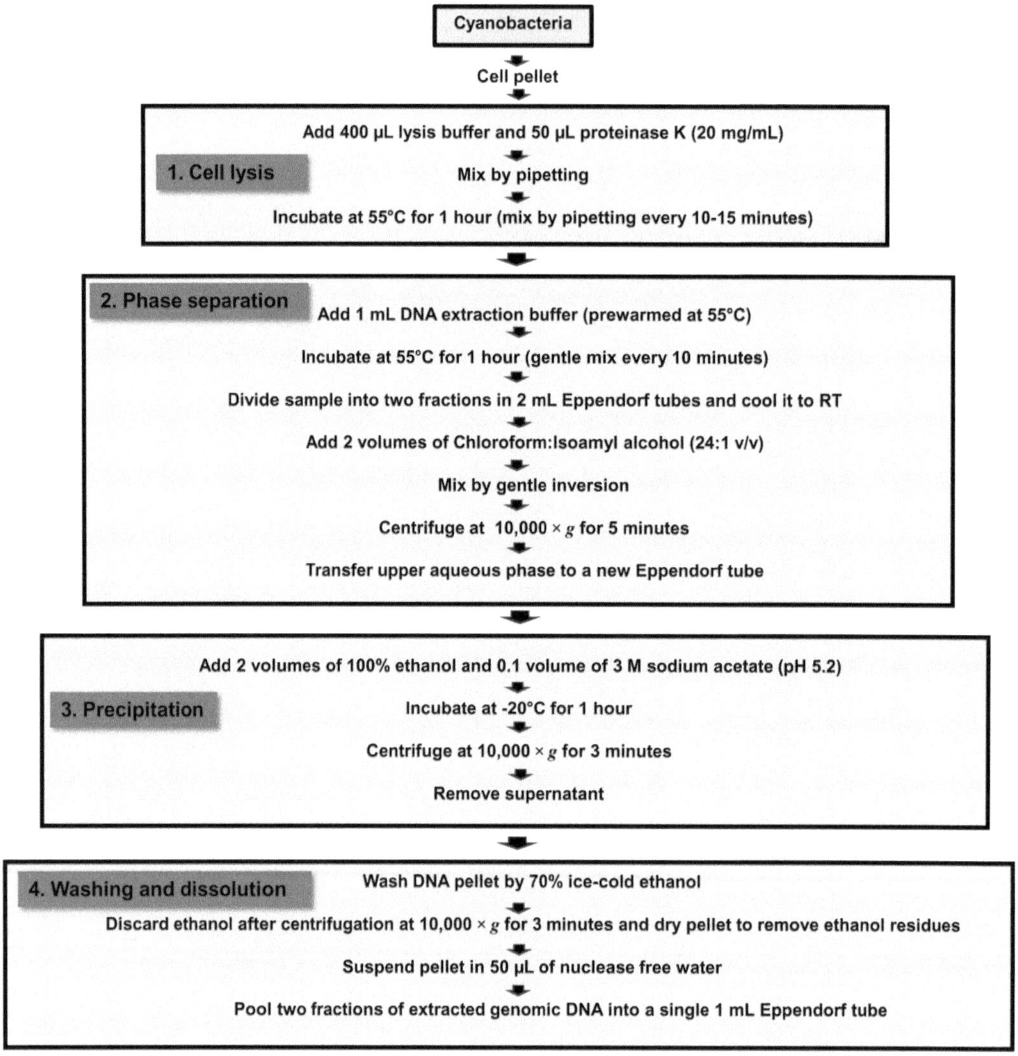

FIGURE 18.1 Flowchart showing different steps involved in the extraction of genomic DNA from cyanobacteria.

and/or mechanical disruption treatment to lyse the cell (Sinha et al. 2001). Also, the protocol given by Singh et al. (2011) does not require additional washing steps to remove the exopolysaccharides for efficient extraction of genomic DNA (Margheri et al. 1999; Fiore et al. 2000; Morin et al. 2010). Application of sarkosyl (0.1%) (Wu et al. 2000) and silica (Saha et al. 2005) has been used earlier to improve cell lysis, but these two treatments may cause shearing of high-molecular-weight DNA.

Similarly, efficient extraction of RNA is needed to study the expression of different genes (qRT-PCR, Northern blotting, microarray, RNA sequencing) under different environmental or developmental conditions in wild-type and/or genetically engineered strains of cyanobacteria.

Several methods have been developed for the extraction of RNA from different cyanobacteria (Table 18.2); however, these methods were developed for individual cyanobacteria, except for a few methods. Here, in addition to the DNA extraction, we also present a modified method that can be used for the extraction of RNA from both unicellular and filamentous cyanobacteria (Figure 18.2) (Singh et al. 2010; Singh and Montgomery 2013).

18.2 EQUIPMENT

1. Autoclave (NSW-227, NSW-India).
2. Refrigerated centrifuge (NEYA16R, Remi, India).

TABLE 18.2
Different Methods Developed for the Extraction of RNA from Cyanobacteria

RNA Extraction Solution	Homogenization Method			Cyanobacteria	References
	Physical	Enzymatic	Heating		
TRIzol	Pestle and mortar	_	_	*Microcystis aeruginosa* PCC 7806, *Planktothrix agardhii* 126/3	Kaebernick et al. 2000; Dittmann et al. 2001; Tonk et al. 2005
XSD (Xanthogenate-SDS-phenol)	_	_	65°C	*M. aeruginosa* PCC 7806, *M. wesenbergii* NIES 107, *M. viridis* NIES 102, *M. incerta* HINDAK 1965/17, *Nodularia spumigena* PCC 73104, *Pseudanabaena* sp. AWT 210 and *Anabaena circinalis* AWT 006, Cyanobacterial blooms	Tillett and Neilan 2000
Phenol-chloroform (BPC)	Zirconia/Silica bead	_	_	*Merismopedia tenuissima* NIES 230, *Synechococcus leopoliensis* UTEX 625, *Arthrospira platensis* NIES 39, *Oscillatoria tenuis* NIES 33, *Phormidium* sp. KCTC AG10164, *Synechocystis* sp. PCC 6803, *Anabaena flos-aquae* UTEX 2557, *Nodularia spumigena* UTEX 2092, *Nostoc* sp. PCC 7120, *Microcystis aeruginosa* UTEX 2666	Kim et al. 2006
PGTX (Phenol, glycerol, guanidine and triton X)	_	_	95°C	*Nostoc punctiforme* ATCC 29133	Pinto et al. 2009
TRIzol	Freeze and thaw	_	_	*Anabaena variabilis* PCC 7937	Singh et al. 2010
Mechanical disruption in PMR1 solution (MO BIOLaboratories, Inc.)	Bead beating	_	_	*Anabaena* sp. 90 and *Microcystis aeruginosa* PCC 7806	Sipari et al. 2010
TRIzol	_	_	95°C	*Planktothrix agardhii* CYA126/8	Tran et al. 2013

(Continued)

TABLE 18.2 *(Continued)*
Different Methods Developed for the Extraction of RNA from Cyanobacteria

RNA Extraction	Homogenization Method				
Solution	**Physical**	**Enzymatic**	**Heating**	**Cyanobacteria**	**References**
TRIzol (Cells transferred after enzymatic digestion)	Bead beating	Lysozyme	_	Cyanobacteria collected from freshwater reservoir	Penn et al. 2014
TRIzol	Bead beating	_	_	*Planktothrix agardhii* LMECYA 256 and *Microcystis aeruginosa* LMECYA 7	Salvador et al. 2016
Acidic phenol	Liquid nitrogen	_	65°C	*Trichodesmium* sp.	Kramer et al. 1996
Hot phenol	Liquid nitrogen using a mortar and pestle	_	65°C	*Synechococcus elongates* PCC 7942, *Anabaena* variabilis PCC 7937	Bovy et al. 1993
Lithium chloride (LiCl)	Liquid nitrogen using a mortar and pestle	_	80°C	*Arthrospira (Spirulina) platensis* PCC 7345	Pathak and Lochab 2010
TRIzol/Chloroform	Liquid nitrogen using a mortar and pestle, ultra-sonication, beads etc.	_	_	*Planktothrix agardhii*	Kim Tiam et al. 2019
Hot phenol	_	_	65°C	*Synechocystis* sp. PCC 6803	Mironov and Los 2015

Source: The table was developed based on the information given by Kim Tiam et al. (2019).

3. Water bath (LABQUEST, BOROSIL, Model: WBS020, India).
4. −20°C refrigerator (Samsung, Korea).
5. −80°C freezer (Haier, China).
6. Icebox (Cello, India).
7. Vortexer (LABQUEST, BOROSIL, Model: VMS, India).
8. UV-VIS Spectrophotometer 1800 (Shimadzu, catalog number: 80626, Japan).
9. Agarose gel electrophoresis (Genetix, catalog number: H815181, India).
10. UV trans-illuminator (Laczene Bioscience, RF: 12S6025H, India).
11. Micropipettes (Eppendorf, Germany).
12. Thermocycler (BIO-RAD, USA, Model: C1000 Touch™).
13. UV tray (BIO-RAD, USA, catalog no: 1708271).
14. Gel dock (BIO-RAD, USA, Model: Gel Dock™ EZ Imager).
15. NanoDrop (Thermofisher Scientific, USA).
16. Stirring plate with heating (Tarson, SPINNOT Model: 02, catalog Number: 6030).

18.3 MATERIALS

1. Lysis buffer (see recipe later).
2. DNA extraction buffer (see recipe later).
3. TAE buffer (see recipe later).

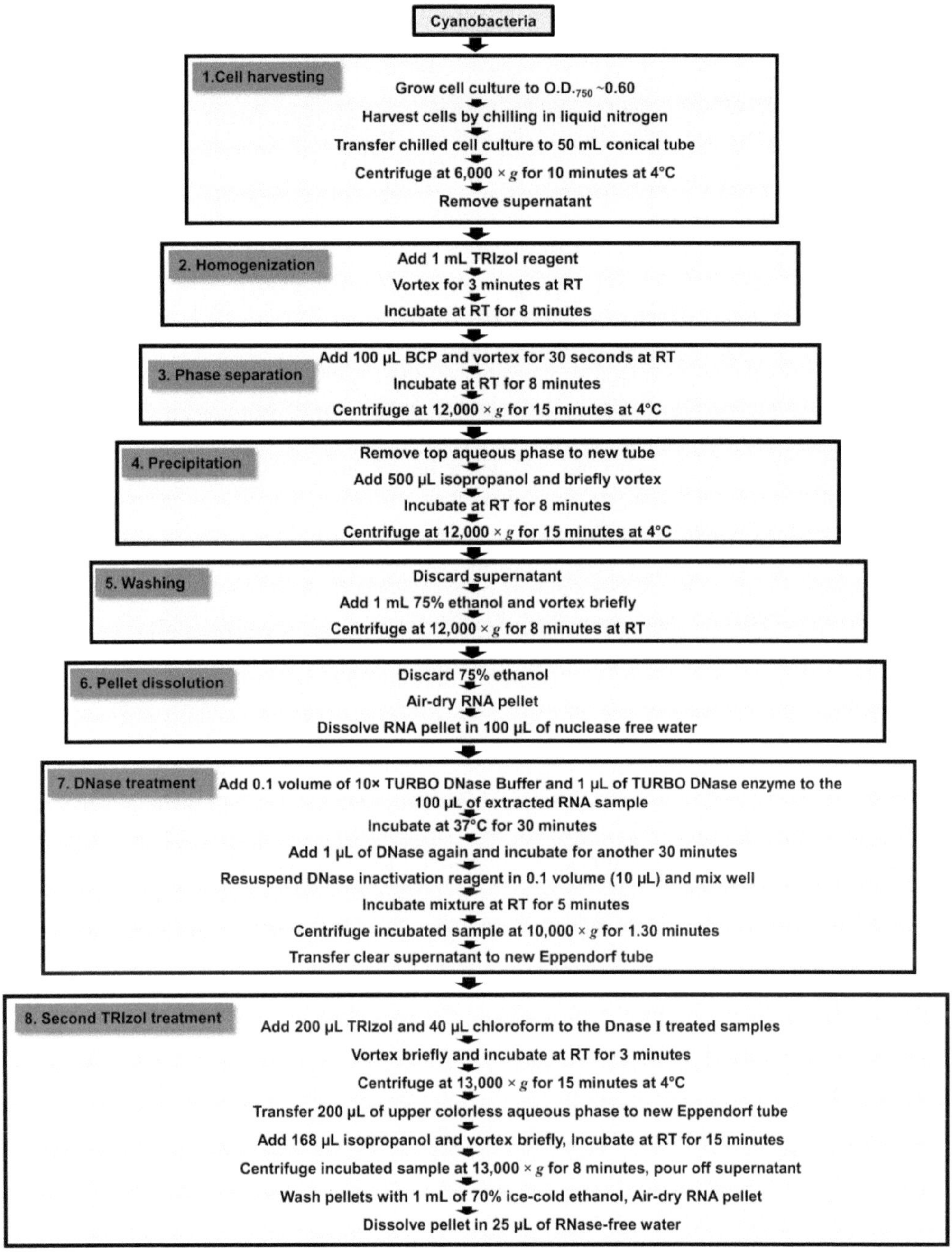

FIGURE 18.2 Flowchart describing different steps involved in the extraction of RNA from cyanobacteria.

4. 0.8%, 1.0% agarose gel (UltaPure™ Agarose, Invitrogen).
5. 1.5 mL microcentrifuge tube, amber (Tarsons, catalog number: 500010X, India).
6. 1 mL micro tip (Abdos, catalog number: P10109, India).
7. 200 µL micro tip (Abdos, catalog number: P10140, India).
8. 10 µL micro tip (Abdos, catalog number: P10116, India).

9. PCR tubes (ABDOS, catalog number: P10701, India).
10. 1 cm path-length quartz cuvette (Quartz cuvette, catalog number: MCQ-254, India).
11. 50 mL plastic conical polypropylene tubes (Abdos, catalog number: P10404, India).
12. 250 mL flask (Borosil, catalog number: 4980021).
13. 100 mL graduated cylinder (Borosil, catalog number: 3021016).
14. 50 mL graduated cylinder (Borosil, catalog number: 3012).
15. 20 mL 1 M HEPES pH 8.0 (Himedia, catalog number: RM380-500G).
16. Absolute ethanol (Merck, catalog number: K48191683632).
17. Isopropanol (SRL, catalog number: 5383952).
18. Reverse transcription system (Promega, catalog number: A3500).
19. Proteinase K (Merck, catalog number: V017025168).
20. TRIzol (Ambion, catalog number: 15596026).
21. TURBO DNA-free™ Kit (Ambion, catalog number: AM1907).
22. 1-Bromo-3-chloropropane (Himedia, catalog number: RM5036).
23. 3 M sodium acetate (pH 5.2) (Merck, catalog number: QC5Q650524).
24. Chloroform:isoamyl alcohol (24:1 v/v) (Himedia, catalog number: MB115).
25. Nuclease-free water (Promega, catalog number: P119E).
26. Loading dye (Thermo Scientific, catalog number: R0611).
27. GeneRuler (Thermo Scientific, catalog number: SM0313).
28. Ethidium bromide (Himedia, catalog number: MB071-1G).
29. RNAse-Off (Takara, catalog number: 9037).
30. Autoclaved double distilled or Milli-Q (MQ) water.
31. Exponentially growing cultures (0.6 OD_{750}) of *Anabaena variabilis* PCC 7937, *Anabaena* sp. PCC 7120, *Synechocystis* sp. PCC 6803, *Synechococcus* sp. PCC 6301, *Rivularia* sp. HKAR-4, *Fremyella diplosiphon* SF33 and *Synechococcus elongatus* PCC 7942.

Note: All cyanobacteria were routinely subcultured in a sterilized BG-11 medium supplemented with 20 mM HEPES buffer (pH 7.1) (Allen 1968) in a culture room at 25 ± 2°C under white fluorescent light (~80 μmol m^{-2} s^{-1} photons). *F. diplosiphon* SF33 was grown under red light (~20 μmol m^{-2} s^{-1} photons). All cultures were grown using continuous light with shaking at 150 rpm. The extraction of DNA and RNA was performed from 50 mL of exponentially growing cyanobacterial cultures (0.6 OD_{750}) in three replicates.

18.4 SOFTWARE

1. Image Lab (Bio-Rad).

18.5 RECIPES

18.5.1 Lysis Buffer for DNA Extraction

Add 4.804 g urea, 0.630 g Tris-HCl, 0.023 g NaCl and 1.169 g EDTA to 10 mL of autoclaved MQ water. Mix until completely dissolved and make the final volume 20 mL using MQ water. Store at 4–8°C.

18.5.2 DNA Extraction Buffer

Dissolve 0.75 g CTAB, 2.045 g NaCl, 0.146 g EDTA and 0.344 g Tris-HCl in 20 mL of autoclaved MQ water. Adjust the pH to 8.0, and thereafter, add 0.25 g sarkosyl and 0.25 mL of mercaptoethanol into the same solution. Adjust the final volume of buffer to 25 mL using MQ water. Store the DNA extraction buffer at 4–8°C and warm it at 55°C with proper mixing before use.

18.5.3 Proteinase K

Dissolve 20 mg of Proteinase K in 1 mL of autoclaved MQ water to get a 20 mg/mL stock solution. Store at –20°C.

18.5.4 Sodium Acetate (pH 5.2)

Dissolve 2.461 g of sodium acetate in 5 mL of autoclaved MQ water. Adjust the pH to 5.2 using HCl or NaOH and make the final volume 10 mL using MQ water. Store at room temperature (RT; 25°C).

18.5.5 70% Ethanol

Add 70 mL absolute ethanol (molecular grade) in 30 mL of MQ water. Store at RT.

18.5.6 75% Ethanol

Add 75 mL absolute ethanol (molecular grade) in 25 mL of MQ water. Store at RT.

18.5.7 50× TAE Buffer (pH 8.0)

Add 57.1 mL of glacial acetic acid into 500 mL of MQ water. Then add 242 g of Tris and 100 mL of 0.5 M EDTA (pH 8.0). Stir until dissolved. Transfer the solution to a 1-liter measuring cylinder. Bring the final volume to 1 liter with MQ water and store it at RT.

18.5.8 1% Agarose Gel

Add 1 g agarose into 100 mL of 1× TAE buffer. Heat it at 50°C until completely dissolved.

18.6 METHODS

18.6.1 DNA Extraction (Figure 18.1)

a. *Cell lysis*:
   1. Harvest 50 mL of exponentially growing cyanobacterial cells (0.6 OD_{750}) by centrifugation at 3000 × *g* for 5 minutes.
   2. Discard the supernatant and transfer cells to a 2-mL nuclease-free Eppendorf tube.
   3. Suspend cells in 400 µL of lysis buffer. Add 50 µL of Proteinase K and mix immediately by pipetting.
   4. Incubate samples for 1 hour at 55°C with mixing every 10–15 minutes by pipetting.

b. *DNA extraction*:
   1. After incubation, add 1 mL of pre-warmed (55°C) DNA extraction buffer, mix gently by inverting the tubes and incubate samples for another hour at 55°C.
   2. Gently invert the Eppendorf tubes four to five times every 10 minutes for proper mixing during incubation.
   3. Divide the sample into two fractions in 2-mL Eppendorf tubes and cool it to RT.
   4. Add two volumes of chloroform:isoamyl alcohol (24:1 v/v) and gently invert tubes until a white emulsion is formed.
   5. After 5 minutes' centrifugation at 10,000 × *g*, transfer 500 µL of the upper water phase to a new 2-mL Eppendorf tube.

c. *Precipitation, washing and DNA elution*:
   1. Add two volumes of 100% ice-cold ethanol and 0.1 volume of ice-cold 3 M sodium acetate (pH 5.2). Mix by inverting the tube once and keep it at –20°C for 1 hour.

2. After 1 hour incubation, centrifuge samples at 10,000 × *g* for 3 minutes and discard the supernatant.
3. Wash the pellet once with 70% ice-cold ethanol (500 μL) and dry the pellet after the complete removal of ethanol.
4. Suspend the pellet in 50 μL of nuclease-free water.

d. Finally, pool the extracted genomic DNA (see Step B3) into a single 1-mL Eppendorf tube.

e. *Storage*:

Extracted DNA samples should be kept at –20°C or –80°C for long-term storage.

f. *Quantification*:

1. Using a spectrophotometer or NanoDrop, quantify the DNA sample. The concentration of DNA is determined by the absorption at 260 nm (1 OD at 260 nm equals 50 μg/mL of DNA).
2. The ratio of absorption at 260 and 280 nm, as well as running DNA samples in 0.8% agarose gel electrophoresis, can tell a lot about the quality of extracted DNA The intact single band close to the loading site in an agarose gel and 260/280 ratio more than 1.6 suggests a good quality of genomic DNA.

18.6.2 RNA Extraction (Figure 18.2)

a. *Cell harvesting*:

1. Exponentially growing cyanobacterial cultures (50 mL) are diluted to OD_{750} ~0.6 and placed in the appropriate growth condition overnight.
2. The next morning, submerge the flasks in liquid nitrogen to bring the temperature of the cell culture to ~4°C.
3. Transfer the cell suspension to labeled 50-mL polypropylene tubes.
4. Centrifuge the cells at 6000 × *g* for 10 minutes at 4°C and pour off as much supernatant as possible.
5. Transfer the cell pellet to a labeled 2-mL Eppendorf tube.
6. Centrifuge the cells at 6000 × *g* for 5 minutes at 4°C to remove the remaining growth medium.

b. *Homogenization*:

1. Add 1 mL TRIzol reagent to the cell pellet.
2. Vortex the tube for 3 minutes at RT.
3. Incubate the sample at RT for 8 minutes.

c. *RNA extraction*:

1. Add 100 μL BCP to each tube and vortex for 30 seconds at RT.
2. Incubate the sample at RT for 8 minutes.
3. After incubation, centrifuge the sample at 12,000 × *g* for 15 minutes at 4°C.

d. *Phase separation and RNA precipitation*:

1. After centrifugation, transfer the upper clear aqueous phase to a new 1.5-mL Eppendorf tube without disturbing the interface.
2. Add 500 μL isopropanol to each tube containing the aqueous phase followed by brief mixing by vortex and incubation at RT for 8 minutes.
3. After incubation, centrifuge the samples at 12,000 × *g* for 15 minutes at 4°C and remove the supernatant. At this point, a jelly-like substance may appear at the bottom of the tubes.

e. *RNA wash and elution*:

1. Add 1 mL of 75% ice-cold ethanol and vortex briefly.
2. Centrifuge at 1200 × *g* for 8 minutes and pour off the supernatant.
3. Dry the pellet by inverting the tube for 5 minutes to remove the ethanol.
4. Add 100 μL of nuclease-free water to each tube to dissolve the pellet.

f. *Quantification and quality check*:
   1. After extraction, store RNA samples at –80°C.
   2. Quantify the concentration of the extracted RNA using a spectrophotometer or NanoDrop. The absorption at 260 nm gives the concentration of RNA (1 OD at 260 nm equals 40 µg/mL RNA).
   3. Check the purity of the RNA by calculating the 260/280 ratio as well as by running RNA samples in 1.5% agarose gel electrophoresis. A 260/280 ratio higher than 1.7 and intact bands of rRNA in agarose gel indicate good quality of RNA.

18.6.3 DNase Treatment

To remove the contamination of genomic DNA, extracted RNA samples are treated with DNase I enzyme (TURBO DNA-free Kit or similar product), as DNA contamination in RNA samples can affect the interpretation of gene expression studies.

1. Add 0.1 volume (10 µL) of 10× TURBO DNase buffer and 1 µL of TURBO DNase enzyme to the 100 µL of RNA sample.
2. Incubate at 37°C for 30 minutes with regular mixing by gently tapping the tubes.
3. Add 1 µL of DNase again and incubate the samples for another 30 minutes at 37°C with gentle mixing at regular intervals.
4. Add 0.1 volume (10 µL) of DNase inactivation reagent and incubate for 5 minutes at RT after proper mixing.
5. After incubation, centrifuge the samples at 10,000 × *g* for 1.5 minutes.
6. Transfer the clear supernatant to a new 1.5-mL Eppendorf tube.

18.6.4 Second-Time TRIzol Treatment

1. Add 200 µL of TRIzol and 40 µL of chloroform to the DNase I–treated RNA sample.
2. Vortex the sample briefly and incubate at RT for 3 minutes.
3. Centrifuge the samples at 13,000 × *g* for 15 minutes at 4°C.
4. Transfer 200 µL of the upper transparent aqueous phase to a new 1.5-mL Eppendorf tube.
5. Add 168 µL of isopropanol and vortex briefly. Incubate at RT for 15 minutes.
6. After incubation, centrifuge the samples at 13,000 × *g* for 8 minutes at 4°C and carefully remove the supernatant without touching the bottom surface of the Eppendorf tube. One may or may not see the pellet at this step and shall move to the next step in any case.
7. Wash the pellet with 1 mL of 70% ice-cold ethanol. Remove the supernatant after centrifugation at 12,000 × *g* for 8 minutes at 4°C. Dry the pellet at RT to remove residues of ethanol.
8. Dissolve the pellet in 25 µL of RNase-free water.
9. Perform qualitative and quantitative analyses of RNA samples as described earlier in the quantification and quality check section (see steps F2 and F3).

18.6.5 Reverse Transcriptase Polymerase Chain Reaction (RT-PCR)

1. Denature 0.5 µg of RNA in 9.8 µL of RNase-free water at 70°C for 10 minutes. Afterward, keep the sample at 4°C until the next step.
2. Add 10.2 µL of master mix containing 2 µL of 10× buffer, 4 µL of 25 mM $MgCl_2$, 2 µL of dNTPs, 1 µL of random primer (100 pmol reverse primer), 0.5 µL of RNasin and 0.7 µL of Amv-RT to the denatured RNA sample. The final volume will be 20 µL at this step.
3. Use water instead of Amv-RT for the No-RT reaction mixture.
4. Incubate 20 µL of the reaction mixture at 24°C for 10 minutes followed by 42°C for 1 hour and 95°C for 5 minutes.

5. After completion of these steps, keep the sample on ice and add 30 μL of nuclease-free water to the cDNA sample. Two μl or any other optimized volume of cDNA can be used in a conventional RT-PCR or qPCR.

18.7 DATA ANALYSIS

The quantification of nucleic acids and their quality check was done by a NanoDrop spectrophotometer. The purity of extracted DNA and RNA was checked by obtaining the A_{260}/A_{280} ratio. A ratio of A_{260}/A_{280} of ~1.8 generally suggests good-quality DNA, and a ratio of A_{260}/A_{280}, which is somewhere between 1.7 and 2.0, suggests good-quality RNA. NanoDrop spectrophotometer data indicated that nucleic acids (DNA and RNA) extracted from different cyanobacteria using the methods presented here are of good quality and quantity (Tables 18.3 and 18.4). Agarose gel electrophoresis analysis suggested that the integrity of genomic DNA and RNA was preserved during their extraction, and their yield was also high (Figures 18.3 and 18.4).

18.8 NOTES/PRECAUTIONS

18.8.1 Quality Control and Precautions to Be Taken during DNA Extraction

1. Samples should not be contaminated with other sources of DNA because this will affect the results.
2. All materials used during extraction should be free of DNase contamination.
3. During the extraction procedure, avoid excessive vortexing or shearing of samples.
4. The DNA extraction buffer and lysis buffer should be freshly prepared.
5. Residuals of 70% ethanol in the final step of washing should be completely removed before dissolving the DNA pellet in nuclease-free water. Overdrying the pellet should be avoided.
6. Barcode labels can be used instead of hand labeling to avoid identification and swapping errors while handling a large number of samples.

18.8.2 Quality Control and Precautions to Be Taken during RNA Extraction

1. The most important quality control parameters considered during RNA extraction are quantity, purity and integrity of the RNA.
2. Before performing the extraction procedure, clean the workspace, pipettes and gloves with 70% ethanol. Also, sterilize conical tubes, Eppendorf tubes and pipette tips using an autoclave before use.

TABLE 18.3
The Concentration and Quality of Genomic DNA Extracted from Different Cyanobacteria

Cyanobacteria	260/280 ratio (A260/A280)	Concentration of DNA (ng/μL)	References
Rivularia sp. HKAR-4	1.93 ± 0.15	478.5 ± 40.31	Singh et al. 2011
Anabaena sp. PCC 7120	1.84 ± 0.14	362.5 ± 17.68	Singh et al. 2011
Anabaena variabilis PCC 7937	1.89 ± 0.07	230 ± 28.28	Singh et al. 2011
Synechocystis sp. PCC 6803	1.84 ± 0.11	433.5 ± 144.96	Singh et al. 2011
Synechococcus sp. PCC 6301	1.91 ± 0.07	259.5 ± 57.28	Singh et al. 2011
Synechococcus elongatus PCC 7942	1.86 ± 0.11	780.53 ± 187.71	This Study
Fremyella diplosiphon SF33	1.88 ± 0.09	249.1 ± 156.24	This Study

TABLE 18.4
Concentration and Quality of RNA Extracted from Different Cyanobacteria

Cyanobacteria	A260/280	Concentration of RNA (ng/μL)	References
Anabaena variabilis PCC 7937	1.973 ± 0.0124	167.6 ± 34.266	Singh et al. 2010
Synechococcus elongatus PCC 7942	1.983 ± 0.0094	247.7 ± 17.140	This Study
Fremyella diplosiphon SF33	1.85 ± 0.035	1127.66 ± 169.01	This Study

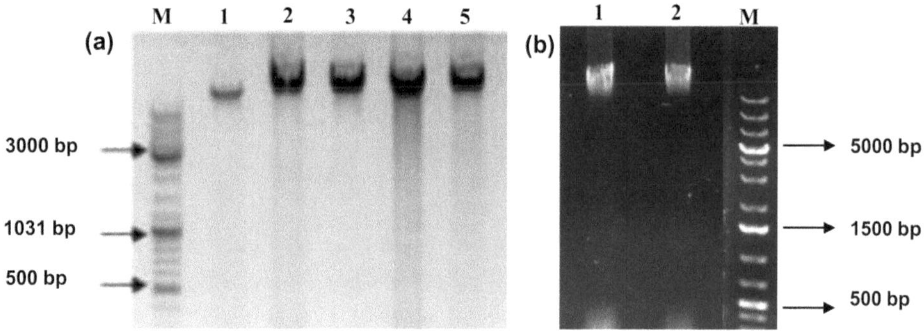

FIGURE 18.3 Agarose gel electrophoresis of extracted genomic DNA. GeneRuler used as a marker (lane M); *Rivularia* sp. HKAR-4 (lane 1); *Anabaena* sp. PCC 7120 (lane 2); *Anabaena variabilis* PCC 7937 (lane 3); *Synechocystis* sp. PCC 6803 (lane 4); *Synechococcus* sp. PCC 6301 (lane 5) (a); *Synechococcus* sp. PCC 7942 (lane 1); *Fremyella diplosiphon* SF33 (lane 2) (b). The representative gel image (a) is based on our previous work (Singh et al. 2011).

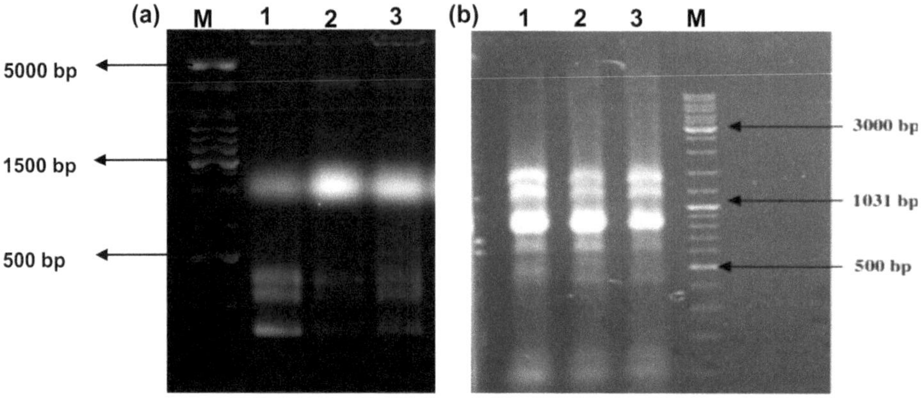

FIGURE 18.4 Agarose gel electrophoresis of total RNA (three replicates) extracted from *Synechococcus elongatus* PCC 7942 (a) and *Anabaena variabilis* PCC 7937 (b). Five μL of extracted RNA was run on 1% agarose gel in 1× TAE buffer. Image b is from our previous work (Singh et al. 2010).

3. Always wear gloves and make sure the working desk, pipettes, pipette tips and solutions are RNase-free. RNase-off solution should be used every time during the extraction procedure.
4. While working with TRIzol, avoid using bleach because the two are incompatible, and TRIzol itself inactivates biological materials.
5. Incomplete homogenization or lysis of cyanobacterial cells gives a very low yield of RNA.
6. Remove all traces of ethanol before dissolving RNA in nuclease-free water.
7. Purified RNA can be stored at −20°C for a short period. However, storing RNA at −80°C in single-use aliquots can help to prevent RNA damage/degradation from multiple freeze-thaw events and accidental RNase contamination.
8. During the 5-minute incubation period of DNase treatment, mixing is required to re-disperse the DNase inactivation reagent. This efficiently removes excess DNase enzyme and divalent cations such as magnesium and calcium that could catalyze RNA degradation.

ACKNOWLEDGMENTS

This work was supported by funding from the Institute of Eminence Incentive Grant, Banaras Hindu University (R/Dev/D/IOE/Incentive/2021-2022/32399) and SERB, New Delhi, India, in the form of a core research grant (SCP/2022/000201) to Shailendra P. Singh. Anjali Gupta and Sapna Tiwari are thankful to the ICMR, New Delhi, India (3/1/3/SRF -2019/HRD-LS), and UGC, New Delhi, India (NTA Ref. No.: 201610202339), respectively.

REFERENCES

Aikawa S, Nishida A, Ho SH, Chang JS, Hasunuma T and Kondo A (2014) Glycogen production for biofuels by the euryhaline cyanobacteria *Synechococcus* sp. Strain PCC 7002 from an oceanic environment. *Biotechnology for Biofuels* 7(1): 8.

Allen M (1968) Simple conditions for growth of unicellular blue green algae. *Journal of Phycology* 4: 1–3.

Angermayr SA, Hellingwerf KJ, Lindblad P and de Mattos MJT (2009) Energy biotechnology with cyanobacteria. *Current Opinion in Biotechnology* 20: 257–263.

Atsumi S, Higashide W and Liao JC (2009) Direct photosynthetic recycling of carbon dioxide to isobutyraldehyde. *Nature Biotechnology* 27: 1177–1180.

Bhardwaj B, Agrawal A, Ledwani L, Vaiphei ST and Jain S (2019) An efficient method for DNA extraction from cyanobacteria isolated from hypersaline and marine environments. *Journal of Phycology* 55: 733–737.

Billi D, Grilli Caiola M, Paolozzi L and Ghelardini P (1998) A method for DNA extraction from the desert cyanobacterium *Chroococcidiopsis* and its application to identification of *ftsZ*. *Applied and Environmental Microbiology* 64: 4053–4056.

Boom RCJA, Sol CJ, Salimans MM, Jansen CL, Wertheim-van Dillen PM and Van der Noordaa JPME (1990) Rapid and simple method for purification of nucleic acids. *Journal of Clinical Microbiology* 28: 495–503.

Bovy A, de Vrieze G, Lugones L, van Horssen P, van den Berg C, Borrias M and Weisbeek P (1993) Iron-dependent stability of the ferredoxin I transcripts from the cyanobacterial strains *Synechococcus* species PCC 7942 and *Anabaena* species PCC 7937. *Molecular Microbiology* 7: 429–439.

Cardellina JH, Munro MH, Fuller RW, Manfredi KP, McKee TC, Tischler M and Boyd MR (1993) A chemical screening strategy for the dereplication and prioritization of HIV-inhibitory aqueous natural products extracts. *Journal of Natural Products* 56: 1123–1129.

Cler L, Bu D, Lewis C and Euhus D (2006) A comparison of five methods for extracting DNA from paucicellular clinical samples. *Molecular and Cellular Probes* 20: 191–196.

Cohen MF, Wallis JG, Campbell EL and Meeks JC (1994) Transposon mutagenesis of *Nostoc* sp. strain ATCC 29133, a filamentous cyanobacterium with multiple cellular differentiation alternatives. *Microbiology* 140: 3233–3240.

De Clerck O, Guiry MD, Leliaert F, Samyand Y and Verbruggen H (2013) Algal taxonomy: A road to nowhere? *Journal of Phycology* 49: 215–225.

de Lorimer R, Bryant DA, Porter RD, Liu WY, Jay E and Stevens SE Jr (1984) Genes for the alpha and beta subunits of phycocyanin. *Proceedings of the National Academy of Sciences USA* 81: 7946––7950.

Dittmann E, Erhard M, Kaebernick M, Scheler C, Neilan BA, von Döhren H and Börner T (2001) Altered expression of two light-dependent genes in a microcystin-lacking mutant of *Microcystis aeruginosa* PCC 7806. *Microbiology* 147: 3113–3119.

Ermakova-Gerdes S and Vermaas W (1999) Inactivation of the open reading frame *slr0399* in *Synechocystis* sp. PCC 6803 functionally complements mutations near the QA niche of photosystem II: A possible role of Slr0399 as a chaperone for quinone binding. *Journal of Biological Chemistry* 274: 30540–30549.

Falkowski PG (1997) Evolution of the nitrogen cycle and its influence on the biological sequestration of CO_2 in the ocean. *Nature* 387: 272–275.

Ferris MJ, Muyzer G and Ward DM (1996) Denaturing gradient gel electrophoresis profiles of 16S *rRNA*-defined populations inhabiting a hot spring microbial mat community. *Applied and Environmental Microbiology* 62: 340–346.

Fiore MF, Moon DH, Tsai SM, Lee H and Trevors JT (2000) Miniprep DNA isolation from unicellular and filamentous cyanobacteria. *Journal of Microbiological Methods* 39: 159–169.

Gaget V, Keulen A, Lau M, Monis P and Brookes JD (2017) DNA extraction from benthic cyanobacteria: Comparative assessment and optimization. *Journal of Applied Microbiology* 122: 294–304.

Giovannoni SJ, DeLong EF, Schmidt TM and Pace NR (1990) Tangential flow filtration and preliminary phylogenetic analysis of marine picoplankton. *Applied and Environmental Microbiology* 56: 2572–2575.

Golden SS, Brusslan J and Haselkorn R (1987) Genetic engineering of the cyanobacterial chromosome. *Methods in Enzymology* 153: 215–231.

Guevara R, Armesto JJ and Caru M (2002) Genetic diversity of *Nostoc* microsymbionts from *Gunnera tinctoria* revealed by PCR-STRR fingerprinting. *Microbial Ecology* 44: 127–136.

Harrison DJ and Thompson EP (2020) A rapid and low-cost method for genomic DNA extraction from the cyanobacterium *Synechocystis*. *Biology Methods and Protocols* 5: bpaa011.

Holm C, Meeks-Wagner DW, Fangman WL and Botstein D (1986) A rapid, efficient method for isolating DNA from yeast. *Gene* 42: 169–173.

Jackman DM and Mulligan ME (1995) Characterization of a nitrogen-fixation (*nif*) gene cluster from *Anabaena azollae* 1a shows that closely related cyanobacteria have highly variable but structured intergenic regions. *Microbiology* 141: 2235–2244.

Jhingan AK (1992) A novel technology for DNA isolation. *Methods in Molecular and Cellular Biology* 3: 15–22.

Joset F (1988) Transformation in *Synechocystis* PCC 6714 and 6803: Preparation of chromosomal DNA. *Methods in Enzymology* 167: 712–714.

Kaebernick M, Neilan BA, Börner T and Dittman E (2000) Light and the transcriptional response of the microcystin biosynthetic genes. *Applied and Environmental Microbiology* 66: 3387–3392.

Kim BH, Oh HM, Lee YK, Choi GG, Ahn CY, Yoon BD and Kim HS (2006) Simple method for RNA preparation from cyanobacteria. *Journal of Phycology* 42: 1137–1141.

Kim Tiam S, Comte K, Dalle C, Duval C, Pancrace C, Gugger M and Bernard C (2019) Development of a new extraction method based on high-intensity ultra-sonication to study RNA regulation of the filamentous cyanobacteria *Planktothrix*. *PLoS One* 14: e0222029.

Kramer JG, Wyman M, Zehr JP and Capone DG (1996) Diel variability in transcription of the structural gene for glutamine synthetase (*glnA*) in natural populations of the marine diazotrophic cyanobacterium *Trichodesmium thiebautii*. *FEMS Microbiology Ecology* 21: 187–196.

Lau AF, Siedleck J, Anleitne J, Patterson GM, Caplan FR and Moore RE (1993) Inhibition of reverse transcriptase activity by extracts of cultured blue-green algae (Cyanophyta). *Planta Medica* 59: 148–151.

Leff LG, Dana JR, McArthur JV and Shimkets LJ (1995) Comparison of methods of DNA extraction from stream sediment. *Applied and Environmental Microbiology* 61: 1141–1143.

Lindberg P, Park S and Melis A (2010) Engineering a platform for photosynthetic isoprene production in cyanobacteria using *Synechocystis* as the model organism. *Metabolic Engineering* 12: 70–79.

Lippke JA, Strzempko MN, Raia FF, Simon SL and French CK (1987) Isolation of intact high-molecular-weight DNA by using guanidinc isothiocyanate. *Applied and Environmental Microbiology* 53: 2588–2589.

Lotti F, Giovannetti L, Margheri MC, Ventura S and Materassi R (1996) Diversity of DNA methylation pattern and total DNA restriction pattern in symbiotic *Nostoc*. *World Journal of Microbiology and Biotechnology* 12: 38–42.

Luo XZJ and Stevens SE Jr (1997) Isolation of full-length RNA from a thermophilic cyanobacterium. *BioTechniques* 23: 904–909.

Mak YM and Ho KK (1991) An improved method for the isolation of chromosomal DNA from various bacteria and cyanobacteria. *Nucleic Acids Research* 20: 4101–4102.

Margheri MC, Bosco M, Giovannetti L and Ventura S (1999) Assessment of the genetic diversity of halotolerant coccoid cyanobacteria using amplified 16S *rDNA* restriction analysis. *FEMS Microbiology Letters* 173: 9–16.

Mironov KS and Los DA (2015) RNA isolation from *Synechocystis*. *Bio-Protocol* 5: e1428.

Morin N, Vallaeys T, Hendrickx L, Natalie L and Wilmotte A (2010) An efficient DNA isolation protocol for filamentous cyanobacteria of the genus *Arthrospira*. *Journal of Microbiological Methods* 80: 148–154.

Motham M, Pumas C and Peerapornpisal Y (2014) Improvement of DNA extraction protocols for *Nostochopsis* spp. *Chiang Mai Journal of Science* 41: 557–567.

Moza MI and Postolache C (2022) Optimized protocol for cyanobacterial 16S rRNA analysis in Danube delta lakes. *Biointerface Research in Applied Chemistry* 12: 1365–1384.

Murray MG and Thompson WF (1980) Rapid isolation of high molecular weight plant DNA. *Nucleic Acids Research* 8: 4321–4325.

Neilan BA (1995) Identification and phylogenetic analysis of toxigenic cyanobacteria by multiplex randomly amplified polymorphic DNA PCR. *Applied and Environmental Microbiology* 61: 2286–2291.

Nishihara H, Miwa H, Watanabe M, Nagashima M, Yagi O and Takamura Y (1997) Random amplified polymorphic DNA (RAPD) analyses for discriminating genotypes of *Microcystis* cyanobacteria. *Bioscience, Biotechnology and Biochemistry* 61: 1067–1072.

Palenik B (1994) Cyanobacterial community structure as seen from RNA polymerase gene sequence analysis. *Applied and Environmental Microbiology* 60: 3212–3219.

Pathak RR and Lochab S (2010) A method for rapid isolation of total RNA of high purity and yield from *Arthrospira platensis*. *Canadian Journal of Microbiology* 56: 578–584.

Penn K, Wang J, Fernando SC and Thompson JR (2014) Secondary metabolite gene expression and interplay of bacterial functions in a tropical freshwater cyanobacterial bloom. *The ISME Journal* 8: 1866–1878.

Pinto FL, Thapper A, Sontheim W and Lindblad P (2009) Analysis of current and alternative phenol-based RNA extraction methodologies for cyanobacteria. *BMC Molecular Biology* 10: 1–8.

Pitcher DG, Saunders NA and Owen RJ (1989) Rapid extraction of bacterial genomic DNA with guanidium thiocyanate. *Letters in Applied Microbiology* 8: 151–156.

Porter RD (1988) DNA transformation. *Methods in Enzymology* 167: 703–712.

Reith ME, Laudenbach DE and Straus NA (1986) Isolation and nucleotide sequence analysis of the ferredoxin I gene from the cyanobacterium *Anacystis nidulans* R2. *Journal of Bacteriology* 168: 1319–1324.

Saba F, Papizadeh M, Khansha J, Sedghi M, Rasooli M, Amoozegar MA and Fazeli SAS (2016) A rapid and reproducible genomic DNA extraction protocol for sequence-based identification of archaea, bacteria, cyanobacteria, diatoms, fungi and green algae. *Journal of Medical Bacteriology* 5: 22–28.

Saha SK, Uma SL and Subramanian G (2005) An improved method for marine cyanobacterial DNA isolation. *World Journal of Microbiology and Biotechnology* 21: 877–881.

Salvador D, Churro C and Valério E (2016) Evaluating the influence of light intensity in *mcyA* gene expression and microcystin production in toxic strains of *Planktothrix agardhii* and *Microcystis aeruginosa*. *Journal of Microbiological Methods* 123: 4–12.

Sharma NK, Rai AK and Stal LJ (2014) Cyanobacteria: An Economic Perspective, First ed. John Wiley and Sons, UK.

Singh SP and Montgomery BL (2013) Distinct salt dependent effects impair *Fremyella diplosiphon* pigmentation and cellular shape. *Plant Signaling and Behavior* 8: e24713.

Singh DP, Prabha R, Kumar M and Meena KK (2012) A rapid and standardized phenol-free method for the isolation of genomic DNA from filamentous cyanobacteria. *Asian Journal Experimental Biological Sciences* 3: 666–673.

Singh SP, Rastogi RP, Häder D-P and Sinha RP (2011) An improved method for genomic DNA extraction from cyanobacteria. *World Journal of Microbiology and Biotechnology* 27: 1225–1230.

Singh SP, Sinha RP, Daiker V and Häder D-P (2010) Quantitative and qualitative extraction of RNA from a filamentous cyanobacterium *Anabaena variabilis* PCC 7937. *Journal of Applied Phycology* 22: 113–116.

Sinha RP, Dautz M and Häder D-P (2001) A simple and efficient method for the quantitative analysis of thymine dimers in cyanobacteria, phytoplankton and macroalgae. *Acta Protozoologica* 40: 187–195.

Sipari H, Rantala-Ylinen A, Jokela J, Oksanen I and Sivonen K (2010) Development of a chip assay and quantitative PCR for detecting microcystin synthetase E gene expression. *Applied and Environmental Microbiology* 76: 3797–3805.

Smoker JA and Barnum SR (1988) Rapid small-scale DNA isolation from filamentous cyanobacteria. *FEMS Microbiology Letters* 56: 119–122.

Srivastava AK, Ara A, Bhargava P, Mishra Y, Rai SP and Rai LC (2007) A rapid and cost-effective method of genomic DNA isolation from cyanobacterial culture, mat and soil suitable for genomic fingerprinting and community analysis. *Journal of Applied Phycology* 19: 373–382.

Tillett D and Neilan BA (2000) Xanthogenate nucleic acid isolation from cultured and environmental cyanobacteria. *Journal of Phycology* 36: 251–258.

Tonk L, Visser PM, Christianse G, Dittmann E, Snelder EO, Wiedner C and Huisman J (2005) The microcystin composition of the cyanobacterium *Planktothrix agardhii* changes toward a more toxic variant with increasing light intensity. *Applied and Environmental Microbiology* 71: 5177–5181.

Tran TDC, Bernar C, Ammar M, Chaouch S and Comte K (2013) Heat shock transcriptional responses in an MC-producing cyanobacterium (*Planktothrix agardhii*) and its MC-deficient mutant under high light conditions. *PLoS One* 8: e73198.

Van der Plas J, Bovy A, Kruyt F, De Vrieze G, Dassen E, Klein B and Weisbeek P (1989) The gene for the precursor of plastocyanin from the cyanobacterium *Anabaena* sp. PCC 7937: Isolation, sequence and regulation. *Molecular Microbiology* 3: 275–284.

West NY and Adams DG (1997) Phenotypic and genotypic comparison of symbiotic and free-living cyanobacteria from a single field site. *Applied and Environmental Microbiology* 63: 4479–4484.

Whitehead L, Long BM, Price GD and Badger MR (2014) Comparing the *in vivo* function of α-carboxysomes and β-carboxysomes in two model cyanobacteria. *Plant Physiology* 165: 398–411.

Wilkins TA and Smart LB (1996). Isolation of RNA from plant tissue. In: A Laboratory Guide to RNA: Isolation, Analysis and Synthesis (Eds. P.A. KriegeKriege). John Wiley and Sons, Inc., New York, NY, pp. 21–42. ISBN: 0-471-12536-9.

Wu X, Zarka A and Boussiba S (2000) A simplified protocol for preparing DNA from filamentous cyanobacteria. *Plant Molecular Biology Reporter* 18: 385–392.

Yilmaz M, Phlips EJ and Tillett D (2009) Improved methods for the isolation of cyanobacterial DNA from environmental samples. *Journal of Phycology* 45: 517–521.

Zehr JP and McReynolds LA (1989) Use of degenerate oligonucleotides for amplification of the *nifH* gene from the marine cyanobacterium *Tricodesmium thiebautii. Applied and Environmental Microbiology* 55: 2522–2526.

19 Detection and Quantification of DNA Damage in Cyanobacteria by Immunodot Blot Assay

Prashant R. Singh, Jainendra Pathak, Rajneesh, Deepak Kumar, Amit Gupta, Ashish P. Singh, Varsha K. Singh, and Rajeshwar P. Sinha

19.1 INTRODUCTION

In recent years, there has been an increase in the intensity of solar ultraviolet radiation (UVR) due to depletion of the ozone layer (Holzinger and Lutz 2006). This increased influx of UVR affects all living organisms ranging from prokaryotes to eukaryotes (Kumari et al. 2008). UVR is classified as UV-A (315–400 nm), UV-B (280–315 nm) and UV-C (100–280 nm). UV-C is biologically irrelevant, as it does not reach the Earth due to absorption by the ozone layer in the stratosphere. Therefore, UV-B is the most deleterious UVR, which induces three common lesions in the DNA: cyclobutane-pyrimidine dimers (CPDs), pyrimidine (6-4) pyrimidone photoproducts (6-4PPs) and Dewar valence isomers (DEWs) (LeClerc et al. 1991; Sinha and Häder 2002; Pathak et al. 2019).

Cyanobacteria, algae and plants are most affected by UVR due to their photosynthetic nature, and therefore their exposure to UVR results in direct damage to DNA and proteins (Sinha and Häder 2002). In addition, indirect effects of UVR involve the formation of reactive oxygen species (ROS), which negatively affect the overall fitness of the organisms by damaging several biological molecules and cellular processes (Pathak et al. 2019). In humans these lesions interrupt DNA replication and cause mutations in cells, aging and cell death (Pfeifer et al. 2005).

Different repair mechanisms work in different organisms such as cyanobacteria, algae, plants and nonplacental mammals. DNA repair utilizes a light-dependent enzyme called photolyase which mediates photoreactivation. In humans, nucleotide excision repair (NER) is the major repair pathway for repairing dimeric photoproducts (Courdavault et al. 2005). UV-induced DNA damage can be used as a biomarker for assessment of the overall health of the organism. Here, we present a method for the detection and quantification of pyrimidine dimer lesions by using an immunodot blot assay. In this method, thymine dimers are detected by primary anti-thymine dimer antibodies and an anti-mouse IgG, Fab-specific, secondary antibody conjugate with peroxidase (Sinha et al. 2001; Singh et al. 2011; Rastogi et al. 2014). This method can detect and quantify pyrimidine dimers in nanogram quantities of DNA in cyanobacteria and macroalgae with high sensitivity and accuracy.

19.2 EQUIPMENT

1. Microcentrifuge tube (1.5 and 2.0 mL) (Abdos Lifesciences, catalogue numbers P10202 and P10203).

DOI: 10.1201/9781003398387-19

2. Glass Petri dishes (Corning, catalogue number 3160-102).
3. Measuring cylinder.
4. 295 nm UV cut-off filter (Ultraphan, Digefra, Munich, Germany) to select the desired wavebands of UV-B (280–315 nm), UV-A (315–400 nm) and photosynthetically active radiation (PAR; 400–700 nm).
5. UV-treatment chamber fitted with UV-B (Ultraviolet-B GT15T8E W:12, Sankyo Denki, Japan), UV-A (Ultraviolet-A TL 15 W:12, Philips, Holland) and PAR (Narva, Fluorescent tube, White cool light, 18 W: T8, Germany).
6. Refrigerated centrifuge (Remi Electrotechnik, model CM-12 PLUS).
7. Water bath.
8. 2–20 µL micropipette (Eppendorf, Eppendorf Research plus, model 3120000038).
9. 20–200 µL micropipette (Eppendorf, Eppendorf Research plus, model 3120000054).
10. 100–1000 µL micropipette (Abdos Lifesciences, catalogue number E11105).
11. Slot dot blot unit (Minifold I, Schleicher and Schuell).
12. NanoDrop 1000 (Thermofisher Scientific).
13. ChemiDoc imaging system (Bio-Rad, catalogue number 17001401).

19.3 MATERIALS

1. Urea powder (Sigma-Aldrich, CAS number 57-13-6).
2. Tris-HCl (Sigma-Aldrich, CAS number 1185-53-1).
3. NaCl Merck Millipore (Merck Millipore, CAS number 7647-14-5).
4. Ethylenediaminetetraacetic acid (EDTA) (Sigma-Aldrich, CAS number 62-33-9).
5. Proteinase K (Sigma-Aldrich, CAS number 39450-01-6).
6. Cetyltrimethylammonium bromide (CTAB) (Sigma-Aldrich, CAS number 57-09-0).
7. Sarkosyl (Merck Millipore, CAS number 137-16-6).
8. Mercaptoethanol (Sigma-Aldrich, CAS number 60-24-2).
9. Chloroform:isoamyl alcohol (24:1) (Himedia, CAS number MB115).
10. 100% ethanol (Merck, CAS number 64-17-5).
11. Sodium acetate (Merck Millipore, CAS number 127-09-3).
12. 70% ice cold ethanol.
13. Sterile double-distilled water.
14. TE buffer (see Recipes).
15. Blotting paper.
16. Nylon membrane (Sigma-Aldrich, catalogue number 15356).
17. PBS-T (see Recipes).
18. Skimmed milk powder (Merck Millipore, catalogue number 1.15363).
19. Primary antibody (anti-thymine dimer antibody, Sigma-Aldrich, catalogue number T1192) (see recipes).
20. Secondary antibody (anti-mouse IgG [Fab-specific]) peroxidase conjugate, Sigma-Aldrich, catalogue number A9044) (see Recipes).
21. Detection reagent for chemiluminescence (Renaissance, NEN Life Science Products, Cologne, Germany).

19.4 RECIPES

19.4.1 Lysis Buffer (50 mL)

Dissolve 4 M urea, 0.2 M Tris-HCl (pH 7.4), 20 mM NaCl and 0.2 M EDTA in DDW and maintain the pH at 7.4

19.4.2 Extraction Buffer (50 mL)

Dissolve 3% CTAB, 1.4 M NaCl, 20 mM EDTA, 0.1 M Tris-HCl (pH 8.0), 1% sarkosyl and 1% mercaptoethanol in 50 mL of DDW.

19.4.3 Proteinase K (20 Mg mL^{-1})

A total of 20 mg of proteinase K is dissolved in 1 mL of DDW.

19.4.4 3 M Sodium Acetate (20 mL)

A total of 4.92 g of sodium acetate is dissolved in 20 mL of DDW and the pH maintained at 5.2.

19.4.5 70% Ice Cold Ethanol

Add 70 mL of 100% ethanol in 30 mL of DDW and keep it on ice.

19.4.6 10 × TE Buffer

A total of 1.57 g of Tris powder and 0.3722 g of EDTA are dissolved in DDW, the pH of the solution is maintained at 8 and the final volume is filled to 100 mL. For 1 × TE buffer, 1 mL of 10 × TE buffer is mixed with 9 mL of DDW.

19.4.7 Phosphate Buffer Saline (PBS-T)

A total of 0.14 M NaCl, 3.4 mM KCl, 10.1 mM Na_2HPO_4 and 1.8 mM KH_2PO_4 are dissolved in DDW and the pH is adjusted to 7.4. Afterward, 0.1% (v/v) Tween 20 is added to 100 mL of solution.

19.4.8 Primary Antibody

The primary antibody is diluted to 1:10,000 in PBS-T.

19.4.9 Secondary Antibody

The secondary antibody is diluted to 1:10,000 in PBS-T.

19.5 METHODS

Figure 19.1 illustrates the different steps involved in thymine dimer detection in cyanobacteria (Sinha et al. 2001; Singh et al. 2011; Rastogi et al. 2014). These steps are as follows.

1. A total of 100 mL of cyanobacterial culture is transferred to sterile glass Petri dishes with the help of a measuring cylinder. The Petri dishes are covered with a 295-nm UV cutoff filter and then transferred to the UV light-treatment chamber.
2. Cyanobacteria are irradiated with PAR, UV-A and UV-B, and the temperature of the UV chamber is maintained at 25 ± 2°C to avoid a heating effect.

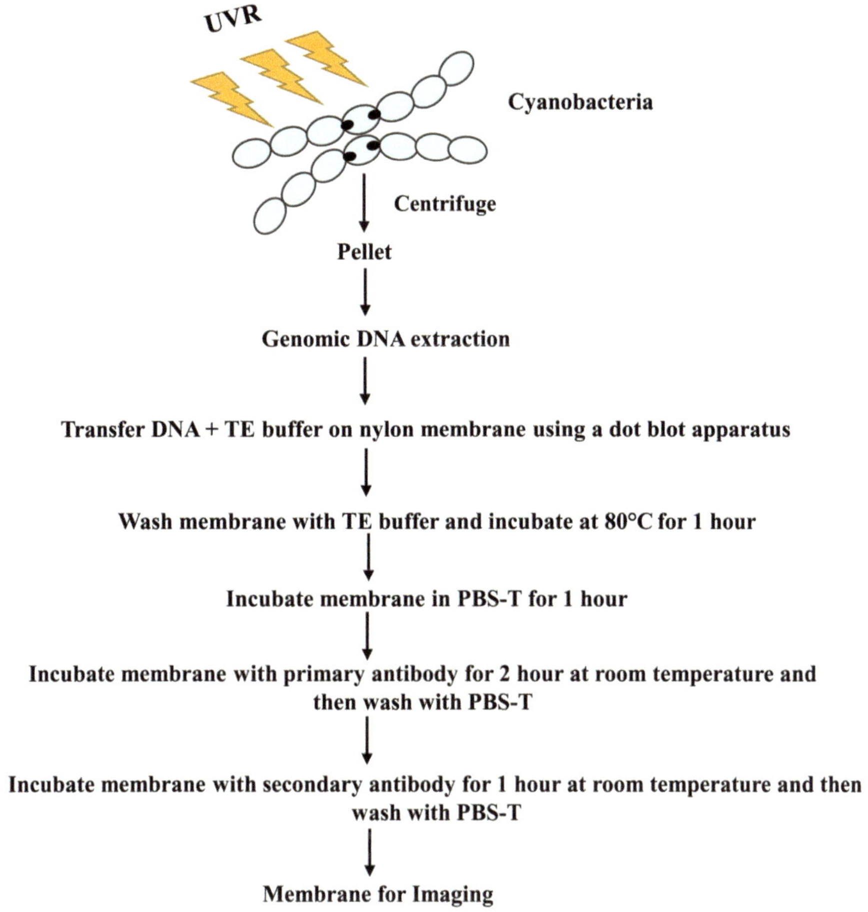

FIGURE 19.1 Flowchart showing steps involved in the thymine dimer detection protocol (color figure in e-book).

3. Genomic DNA isolation:
   a. After 24 hours of the UV treatment, 50 mL of the cyanobacterial cells are harvest and centrifuged at 5000 × *g* for 10 minutes at room temperature.
   b. The pellet is washed with DDW and after that the pellet is transferred to a 2-mL of Eppendorf tube.
   c. The pellet is suspended in 400 µL of lysis buffer.
   d. A total of 50 µL of Proteinase K is added and mixed immediately by pipetting.
   e. The mixture is incubated at 55°C for 1 hour, and every 15 minutes it is mixed by pipetting.
   f. After that, 1 mL of prewarmed (55°C) DNA extraction buffer is added and mixed gently by inverting and again kept in a water bath at 55°C for 1 hour.
   g. After every 10 minutes, the tubes are inverted gently.
   h. After 1 hour of incubation, the mixture is divided into two equal fractions in 2-mL Eppendorf tubes and cooled to room temperature.
   i. Two volumes of chloroform:isoamyl alcohol (24:1) are added to the mixture and mixed gently until a white emulsion appears.
   j. Centrifuge at 10,000 × *g* for 5 minutes.

k. A total of 500 µL of the upper water phase is transferred to the new 1.5-mL Eppendorf tube after centrifugation.
l. Two volumes of 100% ethanol and 0.1 volume of 3 M sodium acetate are added to the tube of the water phase.
m. The mixture is mixed gently by inverting the tube and then the Eppendorf tube is transferred to –20°C for 1 hour.
n. The suspension is centrifuged at 10,000 × *g* for 5 minutes after 1 hour of incubation and the supernatant is discarded.
o. The pellet is washed with 70% ethanol and the excess of ethanol is evaporated after centrifugation at 10,000 × *g* for 5 minutes.
p. The pellet is dissolved in 50 µL of sterile distilled water.
q. DNA is quantified by spectrometry (NanoDrop 1000; Thermofisher Scientific).

4. Detection of thymine dimers:
   a. The blot papers and the nylon membrane are soaked in 3 M NaCl and 0.3 M Na-citrate solution, and the nylon membrane is placed on a slot or dot blot manifold.
   b. The DNA samples are transferred to the nylon membrane and washed in TE buffer.
   c. The membrane is dried at 80°C for 1 hour for DNA immobilization. Afterward, the membrane is incubated for 1 hour in PBS-T containing 5% (w/v) skimmed milk powder to block non-specific sites.
   d. After incubation, the membrane is incubated for 2 hours at room temperature with the primary antibody (anti-thymine dimer antibody), which is diluted in PBS-T + 5% (w/v) skimmed milk powder at a ratio of 1:10,000, and then washed thrice with PBS-T at intervals of 10 minutes.
   e. After incubation with the primary antibody, the membrane is incubated for 1 hour at room temperature with the secondary antibody (anti-mouse IgG [Fab-specific] peroxidase conjugate diluted in PBS-T + 5% (w/v) skimmed milk powder at a ratio of 1:10,000 and washed four times with PBS-T at intervals of 15 minutes.
   f. Finally, the membrane is immersed in a detection reagent for 1 minute before imaging by the ChemiDoc Imaging System. The results are analyzed with the help of the manufacturer's software.
   g. The number of thymine dimers (T^T) is calculated using the following equation (Sinha et al. 2001):

$$y = 1E - 07x^2 + 1E - 05x + 3.6745 \tag{19.1}$$

   h. The frequency F of thymine dimers (T^T)/mega base pair is calculated using the following equation (Sinha et al. 2001):

$$F = T \wedge T / (9.11^{*} 10^{14*} M)^{*} 10^{6} \tag{19.2}$$

where T^T is the number of possible thymine dimers in the irradiated plasmid DNA and M is the total mass of DNA.

19.6 DATA ANALYSIS

The membrane containing DNA – after incubation with primary and secondary antibodies and washing – is analyzed with the ChemiDoc Imaging System for measuring the chemiluminescence intensity representing thymine dimer formation (Figure 19.2).

The frequency of thymine dimers is calculated using the equation given by Sinha et al. 2001. After calculating the thymine dimers, the frequency of thymine dimer/mega base pairs is estimated

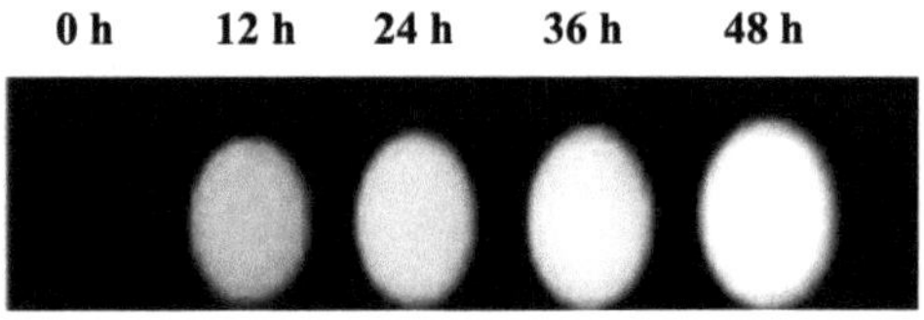

FIGURE 19.2 Dot blot image of DNA of *Nostoc* sp. showing thymine dimer formation after exposure to PAR + UV-A + UV-B radiation for 0, 12, 24, 36 and 48 hours.

using the second equation given by Sinha et al. 2001. The data of the frequency of thymine dimer are exported to Excel and a bar diagram is plotted (Figure 19.3). All measurements are performed at room temperature. All results are presented as mean values of three replicates.

19.7 NOTES/PRECAUTIONS

1. The pH of all reagents should be adjusted before use.
2. UV-treated samples should be used immediately for thymine dimer detection, as freezing of samples at –20°C results in loss of thymine dimers.

ACKNOWLEDGMENTS

Prashant R. Singh (09/013[0795]/2018-EMR-I) is thankful to the Council of Scientific and Industrial Research, New Delhi, for the financial support in the form of a senior research fellowship. Rajneesh is grateful to the Department of Biotechnology, Govt. of India, (DBT-JRF/13/AL/143/2158) for the grant in the form of a senior research fellowship. Deepak Kumar is thankful to the Department of Science and Technology, Inspire Program, New Delhi, India, for the financial assistance in the form of a senior research fellowship (DST/Inspire Fellowship/2015/IF150191). Amit Gupta (09/013[0912]/2019-EMR-I) and Varsha K. Singh (09/013[12862]/2021-EMR-I) are thankful to the Council of Scientific and Industrial Research, New Delhi, for the financial support in the form of junior research fellowships. Ashish P. Singh is thankful to University Grant Commission (UGC), New Delhi, India (NTA Ref. No. 191620014505), for providing a junior research fellowship.

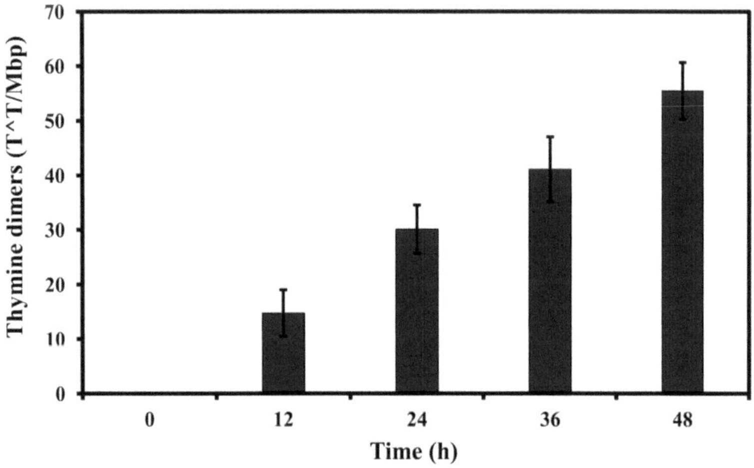

FIGURE 19.3 Frequency of thymine dimers in *Nostoc* sp. after exposure to PAR + UV-A + UV-B radiation for 0, 12, 24, 36 and 48 hours (mean ± SD, n = 3).

REFERENCES

Courdavault S, Baudouin C, Charveron M, Canguilhem B, Favier A, Cadet J and Douki T (2005) Repair of the three main types of bipyrimidine DNA photoproducts in human keratinocytes exposed to UV-B and UV-A radiations. *DNA Repair* 4: 836–844.

Holzinger A and Lütz C (2006) Algae and UV irradiation: Effects on ultrastructure and related metabolic functions. *Micron* 37: 190–207.

Kumari S, Rastogi RP, Singh KL, Singh SP and Sinha RP (2008) DNA damage: Detection strategies. *EXCLI Journal* 7: 44–62.

LeClerc JE, Borden A and Lawrence CW (1991) The thymine-thymine pyrimidine-pyrimidone (6-4) ultraviolet light photoproduct is highly mutagenic and specifically induces 3′ thymine-to-cytosine transitions in *Escherichia coli*. *Proceeding of the National Academy of Sciences USA* 88: 9685–9689.

Pathak J, Rajneesh, Singh PR, Häder D-P and Sinha RP (2019) UV-induced DNA damage and repair: A cyanobacterial perspective. *Plant Gene* 19: 100194.

Pfeifer GP, You YH and Besaratinia A (2005) Mutations induced by ultraviolet light. *Mutation Research/ Fundamental and Molecular Mechanisms of Mutagenesis* 571: 19–31.

Rastogi RP, Singh SP, Incharoensakdi A, Häder D-P and Sinha RP (2014) Ultraviolet radiation-induced generation of reactive oxygen species, DNA damage and induction of UV-absorbing compounds in the cyanobacterium *Rivularia* sp. HKAR-4. *South African Journal of Botany* 90: 163–169.

Singh SP, Rastogi RP, Häder D-P and Sinha RP (2011) An improved method for genomic DNA extraction from cyanobacteria. *World Journal of Microbiology and Biotechnology* 27: 1225–1230.

Sinha RP, Dautz M and Häder D-P (2001) A simple and efficient method for the quantitative analysis of thymine dimers in cyanobacteria, phytoplankton and macroalgae. *Acta Protozoologica* 40: 187–195.

Sinha RP and Häder D-P (2002) UV-induced DNA damage and repair: A review. *Photochemical and Photobiological Sciences* 1: 225–236.

20 Engineering Cyanobacteria for the Synthesis of Novel Products

Rungaroon Waditee-Sirisattha and Hakuto Kageyama

20.1 INTRODUCTION

Cyanobacteria are photosynthetic prokaryotes that show great promise for sustainable technologies. They are capable of converting solar energy, the most abundant form of renewable energy, and efficiently harvesting CO_2 as a carbon source to generate a variety of carbon-based biocompounds. With their natural metabolic diversity, cyanobacteria are prime host cells for industrial applications. They have been used commercially for the production of proteins and nutraceuticals (Spolaore et al. 2006; Garlapati et al. 2019). Furthermore, cyanobacteria offer the benefits of rapid growth, simple cultivation, and easy genetic manipulation. These advantages make them an ideal vehicle for gene manipulation and metabolic engineering.

Metabolic engineering is involved in the construction of genetically engineered cyanobacteria or any target host cells. It plays a pivotal role in the rewiring of basic metabolism to enhance the production of native metabolites or endowing cells with the ability to produce new products. Several strains of cyanobacteria are known to be readily amenable to genetic modification by conventional methods (e.g., natural transformation, conjugation, and electroporation) or by cutting-edge technologies (e.g., clustered regularly interspaced short palindromic repeats [CRISPR]–dependent approaches) (Vijay et al. 2019). Natural transformation refers to the ability of organisms to uptake DNA naturally, which was first observed in 1970 (Shestakov and Khyen 1970). The authors described the uptake of DNA isolated from antibiotic-resistant *Synechococcus elongatus* PCC 7942 (formerly known as *Anacystis nidulans* R2) by non-antibiotic-resistant *Synechococcus*. Since then, this type of DNA uptake has been referred to as natural transformation. Other examples of natural transformation have been seen in *Synechococcus* sp. PCC 7002 (formerly *Agmenellum quadruplicatum* PR-6) (Stevens and Porter 1980), *Synechocystis* sp. PCC 6803, and *Thermosynechococcus elongates* BP-1 (Iwai et al. 2004). Strains able to undergo natural transformation became the preferred strains for cyanobacterial research, the construction of genetically engineered cyanobacteria, and the development of recombinant technology.

The typical workflow for metabolic engineering is shown in Figure 20.1. The design–build–test cycle is generally used for metabolic engineering. The design step is an input of the metabolic pathway of interest. *In silico* techniques can be used to predict potential biosynthesis routes, catalytic enzymes, and bioproduct targets. The second step, build, involves the implementation of the predicted pathway into a host vehicle. The last step, test, can be performed *in vitro*, *in vivo*, or both. This step is important for confirming the production of target compounds and/or metabolites. The output sought from the design–build–test cycle is to achieve a novel pathway for target compound(s) production in the host vehicle.

In this chapter, we present an example of the metabolic engineering approach using a unicellular cyanobacterium, *S. elongatus* PCC 7942. This strain can be readily genetically modified by natural transformation or by other methods. We describe the input step by focusing on mycosporine-like amino acid (MAA) production. MAAs are one type of novel bioproduct derived from cyanobacteria.

DOI: 10.1201/9781003398387-20

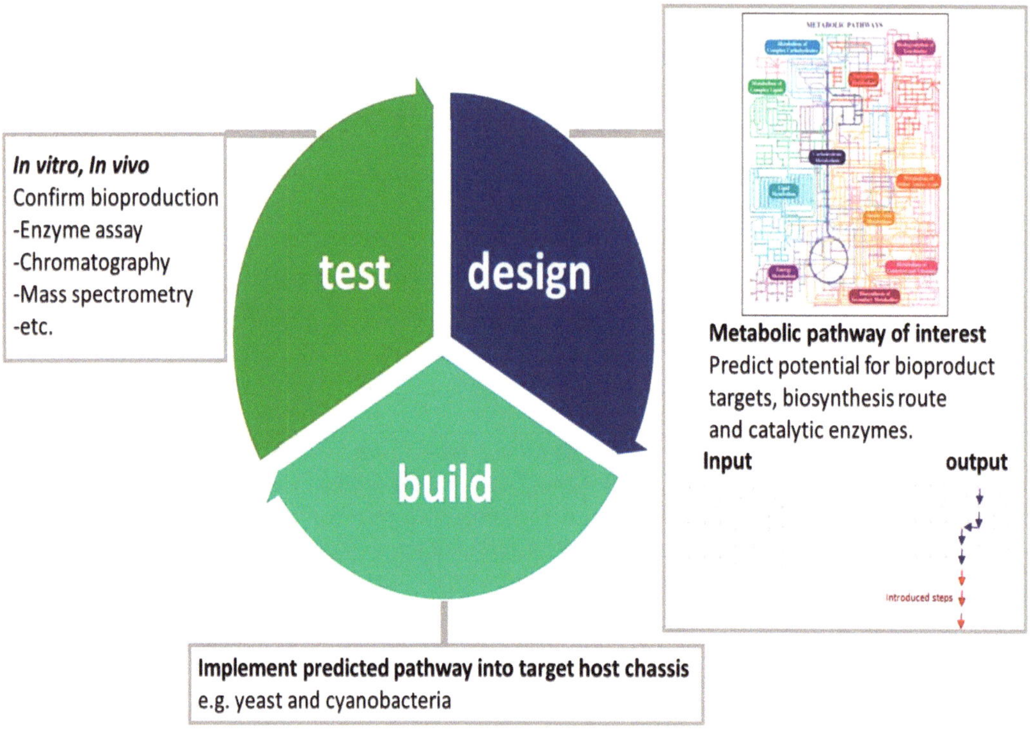

FIGURE 20.1 Typical workflow for metabolic engineering. The design–build–test cycle applied is shown in different colors. The design step is the "input" of the metabolic pathway of interest. Re-routing and/or evaluating the feasibility of a reaction lead to the biosynthetic pathway for a novel compound ("output") (color figure in e-book).

MAAs exhibit a great potential for the development of novel ultraviolet (UV) sunscreens because of their direct and indirect protective effects (Kageyama and Waditee-Sirisattha 2019; Tarasuntisuk et al. 2019; Waditee-Sirisattha and Kageyama 2021). We previously described the construction and expression of an MAA gene cluster (Figure 20.2) from a halophilic species together with its native promotor in *S. elongatus* PCC 7942 (Waditee-Sirisattha et al. 2014). Extremophilic cyanobacteria (e.g., halophiles, thermophiles, and psychrophiles) are active biomolecule producers, endowed with unique biosynthetic pathways. Therefore, we have illustrated metabolic engineering approaches for the production of other possible MAAs using genes sourced from representative extremophilic cyanobacteria whose entire genome sequences are available. A list of putative genes for MAA biosynthesis in extremophilic cyanobacteria is provided in Table 20.1. The protocols necessary for constructing, transforming, and selecting genetically modified *S. elongatus* PCC 7942 are elaborated.

20.2 EQUIPMENT

1. Autoclave (TOMY Digital Biology, Japan).
2. Balance (Mettler Toledo, Switzerland).
3. Bench-top centrifuge (Biosan, Malaysia).
4. Biological safety cabinet (Model MCV-131S, Sanyo, Japan).
5. Gel imaging (Model Gel Doc EZTM, Bio-Rad Laboratories, USA).
6. Micropipette (Eppendorf Research Plus, Eppendorf, Germany).

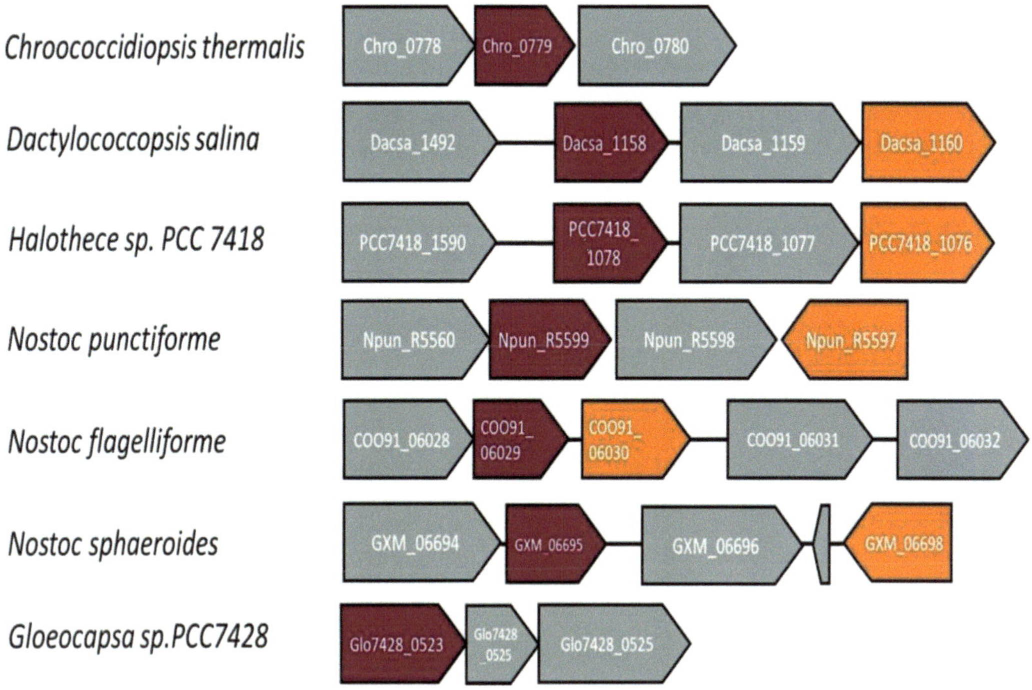

FIGURE 20.2 Diversity of the genetic organization of MAA biosynthetic gene clusters in extremophilic cyanobacterial genomes. The Gene ID retrieved from the KEGG database is shown in the box for each strain. Details of the Gene ID and the definition of KEGG orthology (KO) are shown in Table 20.1 (color figure in e-book).

7. NanoDrop 2000 UV-Vis Spectrophotometer (Thermo Scientific, USA).
8. pH meter (Mettler Toledo, Switzerland).
9. Spectrophotometer (GENESYs 20, Thermo Fisher Scientific, USA).
10. Sonicator VP-5s (TAITEC, Saitama, Japan).
11. Temperature-controlled growth chamber with lights and platform shaker.
12. Thermal cycler (Model C1000 Touch, Bio-Rad Laboratories, USA).
13. Ultralow-temperature freezer (Thermo Scientific, USA).

20.3 MATERIALS

20.3.1 Strains and Cell Cultures

1. *S. elongatus* PCC 7942.
2. *Escherichia coli* DH5α.
3. Blue-green (BG) 11 stock solution.
4. Luria broth (LB).
5. 3 M hydrochloric acid.
6. 3 M sodium hydroxide.
7. Antibiotic solutions.
8. Culture plates.
9. Agar.

TABLE 20.1
A List of Putative Genes for MAA Biosynthesis in Representative Cyanobacteria. Cyanobacterial Strain, Gene ID, and Definition KEGG Orthology (KO) Are Listed. In Case of No Definition by KO, the Definition in Genebank Is Shown and Marked by an Asterisk (*)

Cyanobacterial Strain	Gene ID	Definition KEGG Orthology
Chroococcidiopsis thermalis	Chro_0778	demethyl-4-deoxygadusol synthase
PCC 7203	Chro_0779	caffeoyl-CoA *O*-methyltransferase
	Chro_0780	protein of unknown function DUF201*
		motif: ATP-grasp_3 ATPgrasp_Ter ATP-grasp
Dactylococcopsis salina	Dacsa_1492	3-dehydroquinate synthetase*
	Dacsa_1158	Caffeoyl-CoA *O*-methyltransferase
	Dacsa_1157	hypothetical protein*
		motif: ATP-grasp_3 ATPgrasp_Ter NAGLU_C
	Dacsa_1156	D-alanine-D-alanine ligase
Halothece sp. PCC 7418	PCC7418_1590	3-dehydroquinate synthase*
	PCC7418_1078	Caffeoyl-CoA *O*-methyltransferase
	PCC7418_1077	hypothetical protein*
		motif: ATP-grasp_3
	PCC7418_1076	D-alanine-D-alanine ligase
Nostoc punctiforme	Npun_R5560	3-dehydroquinate synthase
PCC 73102	Npun_R5599	*O*-methyltransferase
	Npun_R5598	conserved hypothetical protein*
		motif: ATP-grasp_3 ATPgrasp_Ter Glyco_transf_4
	Npun_R5597	D-alanine-D-alanine ligase
Nostoc flagelliforme	COO91_06028	demethyl-4-deoxygadusol synthase
	COO91_06029	caffeoyl-CoA *O*-methyltransferase
	COO91_06030	D-alanine-D-alanine ligase
	COO91_06031	putative ATP-dependent carboligase, ATP-grasp superfamily*
	COO91_06032	putative ATP-dependent carboligase, ATP-grasp superfamily*
Nostoc sphaeroides	GXM_06694	demethyl-4-deoxygadusol synthase
	GXM_06695	caffeoyl-CoA *O*-methyltransferase
	GXM_06696	ATP-grasp domain containing protein*
		motif: ATP-grasp_3 ATPgrasp_Ter Glyco_transf_4
	GXM_06697	hypothetical protein
	GXM_06698	D-alanine-D-alanine ligase
Gloeocapsa sp. PCC 7428	Glo7428_0523	hypothetical protein
		motif: Pfam: TP-grasp_3 ATPgrasp_Ter ATPgrasp_ST
	Glo7428_0524	*O*-methyltransferase family 3
	Glo7428_0525	2-epi-5-epi-valiolone synthase
		3-dehydroquinate synthase*

10. Centrifuge tubes, 15–50 mL.
11. Microcentrifuge tubes, 1.5 mL.
12. Gloves.
13. 75% ethanol in spray bottle.
14. Laminar flow hood or biosafety cabinet.

20.3.2 Plasmid Construction

1. Cloning vectors.
2. Expression vectors.
3. Shuttle vectors.
4. Chemically competent *E. coli* strains.
5. Taq polymerase.
6. Nuclease-free water.
7. Restriction enzymes.
8. T4 DNA ligase.

20.3.3 Transformation, Selection, and Verification

1. Falcon tubes, 14 mL.
2. Cell spreaders, sterilized.
3. Inoculation loops or toothpicks, sterilized.
4. Petri dishes.
5. Glass Erlenmeyer flasks, 250 mL.

20.4 RECIPES

20.4.1 BG-11 Medium

BG-11 is one of the most popular growth media for culturing cyanobacteria. For the recipe and method, see details as described earlier (Castenholz 1988).

20.4.2 LB Medium

To prepare the LB medium, an LB tablet is now available, such as Lennox (Sigma-Aldrich, USA). Alternatively, it can be prepared as described earlier (Sambrook et al. 1989).

20.4.3 Antibiotic Solutions

Here, we give examples for preparing two antibiotics, streptomycin and spectinomycin.

20.4.3.1 Streptomycin (100 mg/mL)

Prepare 1000× stocks by dissolving 1000 mg streptomycin salt in 10 mL deionized water. Filter and sterilize using a syringe and 0.2-µm filter into autoclaved 1.5-mL microcentrifuge tubes. Store in 1-mL aliquots at –20°C.

20.4.3.2 Spectinomycin (100 mg/mL)

Prepare 1000× stocks by dissolving 1000 mg spectinomycin salt in 10 mL deionized water. Filter and sterilize using a syringe and 0.2-µm filter into autoclaved 1.5 mL microcentrifuge tubes. Store in 1-mL aliquots at –20°C.

20.5 METHODS

This section describes methods in the following steps.

20.5.1 Inoculating, Maintaining, and Preserving *S. elongatus* PCC 7942

All handling of cyanobacteria and media should be performed using sterile techniques in a laminar flow hood. Wild-type *S. elongatus* PCC 7942 can be obtained from the Pasteur Culture Collection of Cyanobacteria (PCC) (https://webext.pasteur.fr/cyanobacteria/).

Prepare BG-11 medium components as specified in the recipe by dissolving the solid ingredients in deionized water (Castenholz 1988). Autoclave each component on a liquid cycle (121°C, 1.0546 kg/cm^2) for 15 minutes. Trace metals should be filter-sterilized. All components should be stored at 4°C.

To prepare the BG-11 liquid medium, add 600 mL deionized water to a beaker. Then add 1.5 g $NaNO_3$ and stir until completely dissolved. Thereafter, add the other components and finally the trace elements. (Alternatively, components can be prepared as stock solutions, such as 100× or 1000×.) The final volume should be made up to 1000 mL. Adjust the pH to 8.0 using 3 M HCl and 3 M NaOH as needed. Autoclave each component on a liquid cycle (121°C, 1.0546 kg/cm^2) for 15 minutes. Store at 4°C. To prepare the BG-11 agar, add 15 g of agar prior to autoclaving. BG-11 agar can be stored in a water bath or an oven (at a temperature of 60–65°C). To prepare BG-11 liquid agar, pour it into plates and allow it to stand until solidified. Agar plates can be stored at 4°C.

To maintain a liquid cyanobacterial culture, add the liquid BG-11 medium and appropriate antibiotics (for transformants) to an Erlenmeyer flask. Inoculate the liquid culture (2–5% v/v) by pipetting to 50–100 mL of liquid BG-11 medium. Incubate at 28 ± 2°C on a shaker (150 rpm), 50–120 μmol photon m^{-2} s^{-2} for 5 days. Cell growth can be monitored by taking optical density measurements at 730 nm (OD_{730}). For routine culture maintenance, the OD_{730} should remain below ~2.0. The passage number (the number of times a colony is re-streaked or transferred to a new flask) should be kept as low as possible to minimize genetic drift of the strain.

To maintain a solid cyanobacterial culture, pipette 10–100 μL of liquid culture onto a BG-11 agar plate and spread using autoclaved glass beads or a cell spreader. Seal the plate with surgical tape or laboratory film and incubate at 28 ± 2°C, 50–120 μmol m^{-2} s^{-2} for 7–14 days.

To freeze a stock of cyanobacterial culture, grow the culture in BG-11 medium supplemented with appropriate antibiotics to OD_{730} 0.6–1.0. Transfer the culture to a centrifuge tube and centrifuge at 4000 × *g* for 5 minutes. Discard the supernatant and resuspend the cell pellet in a 1/10 volume of BG-11 liquid medium. Add DMSO to a final concentration of 5%. Aliquot into 1.5-mL microcentrifuge tubes and store at –80°C.

20.5.2 Designing and Constructing Plasmids for Transformation

Retrieve the DNA sequence of interest. Native cyanobacterial genes can be accessed from the Kyoto Encyclopedia of Genes and Genomes (KEGG) database or Cyanobase (http://genome.microbedb.jp/cyanobase/).

Preparation of genomic DNA of target organisms. This step is necessary as the template for the amplification of target genes. Alternatively, synthetic genes can be applied, for example, GenScript (https://www.genscript.com/).

Selection of a cloning vector to use for assembling the DNA construct in *E. coli*. Examples of cloning vectors include pBlueScript II vectors (Agilent, USA), pUC19 (Thermofisher, USA), or TA vectors such as pGEM-T (Promega, USA) and pMD20 (Takara, Japan). The vector copy number, selection marker, and multiple cloning sites (MCSs) should be considered to meet for target genes. For example, pUC19, which uses the pMB1 replicon to achieve a high copy number in *E. coli*, contains the ampicillin antibiotic-resistance marker.

Select a cloning method to insert the gene of interest into the cloning vector. Many options are available, including "cut-and-paste" using restriction enzymes and ligase, GoldenGate assembly (New England Biolabs), and others. Commonly, a gene is inserted into the MCS of a vector using a cut-and-paste approach.

Design polymerase chain reaction (PCR) primers to amplify the gene or DNA region of interest. In general, primers should be 15–30 bases long, have a melting temperature (Tm) between 50 and 60°C, and have minimal predicted primer dimers or secondary structure. Primer designing tools are available, such as IDT (https://sg.idtdna.com/pages/tools/primerquest?returnurl=%2FPrimerquest%2FHome%2FIndex). If using cut-and-paste to insert the gene into the MCS, the addition of restriction endonuclease recognition sequences to the 5′ end of both primers can be designed. In this case, at least four bases should be added to the far 5′ end to facilitate enzyme binding.

Follow standard cloning practices for the method of choice to assemble the plasmid, such as the standard cloning protocols described by Sambrook et al. (1989).

Confirmation and verification that the correct clones have been incorporated into cloning vectors can be performed by restriction enzyme analysis to check the presence of the inserted gene in the cloning vector. Nucleotide sequencing is also recommended to verify that the amplified gene matches perfectly with the nucleotide sequence derived from the database.

Follow standard practices for the method of choice to transfer the plasmid into an expression vector or shuttle vector. Examples of expression vectors include pSyn1 and pSyn6 (Invitrogen, USA); an example of a shuttle vector is pUC303 (Kuhlemeier and Van Arkel 1987).

20.5.3 Natural Uptake of DNA

To concentrate *S. elongatus* PCC 7942 cells for the natural uptake of DNA, grow cells to the late-log phase (OD_{730} >1.5). For each transformation, aliquot 10 mL of cell culture into a 15-mL conical centrifuge tube on ice. Centrifuge each tube at 4000 × *g* for 5 minutes at 4°C. Resuspend the cells in the same volume of fresh BG-11 medium, vortex, and centrifuge. Repeat this step at least three times to enhance the natural uptake of DNA. The final step comprises washing and resuspending the cell pellet in 100 μL of BG-11 medium. In addition to the tubes destined for transformation, prepare a negative control (no plasmids).

Add 1 μg of the plasmid being used for transformation to the suspension and incubate cells for at least 5 hours under dim light at 30°C. The tubes should be gently shaken on a rocking platform, after approximately 1 hour, to resuspend the cells.

Pipette 100 μL of the transformation mixture onto a BG-11 agar plate without antibiotics and spread using a cell spreader. Seal the plate with surgical tape and incubate under low light for 48 hours at 30°C for recovery. Note that the transformation mixture can be spread onto a BG-11 agar plate with selected antibiotics. These BG-11 agar plates should be incubated under continuous, cool-white fluorescent light (intensity 50–120 μmol m^{-2} s^{-2}) until colonies appear.

Alternatively, pipette 100 μL of the transformation mixture onto a BG-11 agar plate without antibiotics and spread using a cell spreader. After 48 hours, select for transformants (colonies harboring plasmids) by adding antibiotic solutions (at desired concentrations) to the bottom of the plate. After several days, transformants should retain a green color while non-transformants become pale. Single green colonies should be observed after 7–14 days.

20.5.4 Selecting and Verifying the Transformants

To verify that the gene of interest has been inserted, perform a colony PCR with specific primers. In the case described here, specific primers for MAA-encoding genes should result in positive signals with candidate transformants, as wild-type *S. elongatus* PCC 7942 cells do not possess MAA-encoding genes.

If using vectors that integrate into the chromosome of wild-type *S. elongatus* PCC 7942, for example, pSyn6 (Invitrogen, USA), further steps are necessary to achieve complete segregation.

To achieve full segregation, re-streak single colonies on a fresh BG-11 plate with antibiotics and culture at 30°C in an illuminated incubator. Repeat this process once per week for 3–4 weeks or until full segregation is achieved. Antibiotic resistance can be tested for the complete segregation clone. Also, specific primers for neutral sites (LF and LR) can be used to detect complete segregation. An example of complete segregation using pSyn6 as an expression vector in wild-type *S. elongatus* PCC 7942 can be checked, as previously described (Ngoennet et al. 2020).

20.6 DATA ANALYSIS OF TARGET COMPOUNDS

A convenient method for detecting biosynthesized MAAs in transformed cyanobacteria is the measurement of absorption spectra of cyanobacterial extracts in an aqueous solution containing methanol or ethanol. For example, when using methanol, add an amount of methanol calculated by the formula: methanol volume [mL] = cell fresh weight [g] × 8, as used by Ngoennet et al. (2018). Then, disrupt the cells using a sonicator. When using a sonicator VP-5s, set the output to 7 and repeat for 30 seconds on and 30 seconds off, for a total processing time of 60 seconds. Then, after allowing it to stand at room temperature for about 15 minutes, centrifuge the suspension at 15,000 × *g* for 10 minutes at 25°C. Collect the supernatant and measure the absorption spectrum with a spectrophotometer. For the control, it is better to use a transformant with an empty vector that possesses no target genes. The presence of MAAs in the extracts can be easily determined by confirming whether a peak is detected at a wavelength near the absorption maximum of the target MAA.

If you have authentic MAA standards, the MAA present can be quantified using high-performance liquid-chromatography (HPLC) analysis. An HPLC instrument is sufficient if it has a UV/VIS detector, but ideally a photodiode array (PDA) detector should be used. An octadecylsilyl (ODS) column, which is a reverse-phase column, is generally used. As MAA is a small, hydrophilic molecular compound, it tends to have a shorter retention time when operated in reverse-phase mode. An aqueous solution of between 0.1% and 1% formic or acetic acid is often used as the eluent.

If you wish to characterize the structure of an unknown MAA, more detailed analyses in addition to HPLC analysis are essential. For example, it is necessary to determine the molecular weight of the target MAA using liquid chromatography–mass spectrometry (LC–MS) analysis and to determine its molecular structure using nuclear magnetic resonance (NMR) analysis.

These analysis methods have been described previously (Kageyama and Waditee-Sirisattha 2018; Ngoennet et al. 2018).

REFERENCES

Castenholz RW (1988) Culturing methods for cyanobacteria. In: Cyanobacteria, Methods in Enzymology Vol. 167 (Eds. L. Packer and A.N. Glazer). Academic Press, San Diego, pp. 68–93. ISBN: 978-0-12-182068-8

Garlapati D, Chandrasekaran M, Devanesan A, Mathimani T and Pugazhendhi A (2019) Role of cyanobacteria in agricultural and industrial sectors: An outlook on economically important byproducts. *Applied Microbiology and Biotechnology* 103(12): 4709–4721.

Iwai M, Katoh H and Katayama M and Ikeuchi M (2004) Improved genetic transformation of the thermophilic cyanobacterium, *Thermosynechococcus elongatus* BP-1. *Plant and Cell Physiology* 45(2): 171–175.

Kageyama H and Waditee-Sirisattha R (2018) Mycosporine-like amino acids as multifunctional secondary metabolites in cyanobacteria: From biochemical to application aspects. In: Studies in Natural Products Chemistry Vol. 59 (Ed. Atta-ur-Rahman). Elsevier, Amsterdam, pp. 153–194. ISBN: 978-0-323-91296-9

Kageyama H and Waditee-Sirisattha R (2019) Antioxidative, anti-inflammatory, and anti-aging properties of mycosporine-like amino acids: Molecular and cellular mechanisms in the protection of skin-aging. *Marine Drugs* 17(4): 222.

Kuhlemeier CJ and Van Arkel GA (1987) Host-vector systems for gene cloning in cyanobacteria. In: Recombinant DNA Part D, Method in Enzymology Vol. 153 (Eds. R. Wu and L. Grossman). Academic Press Inc, San Diego, pp. 199–215. ISBN: 978-0-12-182054-1

Ngoennet S, Nishikawa Y, Hibino T and Waditee-Sirisattha R and Kageyama H (2018) A method for the isolation and characterization of mycosporine-like amino acids from cyanobacteria. *Methods and Protocols* 1(4): 46.

Ngoennet S, Sirisattha S, Honda M, Hibino T, Kageyama H and Waditee-Sirisattha R (2020) Morphological plasticity of hyperelongated cells caused by overexpression of translation elongation factor P in *Synechococcus elongatus* PCC 7942. *Applied Microbiology and Biotechnology* 104(20): 8801–8812.

Sambrook J, Fritsch EF and Maniatis T (1989) Molecular Cloning: A Laboratory Manual, Second Ed. Cold Spring Harbor Laboratory Press, Plainview.

Shestakov SV and Khyen NT (1970) Evidence for genetic transformation in blue-green alga *Anacystis nidulans. Molecular and General Genetics* 107: 372–375.

Spolaore P, Joannis-Cassan C, Duran E and Isambert A (2006) Commercial applications of microalgae. *Journal of Bioscience and Bioenergy* 101(2): 87–96.

Stevens SE and Porter RD (1980) Transformation in *Agmenellum quadruplicatum. Proceedings of the National Academy of Sciences USA* 77(10): 6052–6056.

Tarasuntisuk S, Palaga T, Kageyama H and Waditee-Sirisattha R (2019) Mycosporine-2-glycine exerts anti-inflammatory and antioxidant effects in lipopolysaccharide (LPS)-stimulated RAW 264.7 macrophages. *Archives of Biochemistry and Biophysics* 662: 33–39.

Vijay D, Akhtar MK and Hess WR (2019) Genetic and metabolic advances in the engineering of cyanobacteria. *Current Opinion in Biotechnology* 59: 150–156.

Waditee-Sirisattha R and Kageyama H (2021) Protective effects of mycosporine-like amino acid-containing emulsions on UV-treated mouse ear tissue from the viewpoints of antioxidation and antiglycation. *Journal of Photochemistry and Photobiology B: Biology* 223: 112296.

Waditee-Sirisattha R, Kageyama H, Sopun W and Tanaka Y and Takabe T (2014) Identification and upregulation of biosynthetic genes required for accumulation of mycosporine-2-glycine under salt stress conditions in the halotolerant cyanobacterium *Aphanothece halophytica. Applied and Environmental Microbiology* 80(5): 1763–1769.

21 Sample Preparation and Visualization Protocol for Electron Microscopy–Based Analysis of Cyanobacteria

Soumila Mondal, Bhaskar Sharma, Abhisek Majumdar, Sapna Tiwari, Pankaj K. Maurya, Anjali Gupta, Shailendra P. Singh, and Prabhakar Singh

21.1 INTRODUCTION

Electron microscopy (EM) is an effective imaging technique that has revolutionized our knowledge of cellular and subcellular biological structures. It enables the imaging of biological materials with nanometer-scale resolution, providing precise information on cellular morphology, organelle architecture, and cellular activities. In all areas of biology, including cell biology, microbiology, and molecular biology, EM has been used extensively to investigate the ultrastructural properties of biological materials. Electron microscopes use electrons (which have very short wavelengths) as the source of illuminating waves, resulting in high image resolution. To virtualize cellular and subcellular architecture, this resolution is necessary. Cyanobacteria lack apparent structural compartmentalization for cellular functions, necessitating the use of EM to decipher structural organization and interactions between different structures. Numerous ultrastructural details, such as the organization of thylakoids, cell walls, carboxysomes, and nucleoids, are required to comprehend physiological function and evolution (Lang 1968; Mareš et al. 2019). This technique can also disclose the cellular structure of nucleic acids (Rouquette et al. 2009; Ghosh et al. 2021).

The stages of sample preparation have a substantial impact on the final image quality; consequently, a very consistent protocol must be followed when preparing samples for EM-based studies. The scanning electron microscope (SEM) and transmission electron microscope (TEM) are the two types of electron microscopes. SEM is commonly used to examine the surfaces of biological and non-biological specimens, whereas TEM is used to examine the interior and exterior ultrastructure of ultrathin specimen sections (Jaksch and Martin 1995; Baxa 2018). In TEM, electrons are transmitted through the specimen to produce a two-dimensional image of the cell's interior structure. In SEM, an electron beam strikes the surface of an object and generates signals that are surveyed using specific detectors; signal analysis yields a three-dimensional image of the object of study.

SEM and TEM utilize accelerated electrons and electromagnetic optics to produce images in fundamentally distinct ways. Both require a high vacuum so that electrons with low mass can pass down the microscope column unimpeded by air particles. In both instances, the specimen is photographed using either photographic film or a computer interface with digital assistance. Before proceeding with sample preparations and applications of these two EMs, it is essential to comprehend their fundamental construction and operation (Figure 21.1).

 DOI: 10.1201/9781003398387-21

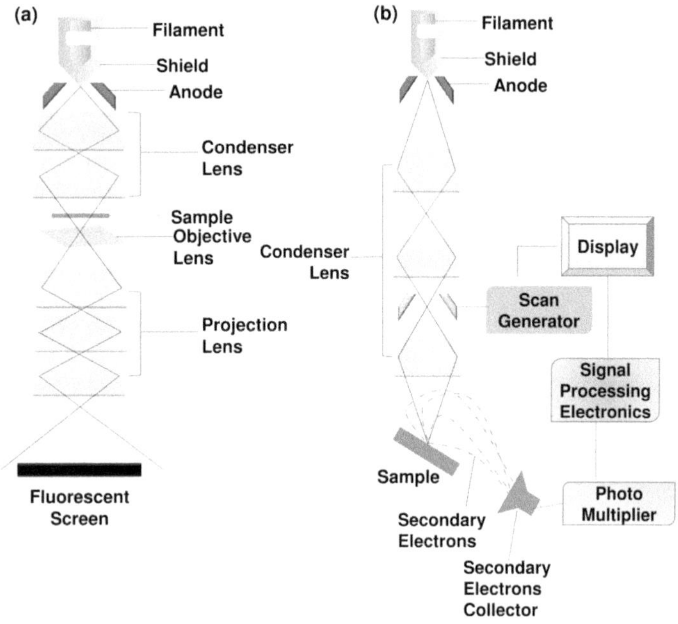

FIGURE 21.1 Schematic diagram of the principle of two types of electron microscopes (EMs). (a) Transmission electron microscope (TEM) and (b) scanning electron microscope (SEM) (color figure in e-book).

A typical TEM employs an electron cannon to generate an electron beam by heating a tungsten filament with a direct current at a very high voltage of negative potential. The negative voltage and heat expel tungsten valence electrons from the V-shaped filament's apex (Crewe et al. 1968). In a vacuum, a highly negative shield drives electrons into a mist at the filament tip. Through an aperture in the shield, some electrons can reach the anode plate. At nearly half the speed of light, electrons are able to penetrate the magnetic fields of electromagnetic lenses (condenser lenses) designed to direct electron beams. Copper wires or solenoids that produce a magnetic field when electricity is sent through a wire are used to construct EM lenses (Shiltsev et al. 2008; Rose 2016). After encountering the magnetic field of the TEM condenser lenses, the electron flow is altered to modify the focal length of the beam and illuminate the specimen's region of interest. As electrons strike various areas of the specimen, their deflection varies depending on the atomic composition of the region. Upon projection onto a viewing screen, certain areas appear gloomy in hue; the region of the specimen with the lowest mass scatters electrons less powerfully, making it appear dazzling on the viewing screen (Unwin and Huxley 1972). The objective lens contributes to the creation of a comprehensive image based on electron scattering. The image is then magnified by three or four projection optics to nearly 1 million times.

Lenses and electron guns in a typical SEM are comparable to those in a TEM. In contrast, the lenses used in SEM concentrate the electron beam onto a very small area of the object. The three condenser lenses utilized in SEM gradually diminish the extent of the electron point (Figure 21.1). When an electron beam impacts the surface of a material, secondary electrons of low energy are ejected from the material's surface (Joy and Pawley 1992). These secondary electrons provide additional information regarding the specimen's surface, which is used to construct the specimen's image. The amount of secondary electrons generated depends on the specimen's topography, the angle of the electron beam's penetration into the material, and the specimen's thickness. When secondary electrons collide with a phosphorescent coating on the detector, they generate a tremendous quantity of light. This light is then transmitted to a photomultiplier, where the feeble signals are amplified and displayed on the screen.

Field emission SEM (FE-SEM) is a type of SEM that provides narrower penetrating beams at both low and high electron energies, producing images with greater clarity, less electrostatic distortion, and higher spatial resolution (Pawley 1997). In this chapter, a protocol for FE-SEM sample preparation is described in detail. The biological samples utilized for SEM are significantly greater than those utilized for TEM. If the organism is properly prepared, SEM can be used to view the entire organism, including spores and small animals. Organisms with a rigid exoskeleton or rigid cell walls can be observed under an SEM using only dry air. However, the majority of biological specimens require adequate chemical fixation to inhibit biological processes and preserve cellular architecture. Sample preparation for TEM is typically more complicated than sample preparation for SEM. The electron gun's high-energy electrons are permitted to pass through the optics and into the tissue/cell section. The preponderance of electrons passes through the narrow section and projects onto the screen to create the image. To enable easy electron transmission (depending on the applied voltage), the section must be exceedingly thin (60–70 nm), which can only be accomplished by adhering to specific procedures for preparing biological materials (Rostgaard and Qvortrup 1989). Before sectioning, the specimen must be chemically fixated and immersed in a matrix to prevent it from disintegrating during microtome processing (Pease and Porter 1981). Since cyanobacteria are microorganisms, it is typically simpler to fix them with an EM fixative, embed them in hardening resins, and section the resin containing dispersed bacteria to visualize their internal structures. Epoxy polymers are the most commonly used embedding media (Finck 1960) for routine room-temperature sample preparation. Sample preparation and visualization of cyanobacteria for TEM and SEM will be discussed in this chapter.

21.2 EQUIPMENT

1. Refrigerated centrifuge (NEYA 16R, Remi, India).
2. Spectrophotometer (Shimadzu, Japan).
3. Zeiss Sigma 300 FE-SEM (Zeiss, Germany).
4. Glass Knife Maker (Leica Microsystems, Vienna, Austria).
5. Ultramicrotome (Ultracut UC7, Leica microsystems, Vienna, Austria).
6. Talos F200 (Thermo Fisher Scientific, USA).

21.3 MATERIALS AND REAGENTS

1. 50 mL exponentially grown cyanobacterial cell culture.
2. 2 mL microcentrifuge tube (Tarson, India).
3. 50 mL Falcon tube (Abdos, India).
4. Micropipettes (Thermo Fisher Scientific India Pvt. Ltd. Mumbai).
5. Paraformaldehyde (Catalog number P6148, Sigma-Aldrich, MO, USA).
6. Glutaraldehyde (Catalog number G002, TAAB Laboratories equipment Ltd, England, UK).
7. Sodium dihydrogen orthophosphate dihydrate (Catalog number 106342, Sigma-Aldrich, MO, USA).
8. Disodium hydrogen orthophosphate dihydrate (Catalog number 137036, Sigma-Aldrich, MO, USA).
9. Milli-Q (MQ) water.
10. Sodium hydroxide pellets (Catalog number Q27815, Thermo Fisher Scientific India Pvt. Ltd. Mumbai).

11. Graded ethanol series (50%, 75%, 95%, and 100%; diluted in Milli-Q water) (Merck Specialties Pvt. Ltd., Mumbai).
12. Acetone (Catalog number Q32007, Thermo Fisher Scientific India Pvt. Ltd. Mumbai).
13. Dry acetone (prepared by adding anhydrous copper sulphate).
14. Toluene (Catalog number 108325, Sigma-Aldrich Chemical Pvt. Ltd., Bangalore, India).
15. Osmium tetroxide (Catalog number 0001, TAAB Laboratories Equipment Ltd, England, UK).
16. Araldite CY212 Resin (Catalog number E006, TAAB Laboratories Equipment Ltd., England, UK).
17. Dodecenyl succinic anhydride (Catalog number D025, TAAB Laboratories Equipment Ltd., England, UK).
18. Methyl nadic anhydride (MNA) (Catalog number M012, TAAB Laboratories Equipment Ltd., England, UK).
19. 2,4,6-Tri (dimethylaminomethyl) phenol-30 (DMP-30) (Catalog number D022, TAAB Laboratories Equipment, Ltd., England, UK).
20. 1.5 mL microcentrifuge tubes (Tarson, India)
21. Glass knife strips, ultramicrotomy grade (Agar Scientific, UK).
22. Toluidine blue O (Catalog number SD211, TAAB Laboratories Equipment Ltd., England, UK).
23. 300-mesh copper grids (Catalog number GG007/C, TAAB Laboratories Equipment Ltd., England, UK).
24. Parafilm M sheets.
25. Tweezers.
26. Uranyl acetate EM (Catalog number U001, TAAB Laboratories Equipment Ltd., England, UK).
27. Lead citrate AR (Catalog number 23605, LADD Research Industries, Inc., VT, USA).
28. Sulfuric acid (Catalog number 258105, Sigma-Aldrich, MO, USA).
29. Hydrogen peroxide (Catalog number 1086009025, Sigma-Aldrich, MO, USA).

21.4 RECIPES

21.4.1 0.2 M Phosphate Buffer (pH 7.4)

1. *Solution A*: Add 3.12 g sodium dihydrogen orthophosphate ($NaH_2PO_4.2H_2O$) to 100 mL of distilled water and vortex to dissolve.
2. *Solution B*: Add 2.84 g disodium hydrogen orthophosphate (Na_2HPO_4) to 100 mL of distilled water and vortex to dissolve.
3. Mix 19 mL of solution A and 81 mL of solution B to get 100 mL 0.2 M phosphate buffer. Adjust the pH to 7.4 using HCl or NaOH.

21.4.2 0.1 M Phosphate Buffer (pH 7.4)

Take 50 mL of 0.2 M phosphate buffer. Add 50 mL distilled water to it and mix thoroughly. Adjust the pH to 7.4 with HCl or NaOH. Keep at 4°C for up to 3 months.

21.4.3 2% Paraformaldehyde Solution

Add 4 g paraformaldehyde to 100 mL of distilled water. Dissolve it at 60°C using a water bath (do not boil). Add a few drops of 0.1 N NaOH solution while dissolving. After dissolving paraformaldehyde completely, cool the solution to room temperature and filter it. Add 100 mL of 0.2 M phosphate buffer to the paraformaldehyde solution to make it 2% solution in 0.1 M phosphate buffer. The solution can be kept at 4°C for up to 3 months.

21.4.4 Primary Fixative Solution for Electron Microscopy

Add 10 mL glutaraldehyde to 90 mL of 2% paraformaldehyde solution. The final concentration of the fixative solution is 2% paraformaldehyde and 2.5% glutaraldehyde in 0.1 M phosphate buffer (pH 7.4). This clear solution can be stored at 4°C for up to 1 month.

21.4.5 Osmium Tetroxide Post-Fixative (Also Called Secondary Fixative)

Combine the following:

1. 2.5 mL of 4% aqueous osmium tetroxide stock from the sealed ampule.
2. 5 mL phosphate buffer stock.
3. 2.5 mL Milli-Q or distilled water.

Mix and keep at 4°C for immediate use. It is better to prepare a fresh batch each time.
Final concentrations: 1% (w/v) osmium tetroxide in 100 mM phosphate buffer (pH 7.4).

21.4.6 Epoxy Resins (Embedding Medium)

Epoxy resins and their components are viscous, and it takes time to make a homogenous mixture of the individual components. The following ingredients are added and stirred vigorously to mix thoroughly to make 100% resin.

1. Araldite CY212	–25 mL
2. Dodecenyl succinic anhydride (DDSA)	–23.2 mL
3. 2,4,6-Tri (dimethylaminomethyl) phenol-30 (DMP-30)	–0.9 mL
4. Methyl nadic anhydride (MNA)	–1.8 mL

21.4.7 Uranyl Acetate Staining Solution

Dissolve 1 g uranyl acetate in 100 mL Milli-Q water by stirring and store overnight at room temperature. The solution is stable for several months at 2–8°C in the dark.

21.4.8 Lead Citrate Staining Solution

1. Dissolve 5 mg of lead citrate in 1 mL of 4.5 mM sodium hydroxide.
2. Centrifuge for 5 minutes at high speed before use. Lead citrate staining solution can be stored at 4°C for 3–6 months.

21.4.9 Piranha Solution

Piranha solution is prepared by mixing 98% sulfuric acid and 35% hydrogen peroxide in a 3:1 (v/v) ratio.

21.5 METHODS

21.5.1 Sample Processing for SEM

Here, we provide a sample processing protocol for FE-SEM. Note that the protocol is only for FE-SEM, which utilizes the freeze-dry method. Room-temperature SEM processing requires a few more steps, including critical point drying and sputter coating. Specialized types of equipment are available in EM labs to perform these additional steps.

21.5.1.1 Fixation of Cyanobacterial Cells for FE-SEM

1. Take 50 mL exponentially growing cyanobacterial cells in a 50-mL Falcon tube. Centrifuge at 6000 × g for 3 minutes at 4°C. Discard the supernatant.
2. Suspend the cell pellet in 0.1 M phosphate buffer (pH 7.4). Centrifuge at 6000 × *g* for 3 minutes at 4°C. Discard the supernatant. Repeat this step twice.
3. Take the pellet in a 2-mL microcentrifuge tube and suspend it in a 2-mL fixative solution (see Recipes). Vortex to mix and keep it at 4°C for 4 hours.
4. Flip the microcentrifuge tube every 30 minutes to avoid settlement of the cells at the bottom of the tube.
5. After 4 hours, centrifuge at 8000 × *g* for 10 minutes at 4°C and discard the supernatant.
6. Add 1 mL 2% paraformaldehyde solution. Do not disturb the pellet.
7. Keep the cells at 4°C until further processing.

21.5.1.2 Sample Preparation for FE-SEM

1. Centrifuge all the samples at 9000 × *g* for 5 minutes to remove the fixative solution. Discard the supernatant.
2. Wash the samples with milli-Q water by centrifugation at 9000 × *g* for 5 minutes.
3. Repeat the washing step five times.
4. After washing, freeze-dry the cyanobacterial samples for 72 hours to obtain solid green samples.
5. Take the samples and keep them in a desiccator under a vacuum to prevent any degradation.
6. Sonicate the solid green samples in water for 5 minutes to make a homogeneously dispersed solution.
7. Drop-cast the prepared solution on a silicon wafer. Pretreat the silicon wafer with piranha solution and wash it with ethanol and water repeatedly to remove the oxide layer from the substrate.
8. After the drop casting, freeze-dry the samples for 72 hours and use them for FE-SEM characterization.

21.5.1.3 FE-SEM Characterization

Visualize the samples under the Zeiss Sigma 300 FE-SEM instrument and capture FE-SEM images of the samples at different magnifications to understand the morphology of the cyanobacterial cells and to compare the change in their cellular structure under different laboratory conditions.

21.5.2 Sample Processing for TEM

21.5.2.1 Sample Preparation for TEM

1. *Fixation of cyanobacterial cells for TEM*
   i. Resuspend the pellet and transfer the cells to a 2-mL microcentrifuge tube. Add 2 mL fixative solution (2.5% glutaraldehyde and 1–2% paraformaldehyde in a 0.1 M sodium phosphate buffer, pH 7.2–7.4). Vortex to mix and keep the samples at 4°C for 4 hours (fixation is generally performed at 4°C for 2–12 hours).
   ii. Flip the microcentrifuge tube every 30 minutes to avoid settlement of the cells at the bottom of the tube.
   iii. After 4 hours of incubation, centrifuge at 700 × *g* for 5 minutes at 4°C and discard the supernatant.
   iv. Keep the cells at 4°C until secondary fixation.
2. *Secondary fixation*
   i. Secondary fixation is done in 1% osmium tetroxide (OsO_4) for 1 hour at room temperature or overnight at 4°C followed by washing three times with 0.1 M phosphate buffer (pH 7.4).
   ii. Centrifuge the samples at 1000 × *g* for 5 minutes.

3. *Dehydration*
   i. Discard the supernatant and add acetone for dehydration.
      Dehydration of pelleted samples is done with three washes of acetone for 15 minutes, 20 minutes, and 30 minutes.
   ii. Add dry acetone and incubate for 10 minutes (repeat three times).
4. *Clearing*
   i. Add toluene to the samples and replace it twice with fresh toluene after 10-minute intervals.
5. *Infiltration of resin and block preparation*:
   i. Add a mixture of 1 part resin:1 part toluene and incubate overnight at 4°C.
   ii. A second change to pure resin at 4°C for 2 hours.
   iii. Change the pure resin two times and incubate at 4°C for 2 hours.
   iv. After incubation, embedding is done in freshly prepared resin followed by polymerization at 45°C for 12–18 hours.
   v. Increase the temperature to 60°C for 12–18 hours to get the blocks.
   vi. Preserve the blocks at room temperature until sectioning using an ultramicrotome.
6. *Ultramicrotomy*
   i. Before sectioning, check the location of the cyanobacteria pellet (black in appearance) in the transparent block.
   ii. Remove extra resin from the sides and trim the EM blocks in the shape of a trapezium.
   iii. Place the blocks in the ultramicrotome's block holder and cut 0.5-mm-thick sections. These sections are called semithin or light microscopy (LM) sections. Sections are cut using a glass or diamond knife. Glass knives for the TEM sectioning are generally made in EM facilities (see the details given after the protocol).
   iv. Place the semithin sections on glass microscopy slides and dry them by placing the slides on a hotplate at 80°C. Stain these sections with 1% toluidine blue (in 1% borax solution) for 20–30 seconds followed by washing with water and drying.
   v. Observe the sections under an optical microscope to check the quality of blocks and the presence of cells.
   vi. Trim further for the area of interest (where cells are present) decided by observing semithin sections under the light microscope.
   vii. Set the section thickness 60 nm and cut ultrathin sections (Figure 21.2a).
   viii. Ultrathin sections floating on the water of the knife boat are stretched by exposing them to chloroform.
   ix. Check the sections exhibit a silver-gray color interference while floating in the knife boat's liquid. These sections are between 60 and 70 nm.
   x. Lift the ultrathin sections on 300-mesh copper TEM grids using fine tweezers (Figure 21.2b).
   xi. Place the grids on a filter paper, keeping the sections right-side up.
   xii. Keep the grids on a filter paper in a covered Petri dish until processing for staining.
7. *Staining*
   i. Keep a parafilm sheet in a Petri dish.
   ii. Put 50 μL of uranyl acetate as a droplet on the clean parafilm surface inside the Petri dish.
   iii. Put the grid on the droplet, bringing the sections down to react with the stain (Figure 21.3).
   iv. Cover the Petri dish. Put an extra layer of black paper on it to protect the grids from light.
   v. Keep for 10–15 minutes.
   vi. Rinse gently in Milli-Q water and dry on Whatman filter paper.

vii. Place the grid onto the lead citrate droplets on parafilm for 10 minutes.
Do this step preferably in a hood with laminar flow, which prevents atmospheric carbon dioxide from reacting with lead citrate and forming precipitates of lead carbonate.
viii. Wash the grids quickly in Milli-Q water and allow to dry.
ix. Keep the grids in a grid box or a well-covered Petri dish.
x. The sections are now ready to be examined using a TEM.

8. *Imaging*
i. Insert the grid in the microscope and scan the sections on the grid on the TEM.
ii. Capture transmission electron micrographs of the cyanobacteria using digital micrograph software.
iii. Images can be taken at different magnifications depending on the objectives of the study.
iv. Transfer the images to another computer for analysis.

21.5.2.2 Details of Preparation Steps for TEM

Fixation: This is the initial step that inhibits cellular autolysis, preserving the backbone of the cells. An excellent EM fixative preserves practically all of the cell's components. Chemical fixation is the most widely used method to preserve biological specimens (Hayat 1981). The most prevalent components in primary fixatives are glutaraldehyde and paraformaldehyde followed by lipid fixation with osmium tetroxide, known as secondary fixation (Graham and Orenstein 2007).

Primary Fixation: Karnovsky's fixative is the standard main fixative used in EM, and it contains 2.5% glutaraldehyde and 1–2% paraformaldehyde in a 0.1 M sodium phosphate buffer (pH 7.2–7.4). For a 2 × 2 mm pelleted sample, fixation is performed at 4°C for 2– 6 hours.

Formaldehyde: Unsaturated lipids' double bonds can react with formaldehyde. According to reports, pH 7.5–8.0 is where maximum protein binding occurs (Hayat 1981). Nucleic acids react with formaldehyde without causing a loss of linear structure. Small carbohydrate molecules are not preserved and are washed away as acid mucopolysaccharides are not fixed, whereas mucoproteins and glycogen can survive formaldehyde fixation if given sufficient time.

Glutaraldehyde: Glutaraldehyde reacts with phospholipids with free amino groups. Most lipids are extracted during dehydration if glutaraldehyde-fixed samples are not post-fixed with osmium. Proteins are the primary targets of glutaraldehyde, which reacts with many free amino groups and cross-links proteins. Nucleic acid preservation is due to the interaction with the proteins associated with nucleic acids. The majority of carbohydrates, except glycogen, are not well fixed and are lost during subsequent rinses (Hayat 1981).

Secondary fixation: It primarily interacts with lipids and also serves as a stain. Specimens are washed in the buffer to remove excess fixative after secondary fixation and can be kept in the refrigerator until dehydration. By directly oxidizing double bonds on unsaturated lipids, osmium treatment has the potential to serve as a preservative for all classes of lipids, whether through a direct chemical reaction or indirectly in the case of saturated lipids. Long-term osmium fixation causes proteins to gradually denature, which eventually allows for the extraction of smaller peptides during buffer washes and the dehydration series.

Dehydration: When resins are utilized, the water content in the sample is replaced with appropriate organic solvent(s), referred to as dehydration agents, before infiltration. The water is replenished by passing the samples through a sequence of increasing concentrations of the dehydrating agent (ethanol) or numerous acetone changes. Acetone is preferable over ethanol because it is more easily miscible with cleaning agents such as xylene and toluene.

Clearing: To enhance resin penetration, the dehydrating agent is substituted with a clearing agent. Some of the well-known excellent cleaning agents utilized in this procedure are xylene, toluene, and propylene oxide. These chemicals are capable of entirely removing the residues of water left by partial dehydration.

Infiltration and embedding: Infiltration is accomplished by progressively lowering the concentration of the clearing agent and increasing the concentration of resin in the infiltration medium in a 3.1:2, 2:2, and eventually 1:1 ratio (clearing agent:resin). A 1-hour treatment at each phase is adequate for resin penetration in most biological samples. Following infiltration, samples are embedded in BEEM capsules or Eppendorf tubes. The samples are put in such a manner that the desired section cutting is possible.

Polymerization: It is the process of allowing the liquid resin to solidify under heat in a temperature-controlled oven, initially at 45°C overnight, to allow the embedding media to become less viscous, permitting simple resin penetration into the sample. The temperature is raised to 60°C the next morning, and the resin hardens after 12–18 hours. The blocks are removed after polymerization and prepped for section cutting (ultramicrotomy).

Making glass knives for sectioning: Belgium glass strips (glass knife strips, ultramicrotomy grade) are used for making glass knives. First, the glass strip is cleaned with acetone to remove any traces of grease. Using a knife maker, glass strips are cut into squares and then each square is cut diagonally into two triangles. Knives with regular, sharp edges are chosen for sectioning. An adhesive tape is used to make a boat to fill with distilled water. The sections float on this water and are then placed onto grids (Figure 21.2).

Ultramicrotomy: Blocks are initially trimmed in the shape of a trapezium to get serial ribbons. The trimmed block is placed in the ultramicrotome's block holder, and portions are cut with glass or diamond blades. Sections that are 0.4–1.0 mm thick are cut before thin sectioning to observe the cells under an optical microscope. This allows the user to decide on the quality of the embedded sample's fixation as well as the region of interest for thin sectioning. Sections are taken on clean glass slides and warmed on a hot plate at around 80°C. After staining the slides with toluidine blue for 20–30 seconds, samples can be visualized under the light microscope to check the presence of cyanobacteria. The blocks are further trimmed and ultrathin sections (thickness 60–70 nm) are cut (Figure 21.2). The sections that exhibit a silver-gray color interference while floating in the knife boat's liquid are between 60 and 70 nm.

Staining: A twofold staining procedure utilizing aqueous uranyl acetate and alkaline lead citrate is widely used to achieve good contrast (Maunsbach and Afzelius 1999). A little

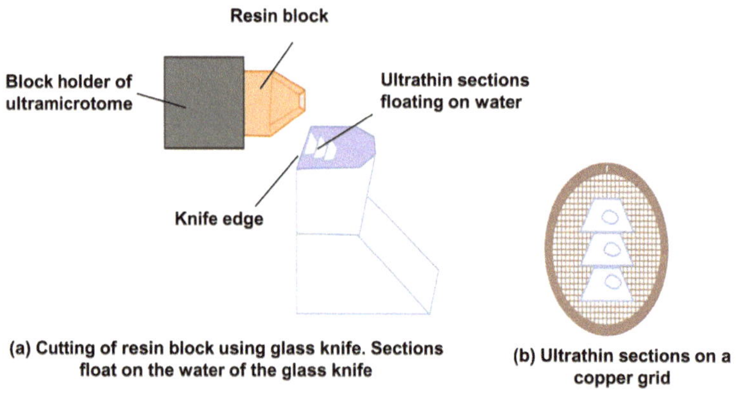

FIGURE 21.2 Schematic diagram shows sectioning of a resin-embedded specimen using a glass knife. (a) The cut sections float on the distilled water surface and (b) are placed onto the TEM grid (color figure in e-book).

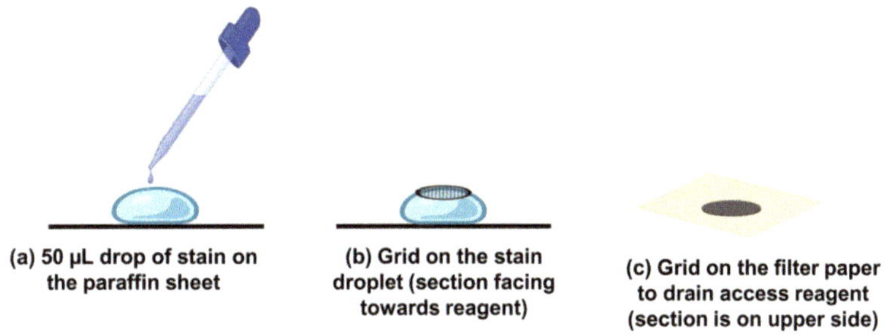

FIGURE 21.3 Representation of the method used to stain ultrathin sections. (a) A droplet is put on the paraffin sheet kept in a Petri dish; (b) the grid containing the ultrathin section is put onto the droplet; (c) to drain the excess reagent, the grid is put on the filter paper (color figure in e-book).

amount of stain is placed in droplets over a clean parafilm surface, the grid is placed onto the drop for an appropriate time, and then the excess stain/water is drained with filter paper (Figure 21.3).

21.6 DATA ANALYSIS

EM is frequently used for qualitatively analyzing the ultrastructure of cyanobacteria. It is utilized for studying the morphological characteristics of a single cell, which enables images of individual cells to be directly captured at a high resolution. SEM has been used for analysis of extracellular topology, cell-cell interactions, and extracellular matrix. TEM offers enormous potential to study structural changes inside the cells. The changes inside the cell can be easily detected by taking images of that area at various magnifications. Images that are taken by EM should be represented with an appropriate scale (Figure 21.4).

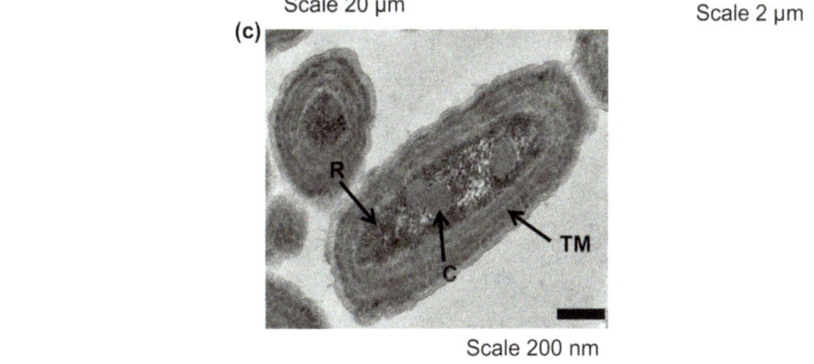

FIGURE 21.4 Electron microscopy (EM) images of cyanobacteria. Images of the surface of (a) *Oscillatoria* sp. and (b) *Synechococcus elongatus* PCC 7942 were captured using a scanning electron microscope (SEM). Transmission electron microscope (TEM) image shows the ultrastructure of *S. elongatus* PCC 7942 (c). TM, Thylakoid membrane; C, Carboxysome; R, Ribosome.

21.7 NOTES/PRECAUTIONS

1. Some of the reagents used for EM are very toxic. Always use appropriate personal protective equipment and ensure adequate ventilation while handling reagents.
2. EM images are at very high magnification, and a very small change due to sample handling and processing may lead to misinterpretation.
3. EM images are generally used for qualitative interpretations. Always use the same magnification while comparing two different images. Images should always be presented with a scale bar.
4. Glutaraldehyde and paraformaldehyde are known to cause oral, dermal, and respiratory toxicity. Always use appropriate personal protective equipment and ensure adequate ventilation while handling these reagents.
5. Osmium tetroxide solution must be prepared a day before use, as it is poorly soluble in water. Osmium tetroxide is toxic if ingested, inhaled, or absorbed through the skin. It should be opened and used in the fume hood. Always use nitrile gloves while handling osmium tetroxide solutions.
6. It is advisable to prepare resins in small batches just before use. Dilutions of complete resin mixtures and dehydrating agents (acetone) should be prepared by dispensing the resin components into new disposable beakers. Components of resin and dehydration agents are kept capped to prevent evaporation. Preparations should be done in a fume hood. Accurate weighing of all components of resins is critical in their preparation to ensure proper characteristics of the resin.
7. Uranyl ions form an insoluble uranyl phosphate, and therefore, uranyl acetate should not be used with phosphate buffer. Samples must be first rinsed in distilled water.
8. Make sure not to breathe on the grid when making lead citrate stain.
9. The resin blocks are very stable and can be kept at room temperature for months. These blocks have to be sectioned using ultramicrotome. Ultramicrotomy requires a specialized technical skill and is generally done only in core EM facilities.
10. An artifact is damage caused to a sample during the preparation procedure that is easily mistaken for the microstructure of the sample. Mechanical, chemical, ionic, or physical activity can all result in artifacts (Mollenhauer 1993; Ayache et al. 2010; Hillman 2011). It is certain that all the samples get altered by the preparation procedures and by interaction with the electron beam. These artifacts should be minimized and identified for proper interpretation of the images. In SEM analysis, anything that falls on the samples can appear to be a fragment of the sample. In addition, fractures are frequently caused by dehydration and the microscope vacuum. Other artifacts may be created by electron beam irradiation during TEM/SEM observation.
    a. The following table lists some of the TEM preparation artifacts:

Preparation-Induced Artifacts	Artifacts Induced during Observation
Contamination, structural change, cracks, inclusion of abrasive grains, deformation, dislocations, selective abrasion, microstructure change, change of natural contrast	Charge effects, microstructure change, sample tearing

ACKNOWLEDGMENTS

This work was supported by funding from an Institute of Eminence incentive grant, Banaras Hindu University (R/Dev/D/IOE/Incentive/2021-2022/32399). The financial support from the Science and Engineering Research Board (SERB), New Delhi (SCP/2022/000201) is greatly acknowledged. ST (NTA Ref. No.-201610202339) is thankful to UGC, New Delhi, India for the junior research fellowship. AG thanks ICMR, New Delhi for the senior research fellowship (3/1/3/JRF-2019/HRD-LS).

Sophisticated Analytical Instrumentation Facility (SAIF), AIIMS, New Delhi, is acknowledged for the TEM experiments.

REFERENCES

Ayache J, Beaunier L, Boumendil J, Ehret G and Laub D (2010) Artifacts in transmission electron microscopy. In: Sample Preparation Handbook for Transmission Electron Microscopy: Methodology (Eds. J. Ayache, L. Beaunier, J. Boumendil, G. Ehret and D. Laub). Springer, New York, NY, pp. 125–170. ISBN: 978-0-387-98182-6

Baxa U (2018) Imaging of liposomes by transmission electron microscopy. *Methods in Molecular Biology* 1682: 73–88.

Crewe AV, Wall J and Welter LM (1968) A high-resolution scanning transmission electron microscope. *Journal of Applied Physics* 39(13): 5861–5868.

Finck H (1960) Epoxy resins in electron microscopy. *The Journal of Biophysical and Biochemical Cytology* 7(1): 27.

Ghosh I, Atsuzawa K, Arai A, Ohmukai R and Kaneko Y (2021) TEM observation of compacted DNA of *Synechococcus elongatus* PCC 7942 using DRAQ5 labelling with DAB photooxidation and osmium black. *Microscopy* 70(3): 316–320.

Graham L and Orenstein JM (2007) Processing tissue and cells for transmission electron microscopy in diagnostic pathology and research. *Nature Protocols* 2(10): 2439–2450.

Hayat MA (1981) Methods of fixation. In: Fixation for Electron Microscopy (Ed. M.A. Hayat). Academic Press, pp 200–208. ISBN: 9780323147293

Hillman H (2011) Artifacts of electron microscopy in cells. *Physiological Chemistry and Physics and Medical NMR* 41: 61–63.

Jaksch H and Martin JP (1995) High-resolution, low-voltage SEM for true surface imaging and analysis. *Fresenius' Journal of Analytical Chemistry* 353: 378–382.

Joy DC and Pawley JB (1992) High-resolution scanning electron microscopy. *Ultramicroscopy* 47(1–3): 80–100.

Lang NJ (1968) The fine structure of blue-green algae. *Annual Review of Microbiology* 22: 15–46.

Mareš J, Strunecký O, Bučinská L and Wiedermannová J (2019) Evolutionary patterns of thylakoid architecture in cyanobacteria. *Frontiers in Microbiology* 10: 277.

Maunsbach AB and Afzelius BA (1999) Section-staining. In: Biomedical Electron Microscopy (Eds. A.B. Maunsbach and B.A. Afzelius). Academic Press, San Diego, pp. 207–228. ISBN: 9780080528090

Mollenhauer HH (1993) Artifacts caused by dehydration and epoxy embedding in transmission electron microscopy. *Microscopy Research and Technique* 26(6): 496–512.

Pawley J (1997) The development of field-emission scanning electron microscopy for imaging biological surfaces. *Scanning* 19: 324–336.

Pease DC and Porter KR (1981) Electron microscopy and ultramicrotomy. *The Journal of Cell Biology* 91(3): 287s–292s.

Rose HH (2016) Optics of high-performance electron microscopes. *Science and Technology of Advanced Materials* 9(1): 014107.

Rostgaard J and Qvortrup K (1989) Ultrathin sectioning for electron microscopy: The distilled water in the knife trough may extract phosphatase reaction products from the sections. *Journal of Microscopy* 156(2): 253–257.

Rouquette J, Kalland KH and Fakan S (2009) Visualisation of RNA by electron microscopic in situ hybridisation. *The Nucleus* 464: 403–413.

Shiltsev V, Bishofberger K, Kamerdzhiev V, Kozub S, Kufer M, Kuznetsov G, Martinez A, Olson M, Pfeffer H, Saewert G, Scarpine V, Seryi A, Solyak N, Sytnik V, Tiunov M, Tkachenko L, Wildman D, Wolff D and Zhang X-L (2008) Tevatron electron lenses: Design and operation. *Physical Review Special Topics-Accelerators and Beams* 11(10): 103501.

Unwin PNT and Huxley HE (1972) Electron microscopy of biological specimens by means of an electrostatic phase plate. *Proceedings of the Royal Society of London. A. Mathematical and Physical Sciences* 329(1578): 327–359.

22 Cyanobacterial Biofertilizer Production Technology

Sonal Mishra and Rajeshwar P. Sinha

22.1 INTRODUCTION

Cyanobacteria form a large and diverse group of oxygenic photoautotrophic prokaryotes, many of which can produce hydrogen, energy, biofertilizers, secondary metabolites, cosmetics and medicines (Mishra et al. 2019a, 2019b Chittora et al. 2020; Mishra et al. 2022a). Cyanobacteria have attracted special attention in recent years due to their significant applications in the field of biotechnology (Mishra and Sinha 2022). Cyanobacteria are regarded as widespread biomass producers in both fresh and marine water ecosystems (Häder et al. 2011; Mishra et al. 2022b). They can also be found in extreme environmental conditions (Anderson 2005). Globally, cyanobacteria are playing an important role in the maintenance of various ecologically and economically important agriculture ecosystems by increasing soil fertility, resulting in increased yield of rice (Song et al. 2005). Cyanobacteria inoculation is often reported as a safe agricultural practice to reduce costs and energy inputs, contributing to economic viability in paddy habitats (Mishra and Pabbi 2004). Cyanobacteria act as biofertilizers, either as free-living organisms or form a symbiotic association with pteridophytes (*Azolla*, the water-fern) and gymnosperms (the coralloid root of *Cycas*), and this is the main principle of many species – that they have a nitrogen-fixing ability which in turn increases soil fertility (Fay 1983).

Several reports on the broad distribution of cyanobacterial forms such as *Anabaena, Nostoc, Oscillatoria, Aphanothes* and *Phormidium* have already been published (Pal and Santra 1982; Garcia-Pichel and Belnap 1996). Some dominating heterocystous nitrogen-fixing cyanobacterial species have been reported from the soil of Cuttack and Orissa, which are *Nostoc, Anabaena, Tolypothrix, Aulosira, Calothrix* and *Cylindrospermum*. Distribution of isolated cyanobacteria from the soil of Howrah, Cuttack and Bhuwaneshwar specified that heterocystous strains prevail in the natural habitat (Saxena et al. 2007). These cyanobacterial species can be used as biofertilizers because of their ability to secrete growth-promoting substances (such as auxin, gibberellin, vitamins and amino acids), augment water-holding capacity, increase soil biomass or prevent the growth of weeds (Roger and Reynaud 1982; Rodriguez et al. 2006; Saadatnia and Riahi 2009). Over the last decade, a widely applicable and cost-effective method for cyanobacterial biofertilizer production technology has been reported with some modifications: (1) Trough or tank method, (2) pit method, (3) field method and (4) rice nursery cum cyanobacterial production method (Figure 22.1) (Vaishampayan et al. 2001; Sahu et al. 2012).

22.2 EQUIPMENT

1. Dry glassware.
2. Sterile N95 mask.
3. Galvanized iron sheet or permanent tank.
4. Troughs or tank pits.
5. Tray (2 m × 1 m × 23 cm).

22.3 MATERIALS

1. Plastic bags.
2. Paddy field soil.

DOI: 10.1201/9781003398387-22

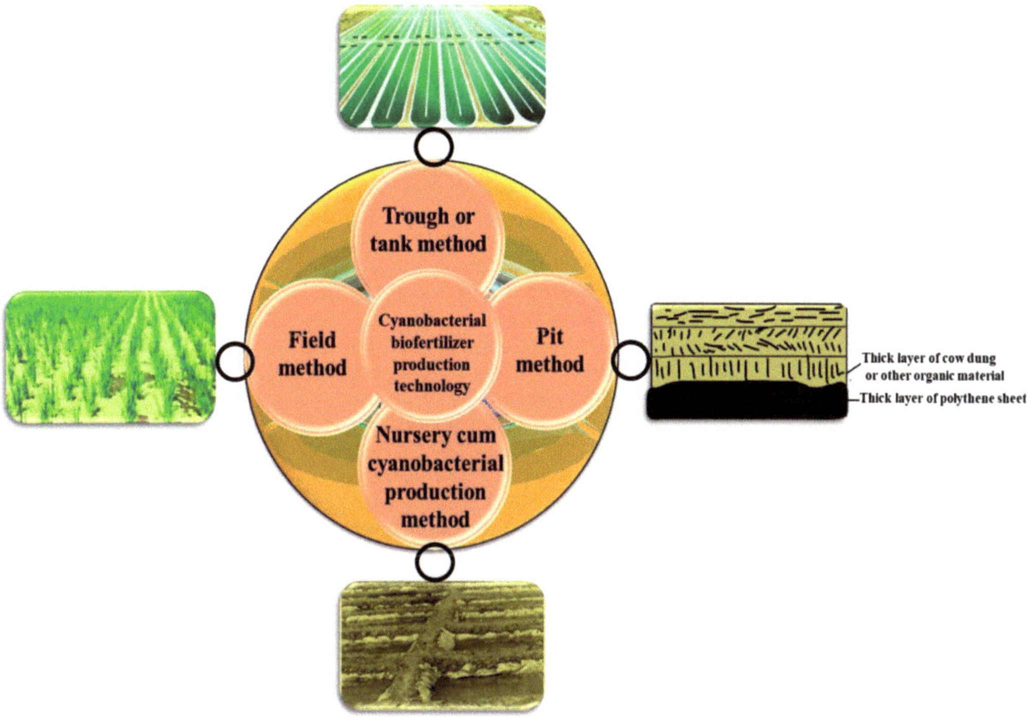

FIGURE 22.1 Four different methods for the production of cyanobacterial biofertilizer (color figure in e-book).

3. Dried cyanobacterial flakes.
4. Superphosphate.
5. Carbofuran.
6. Sodium molybdate.
7. Furadan granules.
8. Malathion.
9. Thick polythene sheet.

22.4 RECIPES

22.4.1 To Avoid Mosquito and Insect Infestation

A total of 10–15 g of furadan granules or malathion were added to the soil and water mixture.

22.4.2 Acidic Soil Neutralized by Adding $CaCO_3$

In the case of acidic soil, $CaCO_3$ can be added to neutralize the pH.

22.5 METHODS

Any technology usually is successful based on its techno-economic viability. The technology can give extra income to farmers by auctioning cyanobacterial biofertilizers. In general, four methods of cyanobacterial biofertilizer production technology have been reported, and the specific experimental steps of this technology are given in (Figure 22.2).

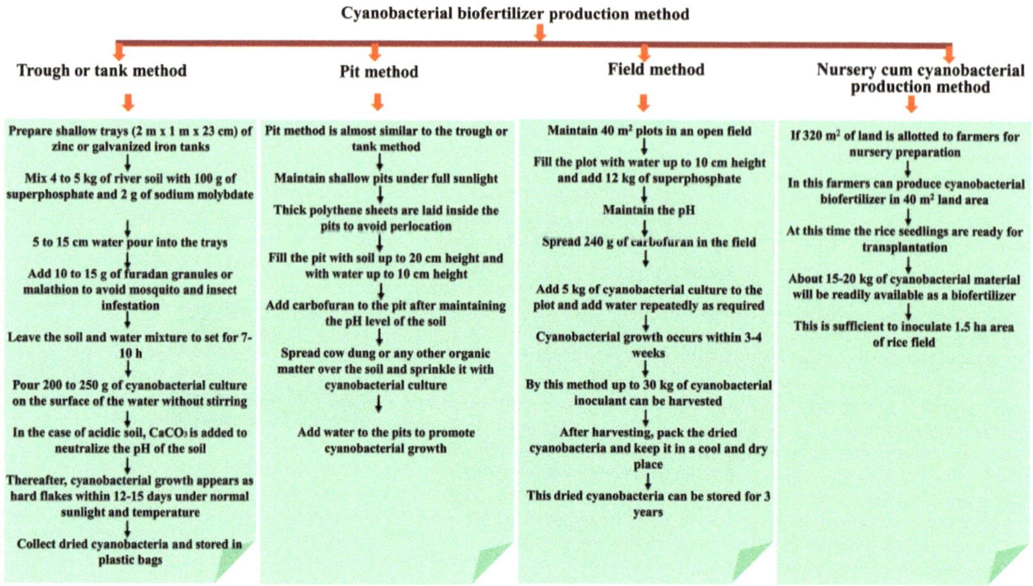

FIGURE 22.2 Flowchart showing the various steps involved in the methods mentioned in the text (color figure in e-book).

22.5.1 Trough or Tank Method

1. Shallow trays (2 m × 1 m × 23 cm) of zinc or galvanized iron tanks are prepared. The size of the tank can be increased depending on the increase in the production rate.
2. After this, thoroughly mix 4–5 kg of river soil with 100 g of superphosphate and 2 g of sodium molybdate.
3. Pour 5–15 cm of water into the trays and mix properly. The amount of water will also depend on the rate of evaporation at that place.
4. To avoid mosquito and insect infestation, 10–15 g of furadan granules or malathion were added.
5. Leave the mixture of soil and water to set for 7–10 hours. Afterward, 200–250 g of cyanobacterial culture was poured on the surface of the water without unsettling it.
6. The pH of the soil should be neutral. In the case of acidic soil, $CaCO_3$ can be added to neutralize.
7. Within 12–15 days under normal sunlight and temperature, the growth of cyanobacteria will be visible as hard flakes on the surface of the soil and water mixture.
8. The cyanobacterial flakes are allowed to dry by evaporation.
9. When the soil dries out, the cyanobacterial flakes are dissociated from the soil. These flakes are collected and stored in plastic bags.
10. Thereafter, water is added to the trays, and the soil is ready for the next round of cyanobacterial growth. Thus, cyanobacterial harvests can be achieved two to three times.

22.5.2 Pit Method

1. This method is almost similar to the trough or tank method. Shallow pits are maintained under full sunlight.
2. Thick polythene layers are laid inside the pits to avoid percolation.

3. The pit is filled with soil up to 20 cm in height and watered up to 10 cm in height.
4. Carbofuran is added to the pit after maintaining the pH of the soil.
5. Cow dung or any other organic matter is then spread over the soil and sprinkled with the cyanobacterial culture.
6. The pits are watered to promote cyanobacterial growth.
7. After this, the soil is allowed to dry.

This method is easy and less expensive to operate by small farmers.

22.5.3 Field Method

1. Small plots of 40 m^2 are maintained in an open field.
2. The plot is watered up to 10 cm in height and then 12 kg of superphosphate is added.
3. After maintaining the pH, 240 g of carbofuran is spread in the field.
4. After this, 5 kg of cyanobacterial culture is added to the plot and watered repeatedly.
5. Cyanobacterial growth is observed over 3–4 weeks and then the soil is allowed to dry. Approximately 30 kg of cyanobacterial inoculant can be harvested by this method.
6. After harvesting, well-dried cyanobacteria are packed. The bags are kept in a cool and dry place. These bags can be preserved for up to 3 years without diminishing their efficacy.
7. The dried cyanobacterial flakes are introduced into the field or crops 10 days after being planted. This process is called algalization. It increases the yield in rice fields by up to 34%.

22.5.4 Rice Nursery Cum Cyanobacterial Production Method

1. Cyanobacterial biofertilizer can be produced by farmers with seedlings in their nurseries if 320 m^2 of land is allotted to farmers for nursery preparation. Farmers can also produce cyanobacterial biofertilizer in 40 m^2 of land.
2. At this time the rice seedlings are ready for transplantation.
3. Approximately 15–20 kg of efficient cyanobacterial material would be readily available as biofertilizer or sufficient to inoculate a 1.5-ha rice field.

If every farmer produced the cyanobacteria needed to inoculate their land, they would reduce the cost of biofertilizers to be purchased. Thus, the cost and side effects of using chemical fertilizers can also be reduced.

22.6 NOTES/PRECAUTIONS

1. When agricultural chemicals like fertilizers, insecticides, weedicides, etc., are used in the field, apply cyanobacterial biofertilizer 3–4 days after applying chemical fertilizer.
2. For a cumulative effect, one should apply cyanobacterial biofertilizers for at least four consecutive seasons.
3. Repeated application of cyanobacterial biofertilizer is not required at the time of application, as once applied it becomes established in the field and grows again when the environment becomes favorable.

ACKNOWLEDGMENTS

Sonal Mishra is thankful to the University Grants Commission (UGC), New Delhi, India, for the financial assistance in the form of a senior research fellowship.

REFERENCES

Anderson RA (2005) Algal Culturing Techniques. Elsevier Academic Press, Burlington, MA, USA.
Chittora D, Meena M, Barupal T, Swapnil P and Sharma K (2020) Cyanobacteria as a source of biofertilizers for sustainable agriculture. *Biochemistry and Biophysics Reports* 22: 100737.
Fay PN (1983) Blue-Greens. Edward Arnold, London.
Garcia-Pichel F and Belnap J (1996) Microenvironments and microscale productivity of cyanobacterial desert crusts. *Journal of Phycology* 32: 774–778.
Häder D-P, Helbling EW, Williamson CE and Worrest RC (2011) Effects of UV radiation on aquatic ecosystems and interactions with climate change. *Photochemical and Photobiological Sciences* 10: 242–260.
Mishra S, Kumari N, Häder D-P and Sinha RP (2022b) Cyanobacterial blooms and their implications in the changing environment. *Advances in Environmental and Engineering Research* 3: 1–41.
Mishra U and Pabbi S (2004) Cyanobacteria: A potential biofertilizer for rice. *Resonance* 9: 6–10.
Mishra S, Pandey A, Ahmed H, Kumar D, Singh V and Sinha RP (2019a) Mycosporine-like amino acids (MAAs) profile of cyanobacteria from different historical kunds of Varanasi, India. *International Journal of Applied Sciences and Biotechnology* 7: 317–326.
Mishra S, Pandey A and Sinha RP (2019b) Characterization of photoprotective and pharmaceutically important compounds from cyanobacteria and algal assemblages from historical kunds of Varanasi, India. *Journal of Chemical and Pharmaceutical Research* 11: 24–36.
Mishra S and Sinha RP (2022) Biotechnological exploitation of cyanobacterial photoprotective metabolites. *Vegetos* 16: 1–7.
Mishra S, Sinha RP, Sahu N and Kumari N (2022a) Identification and characterization of mycosporine-like amino acid in the cyanobacterium *Scytonema* sp. HKAR-16 by HPLC, Raman spectroscopy and ESI-MS. *Songklanakarin Journal of Science and Technology* 44: 836–840.
Pal TK and Santra SC (1982) Contribution to the cyanophyceae of Murashidabad. *Phykos* 21: 150–152.
Rodriguez AA, Stella AA, Storni MM, Zulpa G and Zaccaro MC (2006) Effects of cyanobacterial extracellular products and gibberellic acid on salinity tolerance in *Oryza sativa* L. *Saline Systems* 2: 1–4.
Roger PA and Reynaud PA (1982) Free-living blue-green algae in tropical soils. In: Microbiology of Tropical Soils and Plant Productivity (Eds. Y.R. Dommergues and H.G. Diem). Springer, Dordrecht, pp. 147–168. ISBN: 978-94-009-7529-3
Saadatnia H and Riahi H (2009) Cyanobacteria from paddy fields in Iran as a biofertilizer in rice plants. *Plant Soil and Environment* 55: 207–212.
Sahu D, Priyadarshani I and Rath B (2012) Cyanobacteria – as potential biofertilizer. *CIBTech Journal of Microbiology* 1: 20–26.
Saxena S, Singh BV, Tiwari S and Dhar DW (2007) Physiological characterization of cyanobacterial isolates from Orissa and West Bengal. *Indian Journal of Plant Physiology* 12: 181–185.
Song T, Martensson L, Eriksson T, Zheng W and Rasmussen U (2005) Biodiversity and seasonal variation of the cyanobacterial assemblage in a rice paddy field in Fujian, China. *The Federation of European Materials Societies Microbiology Ecology* 54: 131–140.
Vaishampayan A, Sinha RP, Häder D-P, Dey T, Gupta AK, Bhan U and Rao AL (2001) Cyanobacterial biofertilizers in rice agriculture. *The Botanical Review* 67: 453–516.

23 Biomass Composition Measurement and Intracellular Flux Analyses for Insights into Cyanobacterial Metabolism

Piyush Pachauri, Preeti Rai, and Shireesh Srivastava

23.1 INTRODUCTION

Cyanobacteria are photosynthetic microorganisms that are widely recognized as a promising platform for the production of biofuels and other value-added compounds. In recent years, there has been growing interest in understanding the metabolic pathways and fluxes in cyanobacteria for rational engineering of these organisms for biotechnological applications (Jaiswal et al. 2018). One of the key challenges in this area is the accurate measurement of biomass composition and intracellular fluxes, which provide crucial insights into the underlying metabolic processes.

The progress in genome sequencing technologies and subsequent bioinformatics analyses has made it possible to quickly sequence and annotate cyanobacterial genomes. These can then be used to build genome-scale metabolic networks and models which have applications in systems biological analyses of cyanobacterial metabolism and other integrative analyses (Jaiswal et al. 2018). An important input in cyanobacterial model reconstruction is measurement of biomass composition. Cyanobacterial strains show interesting diversity in their composition that, in turn, plays an important role in variations in the intracellular fluxes of these organisms. Measurement of intracellular fluxes forms an integral component of modern approaches to design metabolic engineering strategies for biotechnological applications (Ahmad et al. 2020). In this chapter, we discuss the methods employed to estimate the biomass composition of cyanobacteria, the variation in biomass composition of certain cyanobacteria reported in the literature and its diurnal variation. Carbohydrates, proteins, glycogen, lipids, nucleic acids and pigments constitute the major components of a cyanobacterial cell. Thus, the study of the biomass composition of a cyanobacterium involves the estimation of total carbohydrate, protein, glycogen, lipids, RNA, DNA, chlorophyll, carotenoids and phycobiliproteins as well as subunit composition of the macromolecules, e.g., analysis of lipid composition (Ahmad et al. 2020). There are several methods for the estimation of all these components, but some methods are preferred due to their sensitivity and simplicity.

Additionally, we discuss the estimation of intracellular flux distribution from the genome-scale metabolic models and the insights gained from them, both in terms of understanding the metabolic properties of these organisms and how to design metabolic engineering strategies. High-throughput metabolic engineering requires a global understanding of metabolism, and metabolic models can provide a framework to decipher the metabolic capabilities of these strains. Genome-scale metabolic models (GSMMs) help in systems-level analysis of metabolic networks of organisms. The models have led to the successful design of metabolic pathway engineering strategies in organisms such as *Escherichia coli* and *Saccharomyces cerevisiae* (Fatma et al. 2018; Chen et al. 2021).

DOI: 10.1201/9781003398387-23

Among the photoautotrophs, cyanobacteria have become the focus of synthetic biology research because of recent discoveries of strains showing very fast growth rates such as those belonging to the genus *Synechococcus* (Jaiswal et al. 2018). These strains can potentially be used as platforms for producing various high-value compounds using CO_2 and sunlight as raw materials. Reconstruction of reliable metabolic models requires integration of prior genomic and biochemical information of the organism being studied. Information about cyanobacterial metabolism is rather scarce as compared with other widely studied organisms like *E. coli* and yeast. Nonetheless, model-directed engineering of cyanobacterial strains has shown considerable progress (Shirai et al. 2016). In this chapter, we report on various metabolic models developed for cyanobacteria in the last decade, and also we discuss the application of model-based metabolic engineering in cyanobacteria. Lastly, we discuss the emerging applications and future directions in the integrative systems biology analyses of cyanobacteria.

23.2 METHODS TO ESTIMATE BIOMASS COMPOSITION

Carbohydrates, proteins, glycogen, lipids, nucleic acids and pigments constitute the major components of a cyanobacterial cell. Thus, the study of the biomass composition of a cyanobacterium involves the estimation of total carbohydrate, protein, glycogen, lipids, RNA, DNA, chlorophyll, carotenoids and phycobiliproteins as well as subunit composition of the macromolecules, e.g., analysis of lipid composition (Ahmad et al. 2020).

23.2.1 Estimation of Total Carbohydrates

Different colorimetric methods of carbohydrate estimation include the anthrone, orcinol, resorcinol and phenol-sulfuric acid methods. The phenol-sulfuric acid method is the most common among these methods because of its simplicity and reliability (Masuko et al. 2005). Several variants of this method have been reported based on their scale.

23.2.1.1 Anthrone-Sulfuric Acid Method

Freshly prepared anthrone in sulfuric acid (2 g anthrone per liter of concentrated H_2SO_4) is used in this process. The carbohydrates are hydrolyzed using concentrated H_2SO_4 at room temperature for 10 minutes followed by 10-fold dilution of the mixture using water. Two volumes of the anthrone reagent are then added to 0.5 mL hydrolyzed carbohydrate in a test tube with quick vortexing (<5 seconds) to prevent overheating followed by incubation at 90°C for 10 minutes. Absorbance of the solution is measured at 625 nm, and the concentration is calculated by comparing against a standards graph (Schneegurt et al. 1994). This is one of the most common methods of carbohydrate estimation.

23.2.1.2 Phenol-Sulfuric Acid

Phenol-sulfuric acid reaction is a quick, simple and accurate approach for identifying the presence and makeup of sugars and their derivatives in a solution. This method involves phenol and concentrated sulfuric acid treatment of the material, which results in an orange-yellow color in the presence of sugars. This is very a sensitive method and can determine the presence of even trace amounts of sugars with stable color and does not require precise control of conditions. This technique can be used to analyze the composition of polysaccharides and their methyl derivatives in conjunction with paper partition chromatography. It is also affordable and reproducible, making it a valuable option for several applications (Deb et al. 2019). The phenol-sulfuric acid method was published by DuBois et al. (1956) to measure total carbohydrate. The procedure comprises mixing the cyanobacterial suspension in strong sulfuric acid (98% w/w) and 5% phenol in a ratio of 1:5:1 (culture:H_2SO_4:phenol). After mixing the content and keeping it still for 10 minutes, the generated color is read at 490 nm. The concentration is determined in mg/L by a calibration curve created using several standards of unknown concentrations of glucose (Markou et al. 2013). The cellular content can be expressed as a percentage of dry cell weight, calculated using following equation:

$$\text{Carbohydrate (\%)} = [\text{Total carbohydrate amount(g)/Dry biomass weight(g)}] \times 100$$

The minimum amount of monosaccharides required in the previously mentioned original method was 50–450 nmol. This process was scaled down to 10–80 nmol by Masuko et al. (2005). The scaled-down method is amenable to 96 well plates for quick estimation of several samples while saving reagents.

23.2.1.3 Orcinol-Sulfuric Acid Method

This is one of the oldest methods of carbohydrate estimation. It uses equal volumes of sugar and orcinol followed by addition of eight volumes of 80% H_2SO_4. This method has several variants, with 1–4 g orcinol and 70–80% H_2SO_4 used in different methods (Fernell and King 1953; Brückner 1955; Churms 2005). This method is the first one to be altered for use with an automated carbohydrate analyzer. The flow-through from the column is heated at 95°C for 15 minutes after being combined with the reagent. Absorbance is typically measured at 420–425 nm, but in the presence of deoxy sugars or uronic acids, absorbance is measured at two different wavelengths: 420 and 510 nm (Churms 2005).

23.2.1.4 Resorcinol-Sulfuric Acid

Monsigny et al. (1988) used resorcinol (1,3-dihydroxybenzene) in a 75% sulfuric acid solution in calorimetric assay for quantification of neutral sugars. This assay is precise and simple. This method is good for the detection of glycoconjugates in chromatographic column effluents because it is sensitive up to very low concentrations of sugar and it is not affected by the presence of proteins. This method can detect up to 1–100 nmol of neutral sugars present in 20-µL samples as compared to earlier described methods like orcinol, phenol or anthrone, which have higher detection limits of 0.1–1 µmol sugar. Dried sugar powder dissolved in acetic acid is used in this method. For a microtiter assay, 20 µL sugar is mixed with an equal volume of resorcinol followed by five volumes of 75% H_2SO_4. The microtiter plate is incubated at 90°C for 30 minutes before reading the optical density at 430 or 480 nm. The quantification is performed using appropriate standards of known concentrations. A 10-fold higher volume of each component is needed for the regular assay of sugars by this method.

23.2.2 Estimation of Glycogen

Glycogen is a hyper-branched polymer of glucose monomers which are linked through $\alpha(1 \rightarrow 4)$, producing linear polymer amylose and $\alpha(1 \rightarrow 6)$ glycosidic bonds, in turn producing a highly branched polymer called amylopectin. Ninety to ninety-three percent of the links in glycogen are α-D-(1–4) glycosidic bonds and 7–10% are α-D-(1–6)glycosidic linkages (Fernandez et al. 2011; Vidal and Venegas-Calerón 2019). The severe branching gives glycogen a compact and amorphous structure in addition to making it highly water-soluble (Meléndez et al. 1998).

There are three major steps in the estimation of glycogen from any cyanobacterial cell: Extraction of glycogen from the cell, hydrolysis of glycogen to monomeric glucose and estimation of the glucose produced. Glycogen represents the storage of carbohydrate, which can be precipitated using KOH, while total carbohydrates can be estimated by using the phenol-sulfuric acid method (Zavřel et al. 2018).

23.2.2.1 Extraction of Glycogen from Cyanobacteria

Two types of hydrolysis are commonly performed for the extraction of glycogen: 1) Acid hydrolysis and 2) alkaline hydrolysis. Hydrolysis of a cyanobacterial biomass with strong acid at high temperature yields glucose monomers due to hydrolysis of glycogen. Alkaline hydrolysis is the more common method of hydrolysis (Vidal and Venegas-Calerón 2019). Hydrolysis with a strong alkali at high temperature yields glycogen in soluble form in the cell hydrolysate (Vidal and Venegas-Calerón 2019). Then 40% (w/v) KOH is added to a cyanobacteria suspension in water and incubated at 90°C for 1 hour. After cooling the solution to room temperature, the carbohydrate is precipitated by adding two volumes of cold absolute ethanol with overnight incubation at –20°C. The carbohydrate is recovered as a pellet after centrifugation of the mixture. In this method, the KOH-soluble and ethanol-insoluble part is carbohydrate, which is recovered as a pellet by centrifugation.

Therefore, acid hydrolysis directly yields glucose monomers which can be quantified using different methods; however, alkaline hydrolysis gives precipitated glycogen which can be further hydrolyzed to glucose monomers for quantification. The following equation is then used to calculate the glycogen concentration in the sample (Roe and Dailey 1966):

$$DU/DS \times 0.1 \times (\text{volume of extract/gram of tissue}) \times 100 \times 0.9 = \text{mg of glycogen per 100 g of cells}$$

where DU = absorbance of the unknown, DS = absorbance of the standard, 0.1 = mg of glucose in 2 mL of standard solution and 0.9 = a factor for converting the glucose value to the glycogen value.

23.2.2.2 Hydrolysis of Glycogen to Monomeric Glucose

This includes enzymatic or acid hydrolysis of glycogen (Huggett and Nixon 1957). The enzymatic hydrolysis involves several steps: Resuspension of the precipitated glycogen in water followed by treatment with an enzyme cocktail of α-amylase and α-amyloglucosidase (de Porcellinis et al. 2017). α-Amylase is an endo enzyme which hydrolyzes the α-1,4-linkages to produce shorter dextrin chains, which are further hydrolyzed to glucose by the exo-acting α-amyloglucosidase (Ernst et al. 1987; Zhang et al. 2013; de Porcellinis et al. 2016).

23.2.2.3 Estimation of Glucose

The glucose in a solution can be estimated either by the colorimetric method or by high-performance liquid chromatography (HPLC). The HPLC method separates glucose using an Aminex column and 5 mM H_2SO_4 as the mobile phase. The glucose is detected using a refractive index detector (RID) and quantified by comparing the peak area of the unknown against the peak area for a standard concentration. This is a simple, quick, accurate, sensitive method with a lower detection limit of (up to 3.5 mg/100 mL for glucose) (Parpinello and Versari 2000).

23.2.2.4 Enzymatic Method

In this procedure, glycogen is extracted using the KOH method, and hydrolysis is carried out using a mixture of amylase and amyloglucosidase. This method uses glucose oxidase, which converts oxygen to hydrogen peroxide and glucose to lactone. Further, peroxidase generates a pink quinonimine dye from hydrogen peroxide and a phenolic compound and 4-aminoantipyrine, allowing the quantitative detection of the released glucose (de Porcellinis et al. 2017).

23.2.3 Estimation of Total Lipid

Lipid can be isolated using the hexane-isopropanol method or the chloroform-methanol method. The hexane-isopropanol method yields lipid in hexane, while the chloroform-methanol method gives the lipid content in a chloroform layer (Ishida et al. 1997; Watson et al. 2014). Freeze-dried cells are suspended in a mixture of chloroform-methanol-water (1:2:0.8) and swirled for 3 hours. The chloroform layer containing the lipid content is then collected after centrifugation. The aqueous layer is extracted several times to increase lipid recovery. The lipid weight is then calculated by weighing after the chloroform layer is evaporated. The lipid content can also be quantified using following equation:

$$\text{Total lipid} = \left(\text{Weight of lipid in aliquot} \times \text{Volume of chloroform layer}\right)/\text{Volume of aliquot}$$

23.2.4 Estimation of RNA

Several methods can be used to quantify RNA, each with its advantages and disadvantages. Ultraviolet (UV) absorbance is a precise method, but it requires pure RNA samples with no DNA contamination. Orcinol colorimetry is another method, but it is not very accurate, and it can be affected by the presence of carbohydrates (Beck et al. 2018). Thiazole orange is a fluorescent dye that can be used to quantify RNA. This method has good precision, but the fluorescence signal is biased toward the GC content, and it is less sensitive for RNA than for DNA. Fluorescent dyes such as RiboGreen and PicoGreen can also be used to quantify extracted RNA, but they are usually used in combination with kit extractions (Beck et al. 2018). Using ethidium bromide as a fluorescent indicator, it is simple to identify native DNA and RNA in crude tissue homogenates. Sequentially adding DNase or RNase to a single reaction mixture causes a quick drop in fluorescence that is proportional to the concentrations of DNA and RNA present. The difference in fluorescence is used to quantify RNA (Bentle et al. 1981). A very common method of RNA quantification is the perchloric acid method. Ahmad et al. (2020) used this method to isolate and quantify RNA from *Synechococcus* sp. BDU 130192. After isolation of RNA, the absorbance is measured at 260 nm and 280 nm (Ahmad et al. 2020).

23.2.5 Estimation of DNA

The measured DNA content from a cell can be interpreted as the bigger or smaller genome size, as well as the ploidy of the cells as compared to related species (Cesarone et al. 1979; Carnicer et al. 2009). Drugs such as mithramycin have been used to quantify DNA in individual cells but have limited sensitivity. Ethidium bromide exhibits a substantial increase in fluorescence upon binding to DNA, but it also strongly reacts with RNA; therefore, it is important to remove the RNA before performing DNA quantification (Labarca and Paigen 1980). Both DAPI (4′,6′-diamidino-2-phenylindole) and Hoechst 33258 (2-[2-(4-hydroxyphenyl)-6-benzimidazolyl]-6-(1-methyl-4-piperazyl)-benzimidazol*3HCl) have been successfully employed to fluorescently stain DNA in cells. Bisbenzimidazole (Hoechst 33258) has been very commonly used to quantify DNA due to several advantages. The linkage of dye with the DNA is not affected by salts or detergents. The dye has an excitation maximum and emission maxima at 356 nm and 468–490 nm, respectively. The binding of DNA with the dye results in the increase in fluorescence (Labarca and Paigen 1980). First, the dried cyanobacterial cells are lysed using lysozyme. Then, 0.5–2 μg/mL working concentration of Hoechst 33258 dye is prepared in TNE buffer (50 mM Tris-HCl [pH 7.4], 100 mM NaCl and 0.1 mM EDTA). Then 200 μL dye in a 10-μL lysed sample is used for quantification. The fluorescence is measured at excitation/emission wavelengths of 356/468 nm. The concentration is calculated using known concentrations of purified DNA as a standard (Ahmad et al. 2020).

23.2.6 Estimation of Pigments

There are several methods of extraction and quantification of pigments in cyanobacteria.

23.2.6.1 Pigment Extraction

Since chlorophyll and carotenoids are soluble in organic solvents while phycobiliproteins (PBPs) are water-soluble, these pigments cannot be extracted together. Most commonly the solvent extraction using 95% methanol or methanol and chloroform is used followed by heating at 60°C for 30 minutes for chlorophyll extraction. Pigments can also be extracted using acetone, which has less efficacy and takes longer and therefore is not recommended for cyanobacteria. The effects of crushing, freeze-drying and extractability using four solvents (acetone, methanol, methanol/acetone and methanol/acetone/N,N-dimethylformamide [DMF]) were evaluated on pigment extraction, with the best results obtained using methanol/acetone/DMF/water (Cysewski and Todd Lorenz 2003;

Sandybayeva et al. 2022). PBPs from different cyanobacterial cultures were extracted using techniques like freezing and thawing (Mac as-Sánchez et al. 2005), ultrasonic (Abalde et al. 1998) and the use of enzymes such as lysozyme (Minkova et al. 2003) alone or in combination with EDTA (Boussiba and Richmond 1979). Sonication is considered a better method compared to tissue grinding. Use of freeze-thaw cycles for PBP extraction from *Spirulina* sp. is considered a most economic and effective method (Doke 2005). PBPs are extracted using phosphate buffer, and several cycles of extractions are performed to achieve maximum extraction (Lawrenz et al. 2013). The concentrations of chlorophyll and carotenoids are then calculated as:

$$\text{Chl } a[\mu\text{g/mL}] = 12.9447(\text{A}_{665} - \text{A}_{720})$$

$$\text{Carotenoids}[\mu\text{g/mL}] = (1000 \times \text{A}_{470} - 44.76 \times \text{A}_{666})/221$$

PBPs are categorized into three groups based on their absorption characteristics: Phycoerythrin (PE), phycocyanin (PC) and allophycocyanin (APC). Phycobilins show absorption peaks at 545 nm for phycoerythrin and 620 nm for phycocyanin (Lawrenz et al. 2013). The spectrum of the extracted PBPs is taken between 400 and 750 nm. The following equations are used to estimate the concentrations of PC, APC and PE in mg/mL (Bennett and Bogorad 1973):

$$[\text{PC}] = \text{OD}_{615} - 0.474(\text{OD}_{652})/5.34$$
$$[\text{APC}] = \text{OD}_{652} - 0.208(\text{OD}_{615})/5.09$$
$$[\text{PE}] = \text{OD}_{562} - 2.41(\text{PC}) - 0.849(\text{APC})/9.62$$

Spectrophotometry is the most common method of quantification of pigments in cyanobacteria, but chromatography and fluorometric methods are also used sometimes. The fluorometric method is sensitive but has several limitations in the case of cyanobacteria. The Kautsky curve is a time series of chlorophyll fluorescence measurements made when samples are illuminated in the dark (Schreiber et al. 1995). The decrease or increase in fluorescence intensity mostly reflects the status of photosystem (PS)II because PSI has much lower chlorophyll fluorescence than PSII at room temperature. The redox state of Q_A, an electron acceptor of PSII, is one of the variables that affect the yield of chlorophyll fluorescence – a higher fluorescence indicates a more reduced Q_A. Because land plants and cyanobacteria have different chlorophyll fluorescence kinetics, it is challenging to determine the state of photosynthesis in cyanobacteria using a typical analytical methodology designed for land plants (Schreiber et al. 1995). Cyanobacteria differ from plants in several factors, like very quick induction of O_2-dependent electron flow, decrease in the plastoquinone pool under the dark, strong interaction between photosynthetic and respiratory metabolism and noticeable shifts of energy distribution between the two photosystems. Chlorophyll fluorescence-based photosynthetic parameter measurements in cyanobacteria require a longer pulse of 25–50 seconds as compared to a 0.5- to 2-second pulse required by higher plants. The shorter pulses are less intrusive in terms of the initiation of energizing electron flow, which can cause non-photochemical quenching and state changes (Schreiber et al. 1995).

23.3 INTRACELLULAR FLUX ANALYSES IN CYANOBACTERIA

High-throughput metabolic engineering requires a global understanding of metabolism, and metabolic models can provide a framework to decipher the metabolic capabilities of these strains. GSMMs help in systems-level analysis of metabolic networks of organisms. The models have led to the successful design of metabolic pathway engineering strategies in organisms such as *E. coli* and *S. cerevisiae* (Fatma et al. 2018; Chen et al. 2021).

Among the photoautotrophs, cyanobacteria have become the focus of synthetic biology research because of recent discoveries of strains belonging to the *Synechococcus* genus showing very fast growth rates (Jaiswal et al. 2018). These strains can potentially be used as platforms for producing various high-value compounds, using CO_2 and sunlight as raw materials. Reconstruction of reliable metabolic models requires integration of prior genomic and biochemical information of the organism being studied. Less information about cyanobacterial metabolism is available as compared with other widely studied organisms like *E. coli* and yeast. Nonetheless, model-directed engineering of cyanobacterial strains has shown considerable progress (Shirai et al. 2016). In this chapter, we report on various metabolic models developed for cyanobacteria during the last decade. Also, we discuss the application of model-based metabolic engineering in cyanobacteria.

Cyanobacteria have the potential to be used as bioproduction hosts due to their ability to use solar energy to capture carbon dioxide and transform it into valuable metabolites. This makes them particularly useful in the pursuit of sustainable biotechnological processes. In addition, cyanobacteria have several advantages over other photosynthetic organisms, such as faster growth rates, higher efficiency in solar energy capture and conversion and the ability to be grown at a large scale without requiring agricultural land and clean water. They can also potentially be used to degrade aquatic pollutants and clean contaminated water supplies. Cyanobacteria produce compounds similar to those obtained from plants for use as food, fiber and fuel. However, until recently, cyanobacteria have not been modified genetically for large-scale cultivation. Like many types of eubacteria, cyanobacteria can produce polyhydroxyalkanoates, a type of biodegradable thermoplastic polyester. Furthermore, many commercially useful compounds can be derived from the pigments, vitamin, fatty acids and amides, etc., produced by cyanobacteria. It is estimated that cyanobacteria have the potential to produce around 1100 different secondary metabolites (Zavřel et al. 2016).

Genome-scale models (GSMs) are large-scale computational models that describe metabolic pathways as the balances of mass and energy within a living cell or organism. These models are used to simulate and predict the behavior of biochemical reactions, transport processes and non-enzymatic reactions. GSMs are widely used in metabolic engineering and strain development, as they provide a powerful tool for generating and testing hypotheses and can help researchers to understand the mechanisms underlying experimental results. GSMs are typically simulated using numerical optimization algorithms and are useful for their ability to provide a comprehensive explanation of metabolic processes (Feist and Palsson 2008).

23.3.1 Structural Metabolic Modeling

Living organisms maintain homeostasis because of underlying intricate metabolic networks, which continuously process external nutrients to synthesize building blocks of cells, energy and cytoprotective molecules. As the microbes adapted to their diverse environments, they developed diverse and specialized metabolic reactions.

Structural metabolic modeling involves modeling the biochemical structure of metabolic reactions in an organism. It does not require the enzymatic or other kinetic parameters to solve the resulting set of equations to generate the flux distribution pattern.

23.3.1.1 Approaches for Model Reconstruction

There are three major approaches to model construction (Kumar and Maranas 2009).

1. *Interaction-based networks*: These networks are used to describe information transfer in large-scale systems like transcriptomics and proteomics. In these models, the stoichiometry of the network is not considered.
2. *Dynamic models*: These models are based on time-dependent kinetic information in the form of ordinary differential equations (ODEs) about the cell system. These models are used to describe the intracellular signaling networks in a quantitative manner.

3. *Stoichiometric models*: They are based on the biochemical metabolic network of a cellular system. They consider the biochemical reaction stoichiometries at steady state; hence, they also describe intracellular mass flow, assuming stationary systems. The stoichiometric models can be as small as about 50 metabolites and 100 reactions (as in central carbon metabolism); in contrast, the GSMMs are large, composed of more than 500 metabolites and 1000 reactions.

Stoichiometric models have been preferred for the computational analysis of large metabolic networks of organisms. Stoichiometric models are less complex, and they provide a platform for mathematical representation of cellular metabolism.

23.4 GENOME-SCALE METABOLIC MODELS AND FLUX BALANCE ANALYSIS

GSMMs are large stoichiometric models. They involve almost every reaction and metabolite reported for the organism and include the reactions for which a gene is available in the genome of that organism, hence, the name genome-scale metabolic model.

To understand how a GSM is constructed, let us consider a toy model of the metabolic network shown in Figure 23.1.

During the exponential phase of growth of microorganisms, it can be assumed that the concentrations of intracellular metabolites remain constant, and this state is known as the steady state.

Under the steady-state assumption, the rate laws for the intracellular metabolites are reduced to linear mass balance equations, which can be solved to determine the rates V_i or fluxes, with unit amount of moles per time per gram dry cell weight [mmol·(gDCW·h)$^{-1}$). V_{Ei} are the exchange fluxes, and they involve the metabolic reactions involved in the exchange of metabolites to and from the environment. The entire flux distribution can be represented mathematically in the form of an array, also known as the flux vector, $V = \{V_1, V_2, V_3, V_4, V_5\ V_{E1}, V_{E2}, V_{E3}, V_{E4}\}$.

Flux balance analysis (FBA) is the mathematical approach to identify the flux distribution for a given metabolic network using the set of linear mass balance equations through linear programming or linear optimizations. In other words, FBA calculates mathematically how metabolites would flow or at what rates if the growth of an organism is maximized with a specified metabolic network and other growth conditions.

Once the metabolic network representing all the reactions in the organism is constructed, the next step is to represent this reconstruction in a mathematical way in the form of a stoichiometric matrix, S (Figure 23.2), in which each row represents a metabolite and each column represents a metabolic reaction. For example, in reaction V_1, 1 mol of A is consumed and 1 mol of B is produced, hence in column V_1, –1 is the coefficient for A. The "–" sign is used by convention when a metabolite is being consumed in a reaction, and 1 is the coefficient for B. The "+" sign is used when

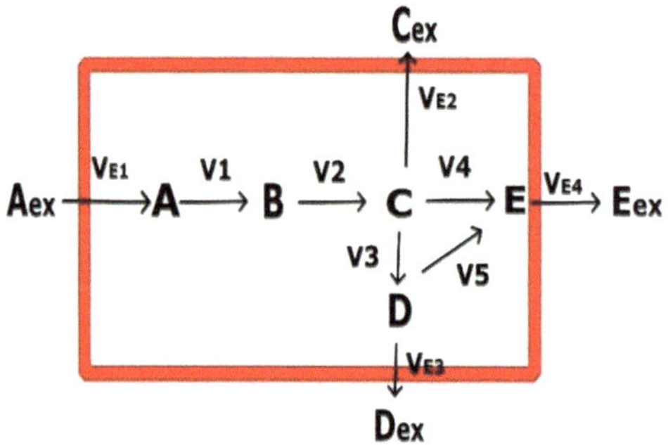

FIGURE 23.1 A toy metabolic network. The network consists of five intracellular metabolites {A, B, C, D, E}, five intracellular reactions rates $\{V_1, V_2, V_3, V_4, V_5\}$ and four exchange reaction rates $\{V_{E1}, V_{E2}, V_{E3}, V_{E4}\}$ (color figure in e-book).

	V_1	V_2	V_3	V_4	V_5	V_{E1}	V_{E2}	V_{E3}	V_{E4}
A	-1	0	0	0	0	1	0	0	0
B	1	-1	0	0	0	0	0	0	0
C	0	1	-1	-1	0	0	-1	0	0
D	0	0	1	0	-1	0	0	-1	0
E	0	0	0	1	1	0	0	0	-1

FIGURE 23.2 Stoichiometric matrix, *S*, for the toy metabolic network (color figure in e-book).

a metabolite is being produced in a reaction. Since the rest of the metabolites are not being used or made in the V_1 reaction, the coefficient for these metabolites is 0.

In GSMs, an additional reaction, known as the biomass reaction, is also incorporated into the model for simulation of the growth of organism. The biomass reaction simulates the consumption or production of metabolites to produce biomass.

At steady state, the dot product of a stoichiometric matrix and metabolic flux vector gives an all-zero vector, signifying that there is no net accumulation or consumption of any intracellular metabolite. That is,

$$\mathbf{S.v = 0}$$

The equation S.v = 0 defines a system of linear mass balance equations, and these equations can be solved to predict the intracellular flux distribution that results in an assumed optimized cellular response, such as maximum growth rate. In order to mathematically solve the resulting system of equations, a cellular flux is chosen as the objective function. For example, for growth maximization, the objective function, Z, is defined as

$$Z = v_{biomass}$$

Where $v_{biomass}$ is the biomass reaction used in the model. Linear optimization or linear programming is used to calculate the flux distribution that either maximizes or minimizes the objective function (Figure 23.3). The solution space of allowed fluxes is defined by the constraints imposed by the reaction bounds of mass balance equations. The resulting optimal solution of FBA is a flux distribution that either maximizes or minimizes the objective function.

23.5 METABOLIC NETWORK RECONSTRUCTION IN CYANOBACTERIA

In order to effectively use cyanobacteria to harness solar energy, it is often necessary to modify their metabolic pathways. A deeper understanding of the metabolic processes occurring during phototrophic growth can help in this task. One way to obtain this understanding is by creating

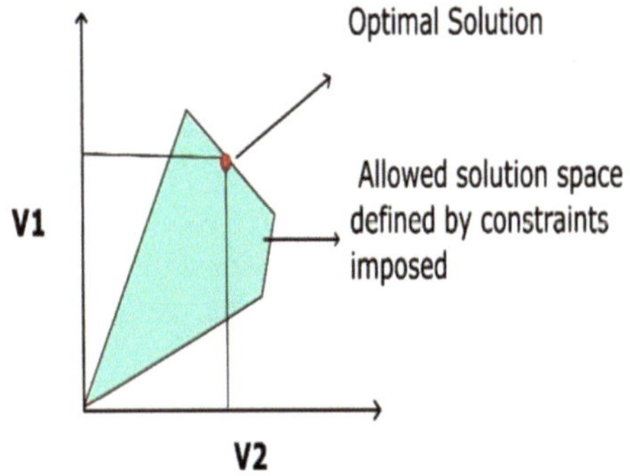

FIGURE 23.3 Identification of flux distribution as an optimal solution (color figure in e-book).

TABLE 23.1
Published Key Genome-Scale Models for Cyanobacteria

Name of the Model	Name of the Species	References
iAnC892	*Anabaena* sp. ATCC 3304	(Hendry et al. 2021)
iSyn706	*Synechococcus* sp. BDU 130192	(Ahmad et al. 2020)
(Model name not available)	*Anabaena* sp. PCC 7120	(Malatinszky et al. 2017)
iTery101	*Trichodesmium erythraeum*	(Gardner and Boyle 2017)
iJB785	*Synechococcus elongatus* PCC 7942	(Broddrick et al. 2016)
(Model name not available)	*Synechococcus* sp. PCC 7002	(Hendry et al. 2016)
iAK888	*Arthrospira platensis*	(Yoshikawa et al. 2015)
(Model name not available)	*Arthrospira platensis* NIES-39	(Yoshikawa et al. 2015)
iSyp708	*Synechococcus* sp. PCC 7002	(Vu et al. 2013)
iSyp611	*Synechococcus* sp. PCC 7002	(Hamilton and Reed 2012)
iAK692	*Spirulina platensis* C1	(Klanchui et al. 2012)
iCce806	*Cyanothece* sp. ATCC 51142	(Vu et al. 2012)
iJN678	*Synechocystis* sp. PCC 6803	(Nogales et al. 2012)
iSyn811	*Synechocystis* sp. PCC 6803	(Montagud et al. 2011)
(Model name not available)	*Synechocystis* sp. PCC 6803	(Yoshikawa et al. 2011)
iSyn669	*Synechocystis* sp. PCC 6803	(Montagud et al. 2010)

detailed, accurate models of the metabolic networks of cyanobacteria based on their genomic information. Recently, several such models have been developed for cyanobacteria (Table 23.1).

FBA was first applied to cyanobacterial metabolism by Shastri and Morgan (2005), followed by an expansion of the model by Hong and Lee (2007). Both of these initial reconstructions were relatively small and primarily focused on the central metabolism. They included an incomplete tricarboxylic acid (TCA) cycle and a potential glyoxylate shunt. The first large-scale model was developed by Fu (2009). Subsequent updated models have incorporated new findings about the cyanobacterial TCA cycle, the potential presence of a glyoxylate shunt, the role of photorespiration in cellular growth and the effects of light-dependent oxidative stress on photosynthetic reactions (Figure 23.4).

23.6 METABOLIC INSIGHTS FROM ANALYSES OF GENOME-SCALE METABOLIC MODELS FOR DIFFERENT CYANOBACTERIAL STRAINS

23.6.1 *Anabaena* sp. ATCC 33047 (iAnC892, Hendry et al. 2021)

Anabaena sp. ATCC 33047 is a marine cyanobacterium that is capable of fixing nitrogen using specialized cells called heterocysts. To simulate the life cycle of this organism, a two-cell model (iAnC892) was developed that included both vegetative cells and heterocysts. The model is composed of unique 953 reactions and 834 metabolites that represent the annotation of 892 metabolic genes. It was structured with two super compartments: A vegetative cell compartment and a heterocyst compartment. These two compartments shared 892 reactions in common but differed in the presence or absence of certain key metabolic processes. For example, the heterocyst compartment lacked the reactions for PSII, RuBisCo and glutamine oxoglutarate aminotransferase, while the vegetative cell compartment lacked the N_2ase reaction.

The intake rate for bicarbonate was constrained to a minimum value of 8.5 mmol·(g DCW·h)$^{-1}$ to reproduce the experimentally observed growth rate of 0.18 h^{-1}. Analysis of this model suggests that the light-dependent electron transport chain (LETC) is critical to produce adenosine triphosphate (ATP) and NADPH in a definite ratio so that nitrogen fixation can occur at an optimal rate. It was

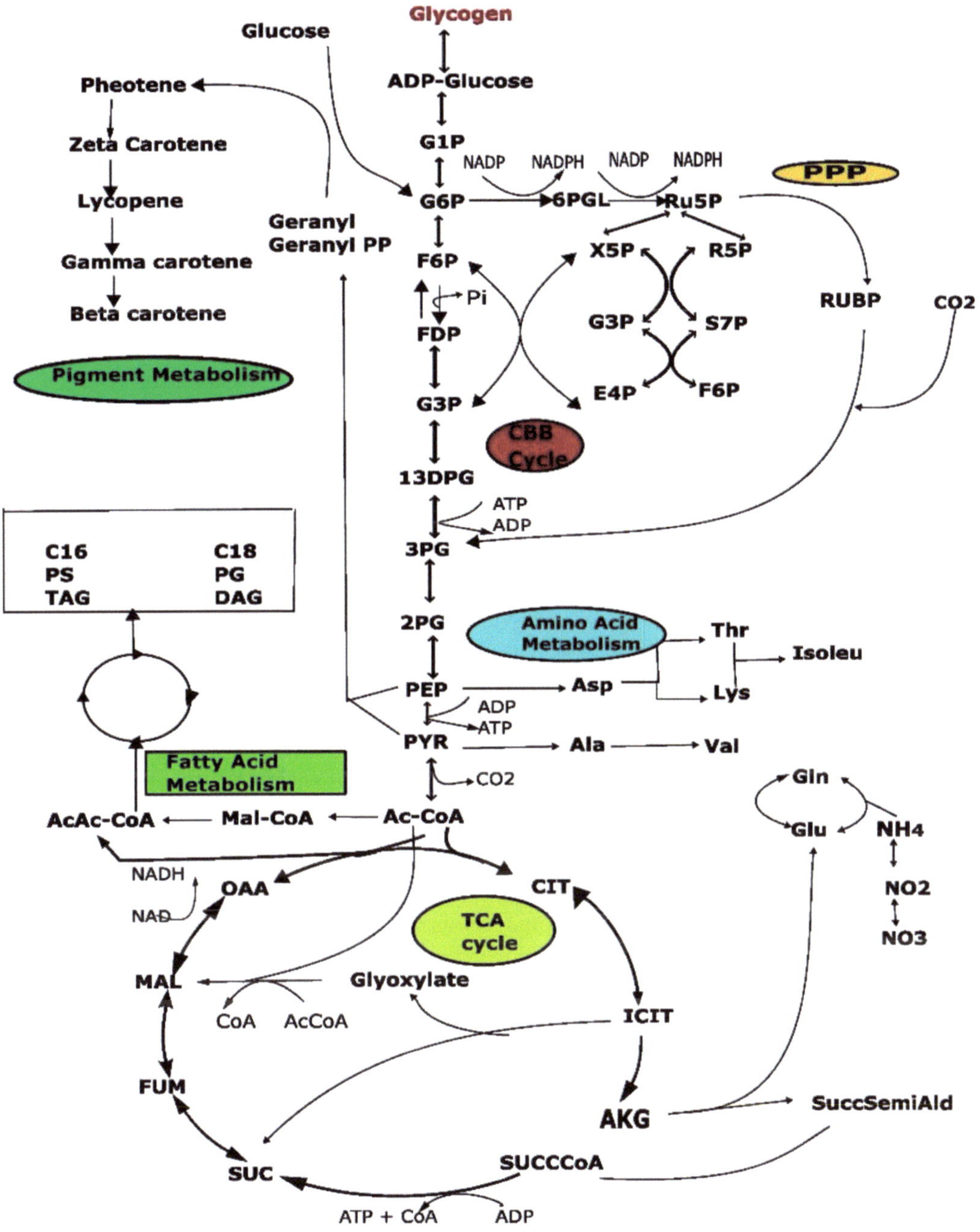

FIGURE 23.4 Core metabolic networks in cyanobacteria (color figure in e-book).

predicted that the TCA and glycolysis can serve as additional sources of reducing equivalents for nitrogen fixation. According to the flux variability analysis (FVA), in this model, the vegetative cell compartment is utilized for fixing CO_2 with the help of NADPH and ATP produced by the LETC, while the heterocyst metabolism is supported by sucrose as an electron source, produced in the vegetative cell. N_2 fixation is confined to the heterocyst compartment. The minimum nitrogenase flux required for optimal growth was found to be 2.5 times lower than the nitrogenase activity measured for this organism using the acetylene reduction assay [1.7 mmol·(gDCW·h)$^{-1}$]. LETC drives the synthesis of ATP in the heterocyst, as suggested from the minimum flux required by heterocyst-PS1. Ammonia is released during nitrogen fixation, which is then used for synthesis of glutamine via the glutamine synthetase reaction. This glutamine is then transported back to the vegetative cell to function as a nitrogen carrier.

One of the potential applications of this model was predicting the genetic modifications for the overproduction of desired chemicals in *Anabaena* sp. ATCC 33047. The model was simulated along with the OptForce algorithm to predict genetic manipulations to produce valerolactam and caprolactam. These chemicals were shown to be produced by incorporating the 5-aminovalerate pathway and the adipyl-CoA pathway into the model.

23.6.2 *Synechococcus elongatus* BDU 130192 (iSyn706, Ahmad et al. 2020)

Synechococcus elongatus BDU 130192 is an Indian euryhaline cyanobacterium. The iSyn706 model was created using an annotated chromosomal genome sequence. This model includes 706 genes, 908 reactions and about 900 different metabolites. Most of the reactions in this model were included in the metabolic pathways for amino acids, carbohydrate, folate and riboflavin. The iSyn706 model includes four compartments: Extracellular, cytosol, periplasm and thylakoid. The cytosol is home to the majority of reactions (797), while 15 reactions related to photosynthesis and respiration are located in the thylakoid and 73 transport reactions are found in the periplasm. During photoautotrophic simulations, the iSyn706 model was set to a fixed growth rate of 0.051 h^{-1}. Based on the measurements, the O_2 evolution rate varied between 1.6 and 1 mmol·(gDCW·h)$^{-1}$. This resulted in a calculated CO_2 uptake, photon absorption and nitrate uptake rates of 1.92, 25.45 and 0.22 mmol·(gDCW·h)$^{-1}$, respectively. The iSyn706 model simulations showed that the predicted O_2 release flux was 2.43 mmol·(gDCW·h)$^{-1}$.

To grow in photosynthetic conditions, about 450 reactions were found to be essential. The highest number of essential reactions was part of the fatty acid metabolism, and the amino acid metabolism constituted the second highest number of essential reactions. According to the model simulations under photoautotrophic conditions, the C3 cycle utilizes ATP and NADPH generated in the light reactions to fix CO_2. An analysis of the flux distributions of central carbon metabolism reactions in both the iSyp708 and iSyn706 models showed that there were differences in the calculated flux distributions when attempting to replicate the measured growth rates, reflecting the variations in the measured biomass compositions. The iSyn706 model predicted almost double the flux through the glycogen synthesis pathway compared to the iSyp708 model, while the fluxes through the TCA cycle were slightly higher in the iSyp708 model compared to iSyn706. The NO_3^- intake in iSyn706 was also about half that of iSyp708, reflecting the lower protein content in BDU 130192.

According to the simulation results, only a part of the TCA cycle is active in the iSyn706 model under simulated photoautotrophic conditions, even though the 2-oxoglutarate decarboxylase (2-OGDC) and succinyl semialdehyde dehydrogenase (SSADH) shunt are present. A part of the total uptake flux is routed into the TCA cycle through AcCoA, while the flux from 2-OG (alpha-KG) is redirected to the glutamine synthase-2-oxoglutarate amidotransferase (GS-GOGAT) cycle. The bacterium uses the enzymes glutamine synthase and GS-GOGAT to convert NO_3^- into NH_4^+ and finally into glutamine and glutamate via the GS-GOGAT cycle. The iSyn706 model shows that synthesis of fatty acids begins with acetyl-CoA and that approximately 3% of total carbon intake goes into fatty acid metabolism, which aligns with the low lipid levels seen in this organism. According to the iSyn706 model, a small portion of flux is directed toward pigment metabolism. FBA was performed on the model to predict the theoretical yields (mol product produced/mol CO_2 consumed) for 10 compounds (4 native and 6 non-native) under phototrophic conditions, with growth rates set at 80% of the wild-type growth rate. The yields of the compounds tested were found to be inversely proportional to the number of carbon atoms in the molecule. Among the native compounds, acetate was produced at the highest yield, while ethanol was the highest-yielding non-native compound. These analyses provided insight into the metabolic capabilities of the organism and revealed the minimum number of gene additions required to produce non-native compounds.

23.6.3 *Arthrospira platensis* C1 (iAK888, Klanchui et al. 2018)

Arthrospira platensis can accumulate large amounts of glycogen when grown under challenging nutrient or environmental conditions. The iAK888 GSMM was developed by improving and updating the previously constructed iAK692 model for *A. platensis* C1. The iAK888 model is a compartmentalized model and completely functional. It consisted of 888 genes, 1096 reactions and 994 metabolites. When compared with published models of *A. platensis*, iAK888 was found to have the highest number of genes, metabolites and reactions. The additional genes and reactions were part of many important metabolic pathways related to carbohydrate, lipid and vitamin metabolism as well as transport.

The iAK888 model accurately predicted growth and glycogen fluxes under various growth conditions. To incorporate glycogen metabolism in *A. platensis*, the iAK888 model included four reactions related to glycogen synthesis and utilization for biomass growth, transport for carbon storage and degradation. In the model, glycogen is treated as a monomer ($C_6H_{10}O_5$) in the synthesis reaction, where one mole of glycogen is produced from one mole of glucose-1-phosphate by the enzyme glycogen synthase. Glycogen phosphorylase, responsible for breaking down glycogen, was set to be inactive during autotrophic growth in the model.

In order to determine the accuracy of the predictions, measured values were compared with the predicted values. For example, maximal growth rate and carbohydrate production fluxes were compared in various conditions like autotrophic, mixotrophic and heterotrophic conditions. The prediction of growth rate was accurate, with an error less than 5% in different conditions. In addition, the model produced oxygen under autotrophic and mixotrophic growth conditions, with a higher oxygen content under autotrophic conditions. Under heterotrophic simulations, the model consumed oxygen and produced CO_2.

Using this model, the effects of nutrient starvation on the growth and production of glycogen were simulated to predict the strategies for increasing the amount of glycogen. The model was simulated with parameters like HCO_3^- uptake flux, 1.6 mmol·(gDCW·h)$^{-1}$; photon uptake, 100 μmol photons m^{-2} s^{-1}. The results showed that limiting the uptake of NO_3^-, PO_4^{3-} and SO_4^{2-} and supplying excess HCO_3^- had a negative effect on growth but a positive effect on glycogen production.

23.6.4 *Synechococcus elongatus* PCC 7942 (iJB785, Broddrick et al. 2016)

The authors have reannotated and reconstructed the metabolic genes in the genome of the unicellular cyanobacterium *S. elongatus* PCC 7942, a freshwater strain which is being developed as a platform for photosynthetic bioproduction of industrial products. They have also designed a method to include light absorption to simulate the cell shading effect. Comparisons of GSM predictions with essentiality data have been used to improve the model and reveal particular aspects of the metabolism of the organism, which resulted in an all-inclusive metabolic model of *S. elongatus*.

Of the 2723 genes in the *S. elongatus* genome, 785 (29%) were finally included, and 118 of these genes (15%) had their functional annotations updated for refinement of gene reaction annotations *in vivo* essentiality data from RB-Tn-Seq. These data were critical to analyze the importance of genes by identifying the genes unable to survive insertion mutants. *In silico* essentiality calls were also made by setting the flux through each enzyme in the model to zero and measuring the impact on growth.

Accurate modeling of obligate phototrophic metabolism required a detailed reconstruction of the photosystem. To resolve discrepancies in the placement of electron transport complexes within cyanobacterial cells, proteomics and fluorescence microscopy techniques were employed. These complexes, previously thought to be located in the cytoplasmic membrane, were actually located at different areas in the cell. The ferredoxin:plastoquinone oxidoreductase complex gene associations were also updated to include additional subunits and the use of ferredoxin as the electron donor. Flux values were in good agreement with ^{13}C metabolic flux analysis (MFA) data for PCC 6803.

The model reveals gaps in our knowledge of *S. elongatus* biology, such as the unknown links of the nucleotide salvage pathway, the limitation of phosphoketolase in bypassing lower glycolysis and non-canonical activity of transhydrogenase. The inclusion of data for essential genes improved the model's accuracy but also highlighted inconsistencies that were not explained in the published literature for *S. elongatus*.

23.6.5 *Synechocystis* sp. PCC 6803 (iJN678, Nogales et al. 2012)

The cyanobacterium *Synechocystis* sp. PCC 6803 is a model photosynthetic prokaryotic organism that can grow in both light and dark conditions. To study its metabolic processes, a GSM was created on the basis of the genomic and biochemical knowledge available for this organism. The final model, called iJN678, contains information on 678 genes, 863 metabolic and transport reactions and 795 non-unique metabolites distributed across 54 subsystems and four different cellular compartments.

Optimal photosynthetic performance in *Synechocystis* requires careful regulation of light and energy conversion through the photosystems and metabolic reactions. One principal factor in this process is the ATP/NADPH ratio in the Calvin cycle, which is responsible for fixing carbon dioxide. The Calvin cycle is the largest consumer of ATP and NADPH and requires a specific ratio of 1.5 ATP to NADPH. However, the process of photosynthetic linear electron flow, which generates ATP and NADPH, only produces a ratio of 1.28 ATP to NADPH, so additional ATP is needed. To address this, *Synechocystis* has developed a range of alternate electron flow (AEF) pathways that can modulate the ATP/NADPH ratio as needed. Although the exact physiological roles of these AEF pathways are not yet fully understood, it is believed that they work in cooperation with linear electron flow (LEF).

The authors of the study made a significant effort to create a comprehensive and compartmentalized model of photosynthesis in *Synechocystis*. The model included reactions for PSII, cytochrome b_6f (cyt BF), PSI and ferredoxin $NADP^+$ oxidoreductase (FNR), which are interconnected through plastoquinone (PQ), plastocyanin/cytochrome C (cyt C) and ferredoxin. The model also includes AEF pathways, such as the ferredoxin PQ reductase (FQR) and NAD(P)H dehydrogenase complexes, as well as photorespiration (PHOTOR) and succinate dehydrogenase (SDH) reactions. These reactions were included in the thylakoid membrane that allowed respiration and photosynthesis to interact with each other. The model was validated by comparing the experimentally measured fluxes to the model-predicted fluxes. The error in prediction was less than 5%. Additionally, the model's predicted flux distributions closely matched experimentally determined fluxes under different growth conditions.

The model also showed that under suboptimal light conditions, maximal growth was only possible with the assistance of the AEF pathways in addition to the LEF pathway, which is consistent with previous findings. Furthermore, the model showed the importance of the AEF pathways and photorespiration during growth with environmental fluctuations such as high light and loss of inorganic carbon. The use of iJN678 allowed researchers to better understand previously unknown aspects of heterotrophic metabolism in this organism. The simulations for heterotrophic conditions predicted the breakdown of glucose to pyruvate via the oxidative line of the pentose phosphate pathway; this was in line with the observed experimental data.

According to the model, autotrophic conditions have the greatest number of essential metabolic genes (350 genes). Heterotrophic and mixotrophic conditions were found to have slightly fewer essential genes, with 261 and 259, respectively. This analysis also found a core group of 259 genes, representing 38% of the total that were essential under all the tested conditions. This result is in close agreement with the 34% of essential genes found in an earlier computational study. The fraction of essential metabolic genes in all growth conditions is higher for *Synechocystis* (38%) compared to other bacteria such as *E. coli* (14%). This result reflects a reduced metabolic robustness of *Synechocystis* in autotrophic conditions.

23.7 EMERGING APPLICATIONS AND FUTURE DIRECTIONS IN THE INTEGRATIVE SYSTEMS BIOLOGY ANALYSES OF CYANOBACTERIA

In recent years, several GSMs have been developed for phototrophic organisms. However, these models are still relatively underrepresented compared to models for heterotrophic organisms. In the case of cyanobacteria, very few GSMMs have been made available, and only a handful of species have had their models refined.

Accurate modeling of phototrophic metabolism requires a detailed understanding of light harvesting and electron transport through various pathways. In recent years, researchers have made considerable progress in this area. For example, Nogales et al. (2012) developed a model that captures the intricate pathways for the electron transport in *Synechocystis* sp. PCC 6803, including cyclic electron transport and accessory pathways. The model made it possible to study entire photosynthetic processes at a systems level (Nogales et al. 2012). Another study by Knoop et al. (2013) expanded on this work by incorporating the role of photorespiration in the growth of cells and the unique characteristics of reactions like light-dependent oxidative stress (Knoop et al. 2013). More recently, an approach was developed that incorporates light absorption and considers the cell shading effects in *S. elongatus* PCC 7942 by modeling light as a metabolite (Hendry et al. 2021). These advances have allowed for a more accurate modeling of phototrophic metabolism and a better understanding of the underlying processes.

One way to evaluate the accuracy of GSMs is by examining how well they can identify essential genes in the metabolic network. This test has been used successfully in bacteria such as *E. coli* and *Pseudomonas putida*. The quality of two GSMs, i.e., iJB785 of *S. elongatus* PCC 7942 and iSynCJ816 of *Synechocystis* sp. PCC 6803, has been assessed using gene essentiality data. In the case of iJB785, 78% of all genes were accurately identified as essential or non-essential genes based on the experimental data. In the second case, the GSM was able to predict gene deletions with 77% accuracy based on a comparison with experimental studies. There are several ways to improve the accuracy of GSMs. One approach is to use data from high-throughput growth phenotyping experiments, such as knockout mutant strains grown in different media conditions, to validate and refine the metabolic network reconstructions. Another way is to integrate transcriptional regulation data to generate better models and improve the capabilities of the reconstructed metabolic networks. Finally, combining multi-omics datasets has also been used to enhance the accuracy of GSMs (Rai and Saito 2016).

REFERENCES

Abalde J, Betancourt L, Torres E, Cid A and Barwell C (1998) Purification and characterization of phycocyanin from the marine cyanobacterium *Synechococcus* sp. IO9201. *Plant Science* 136(1): 109–120.

Ahmad A, Pathania R and Srivastava S (2020) Biochemical characteristics and a genome-scale metabolic model of an Indian euryhaline cyanobacterium with high polyglucan content. *Metabolites* 10(5): 177.

Beck A, Hunt K and Carlson R (2018) Measuring cellular biomass composition for computational biology applications. *Processes* 6(5): 38.

Bennett A and Bogorad L (1973) Complementary chromatic adaptation in a filamentous blue-green alga. *Journal of Cell Biology* 58(2): 419–435.

Bentle LA, Dutta S and Metcoff J (1981) The sequential enzymatic determination of DNA and RNA. *Analytical Biochemistry* 116(1): 5–16.

Boussiba S and Richmond AE (1979) Isolation and characterization of phycocyanins from the blue-green alga *Spirulina platensis*. *Archives of Microbiology* 120(2): 155–159.

Broddrick JT, Rubin BE, Welkie DG, Du N, Mih N, Diamond S, Lee JJ, Golden SS and Palsson BO (2016) Unique attributes of cyanobacterial metabolism revealed by improved genome-scale metabolic modeling and essential gene analysis. *Proceedings of the National Academy of Sciences USA* 113(51): E8344–E8353.

Brückner J (1955) Estimation of monosaccharides by the orcinol–sulphuric acid reaction. *Biochemical Journal* 60(2): 200–205.

Carnicer M, Baumann K, Töplitz I, Sánchez-Ferrando F, Mattanovich D and Ferrer Pand Albiol J (2009) Macromolecular and elemental composition analysis and extracellular metabolite balances of *Pichia pastoris* growing at different oxygen levels. *Microbial Cell Factories* 8(1): 1–14.

Cesarone CF, Bolognesi C and Santi L (1979) Improved microfluorometric DNA determination in biological material using 33258 Hoechst. *Analytical Biochemistry* 100(1): 188–197.

Chen Y, Li F, Mao J, Chen Y and Nielsen J (2021) Yeast optimizes metal utilization based on metabolic network and enzyme kinetics. *Proceedings of the National Academy of Sciences USA* 118(12): e2020154118.

Churms SC (2005) Carbohydrates | sugars – spectrophotometric methods. In: Encyclopedia of Analytical Science (Eds. P. Worsfold, A. Townshend and C. Poole). Elsevier, pp. 408–414. ISBN: 978-0-12-369397-6

Cysewski GR and Todd Lorenz R (2003) Industrial production of microalgal cell-mass and secondary products—Species of high potential: *Haematococcus*. In: Handbook of Microalgal Culture (Eds. A. Richmond). Blackwell Publishing Ltd, Oxford, UK, pp. 281–288. ISBN: 0-632-05953-2

de Porcellinis A, Frigaard NU and Sakuragi Y (2017) Determination of the glycogen content in cyanobacteria. *Journal of Visualized Experiments* 125: e56068.

de Porcellinis AJ, Klähn S, Rosgaard L, Kirsch R, Gutekunst K, Georg J, Hess WR and Sakuragi Y (2016) The non-coding RNA Ncr0700/PmgR1 is required for photomixotrophic growth and the regulation of glycogen accumulation in the cyanobacterium *Synechocystis* sp. PCC 6803. *Plant and Cell Physiology* 57(10): 2091–2103.

Deb D, Mallick N and Bhadoria PBS (2019) Analytical studies on carbohydrates of two cyanobacterial species for enhanced bioethanol production along with poly-β-hydroxybutyrate, c-phycocyanin, sodium copper chlorophyllin, and exopolysaccharides as co-products. *Journal of Cleaner Production* 221: 695–709.

Doke JM (2005) An improved and efficient method for the extraction of phycocyanin from *Spirulina* sp. *International Journal of Food Engineering* 1(5): 1–11.

DuBois M, Gilles KA, Hamilton JK, Rebers PA and Smith F (1956) Colorimetric method for determination of sugars and related substances. *Analytical Chemistry* 28(3): 350–356.

Ernst A, Chen TW and Böger P (1987) Carbohydrate formation in rewetted terrestrial cyanobacteria. *Oecologia* 72(4): 574–576.

Fatma Z, Hartman H, Poolman MG, Fell DA, Srivastava S, Shakeel T and Yazdani SS (2018) Model-assisted metabolic engineering of *Escherichia coli* for long chain alkane and alcohol production. *Metabolic Engineering* 46: 1–12.

Feist AM and Palsson B (2008) The growing scope of applications of genome-scale metabolic reconstructions using *Escherichia coli*. *Nature Biotechnology* 26(6): 659–667.

Fernandez C, Rojas CC and Nilsson L (2011) Size, structure and scaling relationships in glycogen from various sources investigated with asymmetrical flow field-flow fractionation and 1H NMR. *International Journal of Biological Macromolecules* 49(4): 458–465.

Fernell WR and King HK (1953) The simultaneous determination of pentose and hexose in mixtures of sugars. *The Analyst* 78(923): 80–83.

Fu P (2009) Genome-scale modeling of *Synechocystis* sp. PCC 6803 and prediction of pathway insertion. *Journal of Chemical Technology and Biotechnology* 84(4): 473–483.

Gardner JJ and Boyle NR (2017) The use of genome-scale metabolic network reconstruction to predict fluxes and equilibrium composition of N-fixing versus C-fixing cells in a diazotrophic cyanobacterium, *Trichodesmium erythraeum*. *BMC Systems Biology* 11(1): 4.

Hamilton JJ and Reed JL (2012) Identification of functional differences in metabolic networks using comparative genomics and constraint-based models. *PLOS ONE* 7(4): e34670.

Hendry JI, Dinh HV, Sarkar D, Wang L, Bandyopadhyay A, Pakrasi HB and Maranas CD (2021) A genome-scale metabolic model of *Anabaena* 33047 to guide genetic modifications to overproduce nylon monomers. *Metabolites* 11(3): 168.

Hendry JI, Prasannan CB, Joshi A, Dasgupta S and Wangikar PP (2016) Metabolic model of *Synechococcus* sp. PCC 7002: Prediction of flux distribution and network modification for enhanced biofuel production. *Bioresource Technology* 213: 190–197.

Hong SJ and Lee CG (2007) Evaluation of central metabolism based on a genomic database of *Synechocystis* sp. PCC 6803. *Biotechnology and Bioprocess Engineering* 12(2): 165–173.

Huggett ASG and Nixon DA (1957) Use of glucose oxidase, peroxidase, and o-dianisidine in determination of blood and urinary glucose. *Lancet* 270(6991): 368–370.

Ishida T, Hasegawa N, Hayashi NR, Peerapornpisal Y, Ishii M, Igarashi Y and Kodama T (1997) Growth characteristics and dense culture of a thermophilic cyanobacterium, *Chroococcidiopsis* sp. Strain TS-821. *Journal of Fermentation and Bioengineering* 83(6): 571–576.

Jaiswal D, Sengupta A, Sohoni S, Sengupta S, Phadnavis AG, Pakrasi HB and Wangikar PP (2018) Genome features and biochemical characteristics of a robust, fast growing and naturally transformable cyanobacterium *Synechococcus elongatus* PCC 11801 isolated from India. *Scientific Reports* 8(1): 1–13.

Klanchui A, Dulsawat S, Chaloemngam K, Cheevadhanarak S, Prommeenate P and Meechai A (2018) An improved genome-scale metabolic model of *Arthrospira platensis* C1 (iAK888) and its application in glycogen overproduction. *Metabolites* 8(4): 84.

Klanchui A, Khannapho C, Phodee A, Cheevadhanarak S and Meechai A (2012) i AK692: A genome-scale metabolic model of *Spirulina platensis* C1. *BMC Systems Biology* 6(1): 71.

Knoop H, Gründel M, Zilliges Y, Lehmann R, Hoffmann S, Lockau W and Steuer R (2013) Flux balance analysis of cyanobacterial metabolism: The metabolic network of *Synechocystis* sp. PCC 6803. *PLOS Computational Biology* 9(6): e1003081.

Kumar VS and Maranas CD (2009) GrowMatch: An automated method for reconciling in silico/in vivo growth predictions. *PLOS Computational Biology* 5(3): e1000308.

Labarca C and Paigen K (1980) A simple, rapid, and sensitive DNA assay procedure. *Analytical Biochemistry* 102(2): 344–352.

Lawrenz E, Fedewa EJ and Richardson TL (2013) Erratum to: Extraction protocols for the quantification of phycobilins in aqueous phytoplankton extracts. *Journal of Applied Phycology* 25(4): 1269.

Mac as-Sánchez MD, Mantell C, Rodrguez M, Mart nez de la Ossa E, Lubián LM and Montero O (2005) Supercritical fluid extraction of carotenoids and chlorophyll *a* from *Nannochloropsis gaditana. Journal of Food Engineering* 66(2): 245–251.

Malatinszky D, Steuer R and Jones PR (2017) A comprehensively curated genome-scale two-cell model for the heterocystous cyanobacterium *Anabaena* sp. PCC 7120. *Plant Physiology* 173(1): 509–523.

Markou G, Angelidaki I, Nerantzis E and Georgakakis D (2013) Bioethanol production by carbohydrate-enriched biomass of *Arthrospira (Spirulina) platensis. Energies* 6(8): 3937–3950.

Masuko T, Minami A, Iwasaki N, Majima T, Nishimura SI and Lee YC (2005) Carbohydrate analysis by a phenol–sulfuric acid method in microplate format. *Analytical Biochemistry* 339(1): 69–72.

Meléndez R, Meléndez-Hevia E, Mas F, Mach J and Cascante M (1998) Physical constraints in the synthesis of glycogen that influence its structural homogeneity: A two-dimensional approach. *Biophysical Journal* 75(1): 106–114.

Minkova KM, Tchernov AA, Tchorbadjieva MI, Fournadjieva ST, Antova RE and Busheva MC (2003) Purification of C-phycocyanin from *Spirulina* (*Arthrospira*) *fusiformis. Journal of Biotechnology* 102(1): 55–59.

Monsigny M, Petit C and Roche AC (1988) Colorimetric determination of neutral sugars by a resorcinol sulfuric acid micromethod. *Analytical Biochemistry* 175(2): 525–530.

Montagud A, Navarro E, Fernández de Córdoba P, Urchueguía JF and Patil KR (2010) Reconstruction and analysis of genome-scale metabolic model of a photosynthetic bacterium. *BMC Systems Biology* 4(1): 1–16.

Montagud A, Zelezniak A, Navarro E, de Córdoba PF, Urchueguía JF and Patil KR (2011) Flux coupling and transcriptional regulation within the metabolic network of the photosynthetic bacterium *Synechocystis* sp. PCC 6803. *Biotechnology Journal 6*(3): 330–342.

Nogales J, Gudmundsson S, Knight EM, Palsson BO and Thiele I (2012) Detailing the optimality of photosynthesis in cyanobacteria through systems biology analysis. *Proceedings of the National Academy of Sciences USA* 109(7): 2678–2683.

Parpinello GP and Versari A (2000) A simple high-performance liquid chromatography method for the analysis of glucose, glycerol, and methanol in a bioprocess. *Journal of Chromatographic Science* 38(6): 259–261.

Rai A and Saito K (2016) Omics data input for metabolic modeling. *Current Opinion in Biotechnology* 37: 127–134.

Roe JH and Dailey RE (1966) Determination of glycogen with the anthrone reagent. *Analytical Biochemistry* 15(2): 245–250.

Sandybayeva SK, Kossalbayev BD, Zayadan BK, Sadvakasova AK, Bolatkhan K, Zadneprovskaya EV, Kakimov AB, Alwasel S, Leong YK, Allakhverdiev SI and Chang JS (2022) Prospects of cyanobacterial pigment production: Biotechnological potential and optimization strategies. *Biochemical Engineering Journal* 187: 108640.

Schneegurt MA, Sherman DM, Nayar S and Sherman LA (1994) Oscillating behavior of carbohydrate granule formation and dinitrogen fixation in the cyanobacterium *Cyanothece* sp. Strain ATCC 51142. *Journal of Bacteriology* 176(6): 1586–1597.

Schreiber U, Endo T, Mi H and Asada K (1995) Quenching analysis of chlorophyll fluorescence by the saturation pulse method: Particular aspects relating to the study of eukaryotic algae and cyanobacteria. *Plant and Cell Physiology* 36(5): 873–882.

Shastri AA and Morgan JA (2005) Flux balance analysis of photoautotrophic metabolism. *Biotechnology Progress* 21(6): 1617–1626.

Shirai T, Osanai T and Kondo A (2016) Designing intracellular metabolism for production of target compounds by introducing a heterologous metabolic reaction based on a *Synechosystis* sp. PCC 6803 genome-scale model. *Microbial Cell Factories* 15(1): 13.

Vidal R and Venegas-Calerón M (2019) Simple, fast and accurate method for the determination of glycogen in the model unicellular cyanobacterium *Synechocystis* sp. PCC 6803. *Journal of Microbiological Methods* 164: 105686.

Vu TT, Hill EA, Kucek LA, Konopka AE, Beliaev AS and Reed JL (2013) Computational evaluation of *Synechococcus* sp. PCC 7002 metabolism for chemical production. *Biotechnology Journal* 8(5): 619–630.

Vu TT, Stolyar SM, Pinchuk GE, Hill EA, Kucek LA, Brown RN, Lipton MS, Osterman A, Fredrickson JK, Konopka AE, Beliaev AS and Reed JL (2012) Genome-scale modeling of light-driven reductant partitioning and carbon fluxes in diazotrophic unicellular cyanobacterium *Cyanothece* sp. ATCC 51142. *PLOS Computational Biology* 8(4): e1002460.

Watson J, Degnan B, Degnan S and Krömer JO (2014) Determining the biomass composition of a sponge holobiont for flux analysis. In: Metabolic Flux Analysis (Eds. J.O. Kromer, L.K. Nielsen and L.M. Blank). Springer, New York, pp. 107–125. ISBN: 978-1-4939-1170-7

Yoshikawa K, Aikawa S, Kojima Y, Toya Y, Furusawa C, Kondo A and Shimizu H (2015) Construction of a genome-scale metabolic model of Arthrospira platensis NIES-39 and metabolic design for cyanobacterial bioproduction. *PLOS ONE* 10(12): e0144430.

Yoshikawa K, Kojima Y, Nakajima T, Furusawa C, Hirasawa T and Shimizu H (2011) Reconstruction and verification of a genome-scale metabolic model for *Synechocystis* sp. PCC 6803. *Applied Microbiology and Biotechnology* 92(2): 347–358.

Zavřel T, Červený J, Knoop H and Steuer R (2016) Optimizing cyanobacterial product synthesis: Meeting the challenges. *Bioengineered* 7(6): 490–496.

Zavřel T, Očenášová P, Sinetova M and Červený J (2018) Determination of storage (starch/glycogen) and total saccharides content in algae and cyanobacteria by a phenol-sulfuric acid method. *Bio-Protocol* 8(15): e2966.

Zhang B, Dhital S and Gidley MJ (2013) Synergistic and antagonistic effects of α-amylase and amyloglucosidase on starch digestion. *Biomacromolecules* 14(6): 1945–1954.

24 *In Silico* Genome Assembly and Functional Annotation

Pannaga P. Jutur, Ashish Kumar, Mohammed Rehmanji, Rabinder Singh, and Asha A. Nesamma

24.1 INTRODUCTION

Over the last 30 years, technical advances in genome sequencing have greatly enlarged the dimensions of microalgal and cyanobacterial biotechnology (Goodwin et al. 2016; Shendure et al. 2017), increasing the economic viability of the same and resulting in a fast development of genomic resources. Microbial genomics has been advancing at an exponential rate over the last decades, and the pool of publicly available genome assemblies of microalgae is growing. Microalgae are important because of their variety; they are a very diverse group of eukaryotes that arose from initial endosymbiosis and many subsequent endosymbiotic processes (Gray and Doolittle 1982; Cavalier-Smith 1999; Delwiche 1999; McFadden 2001; Palmer 2003; Andersen 2004; Keeling 2004; Brodie and Lewis 2007). These photosynthetic eukaryotes, together with cyanobacteria, are responsible for half of all CO_2 fixation and photosynthesis on the planet. They also play an important role in carbon cycling, accounting for half of global primary production (Pierella Karlusich et al. 2020).

The genetic diversity of microalgae, including cyanobacteria, is demonstrated by diverse phenotypic characters, including size, i.e., 2–60 μm; diverse habitats, e.g., marine and freshwater environments; extreme temperature variations; and adaptation to symbiotic environments (lichens) (Zhang et al. 2020). Although it complicates assembly and sequencing processes, examining this variety aids in understanding evolutionary ideas, metabolism, physiology, discovery of new products and the ability to manipulate them genetically.

The number of sequenced algal genomes, which are crucial to understanding their complicated biology, is expanding, necessitating the development of efficient analytic methods (Table 24.1). The algal genome resource PhycoCosm, for example, contains more than 100 algal genomes assembled and annotated in their entirety, and it is routinely updated. Furthermore, it includes web-based interactive tools that help in environmental and bioenergy studies using microalgae (Grigoriev et al. 2021). These platforms promote scientific community interaction and data exchange, as well as the development of new sequencing initiatives to support research aims. The Joint Genome Institute (JGI) sequenced the first algal genome, that of the diatom *Thalassiosira pseudonana* (Armbrust et al. 2004), which was followed by several others (Merchant et al. 2007; Curtis et al. 2012; Read et al. 2013).

Initially capillary sequencing (as Sanger sequencing) (Sanger et al. 1977) was employed for eukaryotic genome assembly of numerous small DNA fragments. Reading 1 million bases of DNA (1 Mb) with automated Sanger sequencing costs over $1500 and takes more than a day (Rice and Green 2019). Similarly, reading the human genome and precisely assembling it were expensive and time-consuming, for this reason genome initiatives were restricted to a few big universities and concentrated on model species frequently employed in scientific study.

Advancement in the sequencing technologies reduced the costs, saved time, and increased accuracy (Compeau et al. 2011; Li et al. 2012; Simpson and Pop 2015). Platforms such as Pyrosequencing (Margulies et al. 2005), SOLiD (McKernan et al. 2009), Ion Torrent (Rothberg et al. 2011) and Solexa

DOI: 10.1201/9781003398387-24

TABLE 24.1
Selected Microalgal Genomes from Various Classes with Genome Size and Sequencing Platform

Microalgae	Genome Size	Number of Predicted Genes	Year of Completion	Platform of Assembly and Annotation	References
Asterochloris sp. Cgr/DA1pho	55.8 Mb	10,025	2019	Illumina Velvet	(Armaleo et al. 2019)
Auxenochlorella protothecoides	22.9 Mb –	7039 –	2014 –	Illumina HiSeq 2000 platform AUGUSTUS	(Gao et al. 2014)
Bathycoccus prasinos	15.1 Mb	7900	2012	Sanger's Arachne 3.0	(Moreau et al. 2012)
Chlamydomonas reinhardtii CC-503 cw92 mt+	111.1 Mb	17,741	2017	Whole-genome Shotgun sequencing	(Merchant et al. 2007)
Chlorella sorokiniana str. 1228	61.4 Mb	–	2018	PacBio Phrap (SPS-4.24)	(Schoch et al. 2020)
Chlorella sorokiniana UTEX 1230	58.5 Mb	–	2018	Illumina and PacBio Phrap (SPS-4.24)	(Schoch et al. 2020)
Chlorella sorokiniana DOE1412	57.8 Mb	–	2018	PacBio Phrap (SPS-4.24)	(Schoch et al. 2020)
Chlorella variabilis NC64A	46.2 Mb	9791	2010	Whole-genome sequencing Jazz	(Blanc et al. 2010)
Chlorella vulgaris	37.3 Mb	10,724	2019	Illumina and PacBio FALCON	(Schoch et al. 2020)
Coccomyxa subellipsoidea C-169	48.8 Mb	9839	2012	Whole-genome sequencing Jazz	(Schoch et al. 2020)
Dunaliella salina CCAP19/18	343.7 Mb	16,697	2017	Illumina/Pa cBio ALLPATHS-LG	(Polle et al. 2017)
Gonium pectorale	148.8 Mb	–	2016	Illumina AUGUSTUS	(Schoch et al. 2020)
Micromonas commoda NOUM17 (RCC 299)	21.0 Mb	10,262	2009	Whole-genome sequencing	(Worden et al. 2009)
Micromonas pusilla CCMP-1545	21.9 Mb	10,575	2009	Whole-genome sequencing	(Schoch et al. 2020)

(Continued)

TABLE 24.1 ***(Continued)***
Selected Microalgal Genomes from Various Classes with Genome Size and Sequencing Platform

Microalgae	Genome Size	Number of Predicted Genes	Year of Completion	Platform of Assembly and Annotation	References
Micromonas pusilla RCC299/NOUM17	20.9 Mb	10,056	2009	Whole-genome sequencing	(Schoch et al. 2020)
Monoraphidium neglectum	69.7 Mb	16,755	2013	Illumina MiSeq Newbler assembler version 2.6 (Roche)	(Bogen et al. 2013)
Ostreococcus lucimarinus CCE9901	13.2 Mb	7603	2007	Whole-genome sequencing Phred/Phrap/Consed	(Palenik et al. 2007)
Ostreococcus tauri OTH95	12.9 Mb	7699	2014	Illumina	(Blanc-Mathieu et al. 2014)
Ostreococcus sp. RCC809	13.3 Mb	7492	2009	Whole-genome sequencing	(Palenik et al. 2007)
Picochlorum soloecismus DOE101	15.2 Mb	7844	2017	Illumina and pyrosequecing Phrap	(Gonzalez-Esquer et al. 2018)
Scenedesmus obliquus DOE0152Z	210.3 Mb	–	2017	Whole-genome sequencing	(Starkenburg et al. 2017)
Tetraselmis sp.	232.0 Mb	46,696	2019	PacBio FALCON	(Schoch et al. 2020)
Volvox carteri	131.2 Mb	13,076	2010	Whole-genome sequencing	(Prochnik et al. 2010)
Yamagishiella unicocca	140.0 Mb	–	2018	PacBio and Illumina Sequencing	(Hamaji et al. 2018)
Chrysochromulina parva	65.8 Mb	–	2018	Whole-genome sequencing	(Schoch et al. 2020)
Chrysochromulina tobinii CCMP291	59.1 Mb	16,765	2015	Illumina and 454 sequencing Newbler 3.3	(Schoch et al. 2020)
Emiliania huxleyi	167.7 Mb	38,554	2013	Illumina	(Read et al. 2013)
Aureococcus anophagefferens	50.1 Mb	11,522	2011	Sanger's sequencing Jazz	(Gobler et al. 2011)
Ectocarpus siliculosus	195.8 Mb	16,271	2012	Illumina	(Cock et al. 2010)
Fragilariopsis cylindrus CCMP1102	61.1 Mb	21,066	2017	Sanger's and PacBio FALCON	(Mock et al. 2017)

(Continued)

TABLE 24.1 ***(Continued)***
Selected Microalgal Genomes from Various Classes with Genome Size and Sequencing Platform

Microalgae	Genome Size	Number of Predicted Genes	Year of Completion	Platform of Assembly and Annotation	References
Nannochloropsis gaditana	28.5 Mb	10,486	2014	Applied Biosystems SOLiD 3 PASS	(Corteggiani et al. 2014)
Nannochloropsis salina CCMP1766	27.7 Mb	8927	2019	Illumina BRAKER	(Ohan et al. 2019)
Phaeodactylum tricornutum	27.5 Mb	10,408	2008	Illumina JaZZ	(Bowler et al. 2008)
Saccharina japonica	543.4 Mb	–	2015	Illumina Hi-Seq SOAPdenov o2	(Ye et al. 2015)
Thalassiosira oceanica CCMP 1005	92.2 Mb	34,642	2012	Whole-genome sequencing AUGUSTU S	(Lommer et al. 2010)
Thalassiosira pseudonana	32.4 Mb	11,390	2004	Whole-genome sequencing; JAZZ	(Armbrust et al. 2004)
Chondrus crispus	105 Mb	9606	2013	Sanger's sequencing ARACHNE	(Collén et al. 2013)
Cyanidioschyzon merolae 10D	16.5 Mb	4775	2007	Whole-genome sequencing ARACHNE	(Matsuzaki et al. 2004)
	–	–	–	–	–
Galdieria sulphuraria	14.1 Mb	6125	2019	PacBio Canu	(Jain et al. 2015)
Gracilariopsis chorda	92.1 Mb	10,806	2018	Illumina and PacBio HGAP	(Lee et al. 2018)
Porphyridium purpureum	19.7 Mb	8355	2013	Illumina CLC Genomics workbench	(Bhattacharya et al. 2013)

(Bentley et al. 2008) sequencing are a few of the examples. These methods made genome sequencing accessible to smaller labs researching non-model species. Illumina HiSeq 2000 (Korostin et al. 2020) can sequence DNA 10,000 times faster than automated Sanger sequencing at a quarter of the cost (Metzker 2010; Glenn 2011; Heather and Chain 2016).

There is presently no device capable of reading DNA from an entire chromosome. All current genome assembly methods read short segments, for example, 100s bp for Illumina (Minoche et al. 2011), 1000–10,000 bp for PacBio (Minoche et al. 2011; Rhoads and Au 2015), and occasionally 100–1000 bp on the rapidly growing Oxford Nanopore (Jain et al. 2016). Combining long and short readings appears to be a potential method for enhancing genome coverage. It is known as hybrid sequencing because it employs high-quality smaller reads with high precision to fill in gaps together with longer reads to define huge genome assemblies that have been rejected by shorter reads (Miller et al. 2017; Minei et al. 2018), resulting in errorless predictions. Hybrid genome sequencing was recently used to reveal the genome of the marine microalga *Parachlorella kessleri* (Shaikh et al. 2020). We discuss hybrid genome assembly in a straightforward manner in this chapter.

Another important bottleneck in genome sequencing research is genome annotation. Due to a lack of availability of annotation tools, many research projects are finding it difficult to annotate genomes and make them available online. To address this issue, various public repositories are created; however, this is not the primary aim, since these facilities often use in-house bioinformatics assistance/pipelines, again making the data unavailable (Lawson et al. 2007). Hence, enormous amounts of unannotated genomes continue to amass, highlighting the critical need for simpler, more portable annotation procedures. An annotation pipeline should make it simple to annotate, view, and change individual contigs (a set of DNA segments or sequences that overlap in a way that provides a contiguous representation of a genomic region) in order to be considered in a user-friendly manner. This enables users to evaluate partial genome assemblies and annotate regions of interest independently using their own datasets, preferably without the expense of a database and with minimum computational resources.

In this chapter, we describe one such annotation tool, i.e., MAKER (Cantarel et al. 2008). This tool finds repetitions, aligns ESTs and proteins to a genome, predicts gene functions, and incorporates this information into protein-coding gene annotations. Furthermore, its results may be immediately fed into Generic Model Organism Database (GMOD)/GBrowse: the generic genome browser and databases without any post-processing. MAKER is an easy-to-use, yet powerful tool for annotating individual contigs or bacterial artificial chromosomes (BACs) or converting a complete genome sequence into a community-accessible genome database. MAKER is intended to kick-start genomics in developing model organisms by delivering a comprehensive first wave of database-ready protein-coding gene annotations. It is easily trainable: Early run results may be utilised to automatically retrain its gene prediction algorithm, resulting in higher-quality gene models on subsequent runs.

This chapter provides a detailed discussion of genome assembly via hybrid genome sequencing and annotation using MAKER. We provide a systematic procedure for DNA extraction, genome assembly, and annotation. Also, we have emphasised the analysis section as well as numerous criteria and threshold configuration for greater coverage and annotation. This work offers a one-of-a-kind and unusual picture of the present condition of genomic resources. Furthermore, it provides recorded proof that advanced tools and approaches can help in genome assembly. We address many key contributions and advancements that will allow future research to fully harness genomic capabilities.

24.2 EQUIPMENT

Different instruments used in genomics study are shown in Table 24.2, while Figure 24.1 illustrates different DNA sequencing platforms.

TABLE 24.2
List of Instruments Used in Genomic Study

S. No.	Equipment Used	Specifications	Vendor/Manufacturer
1.	Vortexer	Vortex mixers are one of the most used methods for mixing laboratory materials in test tubes, well plates, or flasks.	Fisher Scientific, CAPPRondo
2.	Centrifugation machine	A centrifuge is a device that separates different components of a fluid using centrifugal force. This is accomplished by spinning the fluid at high speeds within a container, separating fluids of varying densities.	Fisher Scientific, Eppendorf®
3.	Thermal cyclers	Thermocyclers, also known as polymerase chain reaction (PCR) equipment, are lab tools used to amplify DNA and RNA samples using the PCR by controlling temperature during cyclical cycles.	Bio-Rad Laboratories, Thermo Fisher Scientific
4.	Nanodrop spectrophotometer	One-microliter-sample quantities can be used with the Nanodrop spectrophotometer to measure nucleic acid contents.	Bio-Rad Laboratories®, Thermo Fisher Scientific®
5.	Gel electrophoresis	A typical laboratory method for identifying, measuring, and purifying nucleic acid fragments is electrophoresis. The negatively charged nucleic acids are made to migrate toward the positive electrode when samples are placed into wells of an agarose or acrylamide gel and exposed to an electric field.	Thermo Fisher Scientific®, Bio-Rad Laboratories®
6.	NextSeq 550 Instrument (Illumina)	Exome sequencing, targeted gene sequencing (amplicon, gene panel), whole-transcriptome sequencing, and cytogenomic arrays. Flexible benchtop sequencing for a variety of research applications.	Illumina®
7.	NextSeq 550 Dx Instrument	Exome sequencing, targeted gene sequencing (amplicon, gene panel), and whole-transcriptome sequencing.	Illumina®
8.	PromethION 2 (P2)	Larger genomes or projects, population-scale sequencing, highly multiplexed small genomes, or targeted regions whole transcriptomes (direct RNA or cDNA).	Oxford Nanopore Technologies®
9.	VolTRAX	Automated sample extraction and library preparation.	Oxford Nanopore Technologies®

24.3 RECIPES

24.3.1 DNA Isolation (Quick-DNA Fungal/Bacterial Microprep Kit; Navarro-Díaz et al. 2022)

1. *Bashing Bead Buffer*: Mostly composed of Tris and EDTA; plays a key role in dissolving the DNA precipitate during DNA extraction. A suitable buffer solution is necessary during isolation in order to preserve the reaction's pH and save nucleic acid. Tris is a chemical compound with the molecular formula $(HOCH_2)_3CNH_2$. The main function of Tris in any biological procedure is to keep the pH of the solution stable and make the cell membrane permeable.

FIGURE 24.1 Different DNA sequencing equipment used by Illumina® and Oxford Nanopore® Technologies in genomics study (colour figure in e-book).

2. *Genomic Lysis Buffer*: A lysis buffer solution used to break the cell wall and nuclear membranes to release nucleic acids. Commonly composed of sodium dodecyl sulphate.
3. *DNA Pre-Wash Buffer*: Wash buffers, which are often alcohol-based, can be used to remove proteins, salts, and other contaminants from the sample. Mostly composed of 80% ethanol, 20 mM NaCl, and 2 mM Tris-HCl (pH of 7.5).
4. *DNA Elution Buffer*: The DNA elution buffer contains 10 mM Tris (pH 8.5) and 0.1 mM EDTA. It can be used in DNA purification methods to elute DNA from columns, plates, and magbeads.

24.4 METHODS

24.4.1 Library Preparation

The preparation of a library is critical for the success of the next-generation sequencing (NGS) process. This procedure prepares DNA or RNA samples to be sequencer compatible. Depending on the lengths of the individual sequences that may be retrieved by the next-generation technology being used, the fragments are typically between 100 and 500 bp in length. The most common method for slicing genomic DNA into pieces of different size is sonication, which involves randomly cutting DNA molecules with high-frequency sound waves (Head et al. 2014).

In NGS, often two main immobilization techniques are employed. In the first technique, a glass slide covered with several copies of a short oligonucleotide serves as the solid foundation. The ends of the DNA fragments are ligated to adaptors, brief snippets of double-stranded DNA whose sequences match those of the oligonucleotide, which are subsequently denatured. Small metallic beads coated in the protein streptavidin serve as the firm support in the second immobilization technique. The DNA fragments are once more ligated to adaptors, but this time, the adaptors have a biotin label on their 5' ends. The fragments are joined to the metallic beads by biotin-streptavidin linkages, a tiny organic molecule that binds powerfully to streptavidin. Only one DNA fragment, on average, attaches to each bead when the DNA fragment–to–bead ratio is established. The beads are then agitated in an oil-water combination to create an emulsion, with the parameters selected so that each aqueous droplet in the emulsion contains only one bead. Then, on a plastic strip, each water droplet is placed into a different well of a multiple array. Amplification of the immobilized DNA fragments by polymerase chain reaction (PCR) is the last stage in the library construction. This creates a sufficient number of identical copies that can be sequenced (Hodges et al. 2009). The adaptors now serve a second purpose by acting as the annealing sites for the PCR primers. Therefore, even if the fragments themselves have a variety of distinct sequences, the same set of primers may be used to amplify all of the fragments. Further, in the glass slide technique, each beginning fragment is amplified into an immobilized cluster of identical fragments because the PCR products adhere to nearby oligonucleotides (Voelkerding et al. 2010). If metallic beads are employed, the PCR is conducted in an oil emulsion such that each PCR result is kept in its own water droplet before being deposited into the wells on the plastic strip.

24.4.2 Sequencing

In recent years, there has been intense competition among several companies to create NGS systems that combine the fastest rate of reliable sequence data creation with the most affordable price. Currently, the most widely used technique relies on reversible terminator sequencing, which like the chain-termination technique, uses modified nucleotides that, when added to the end of a polynucleotide being made by a DNA polymerase, prevent strand formation (Slatko et al. 2018). The distinction is that the termination step is reversible since once the identification of the changed nucleotide has been established, the chemical group linked to its 3-carbon may be removed. The most straightforward way uses a fluorescent label that is unique for each of the four nucleotides as this detachable blocking group (Slatko et al. 2018). Each step of strand synthesis requires a break since the reaction mixture does not include natural deoxynucleotides. During this time, an optical instrument may detect the fluorescent label and determine the terminal nucleotide. The label is subsequently removed by an enzyme, allowing the addition of the subsequent terminator nucleotide and a subsequent round of detection.

After sequencing, the instrument's software recognizes nucleotides (a procedure known as base calling), and it predicts the accuracy of those base calls. When employed in NGS, a primer anneals to the adaptor sequences added to the ends of the DNA fragments during library preparation to start the process (Grumbt et al. 2013). As a result, the library fragment clusters are all sequenced at the same time. With a maximum read length of 300 bp, the approach produces relatively small sequence reads, but because of its extreme parallelism, it may produce up to 2000 Mb of sequence in a single run.

Prior to the development of Illumina technology, pyrosequencing was the foundation of the most widely used next-generation technique. This approach does not artificially end the template copying, since the reaction mixture only comprises deoxynucleotides. When DNA polymerase adds a deoxynucleotides to the 3' end of the developing strand, sulfurylase releases a molecule of pyrophosphate that causes flashes of chemiluminescence that may be used to monitor the progress of strand synthesis. Therefore, the successful copying of one place in the template molecule is shown by a chemiluminescence flash. Further, the 454 sequencing technique is a next-generation approach that uses pyro-sequencing of fragment libraries immobilized on magnetic beads (Chen 2014).

One of the most recent contributions to NGS technology is nanopore sequencing. This technology was created and made commercially available by Oxford Nanopore Technologies (ONT), which also sells portable and desktop sequencers. Single strands of DNA or RNA are routed from the negatively charged '*cis*' side to the positively charged '*trans*' side via a small protein channel called a nanopore that is encased in an electrically impermeable membrane in order to perform nanopore sequencing. The ion current varies as the DNA molecule moves through the nanopore, providing information on the sequence and base modifications (Jurkowski 2020). This method formerly had rather low sequencing read quality. However, the read quality considerably increased after the introduction of the new, better Nanopore proteins and chemistries. The addition of processive enzymes to slow DNA translocation through the Nanopore represented a significant advancement in terms of raising the signal-to-noise ratio (Benner et al. 2007). In particular, it was found that motor protein (phi 29 DNA polymerase) performed better at ratcheting DNA through the Nanopore (Lieberman et al. 2010).

24.5 DNA SEQUENCING TECHNIQUES

The ultimate goal of a genome study is to obtain the whole DNA sequence for the organism being examined. Techniques for DNA sequencing are extremely important in the current scenario. A variety of technologies for DNA sequencing have been developed throughout the years, and more are expected to become essential in the future. The approaches described here are hybrid DNA sequencing techniques that use two NGS platforms: Illumina and Nanopore sequencing. We also explain the data processing of raw reads received from hybrid sequencing with an emphasis on assembly and annotation. NGS is a group of technologies that use a massively parallel strategy to create millions of sequences simultaneously.

24.5.1 Illumina Sequencing

This technique is named after the corporation that sells the required equipment. This approach is based on reversible terminator sequencing, which like the chain-termination method, employs modified nucleotides that, when inserted at the end of a polynucleotide being produced by a DNA polymerase, prevent strand formation (Minoche et al. 2011; Korostin et al. 2020). The distinction is that the termination step is reversible because the chemical group connected to the changed nucleotide's 3-carbon may be removed after the identification of this nucleotide has been validated. In the most basic case, this detachable blocking group is a fluorescent label, one for each of the four nucleotides. Because there are no regular deoxynucleotides in the reaction mixture, each step-in strand synthesis is preceded by a pause during which an optical device detects the fluorescent label, allowing the terminal nucleotide to be identified. The label is subsequently removed by an enzyme, allowing the next terminator nucleotide to be added and the detection procedure to be repeated. When employed in NGS, a primer anneals to the adaptor sequences added to the ends of the DNA fragments during library preparation. As a result, every cluster of fragments in the library is sequenced simultaneously. The approach creates very small sequence reads with a maximum length of 300 bp, but it is so massively parallel that it may generate up to 2000 Mb of sequence in each run.

24.5.2 Nanopore Sequencing

This is fourth-generation sequencing which directly reads the sequence of a DNA molecule without replicating it in any manner. A synthetic membrane with microscopic holes only large enough for a DNA molecule to get through is used in this procedure. A positive electrical current is put up on one side of the membrane and a negative current on the other, causing electrophoresis to induce the DNA molecule to approach one of the nanopores (Jain et al. 2016). This molecule is double-stranded, but the presence of a helicase enzyme near the Nanopore causes the base pairs to be broken, causing the DNA to unwind and just one strand to pass through the pore. Because each of the four nucleotides has a distinct shape and thus occludes the Nanopore in a different way; the sequence of this strand may be read. This results in a slightly different disruption of the flow of ions going through the membrane. These disturbances are monitored in order to determine the polynucleotide sequence.

Because there is no synthesis, the sequence length is not restricted by polymerase processivity, and readings of up to 50 kb have been observed. These systems enable hybrid sequencing to yield 10-fold more data than standard Sanger sequencing does. The higher amount of experimental data generated by NGS has necessitated the development of new prototypes for data computing and information extraction.

Before delving into the technicalities of data analysis, it is crucial to grasp a few concepts related to NGS data analysis.

24.5.3 Raw Data Output of NGS

Illumina-based NGS equipment generates files with an output format. bcl (Binary Base call) (Holtgrewe et al. 2020). These files are binary files containing both the base call and the quality of that base call for each 'tile' in each cycle. The. bcl file type has the benefit of recording each base call as the machine actually calls that base. In contrast to the. fastq file format, where base calls are recorded after the full sequence is read, calling the bases for each cluster in a specific tile for a certain cycle number in. bcl format is a far more efficient operation for the sequencing machine. The bcl format, however, is not useful for any other reason other than being convenient and effective for the sequencing machine.

For running in the assembler Masurca (discussed further), a config file is required where all the input parameters are specified.

24.5.4 Fastq Format

In the bioinformatics field, the. fastq file format is generally used to record raw sequencing data. For each read, this format comprises four lines. The first line begins with a '@' character and is followed by the sequencer identifying name that the sequencer assigned to the read. The second line is the read sequence, the third line is just a '+' sign functioning as a spacer, and the fourth and final line is the Phred quality scores for the bases in the second line (Holtgrewe et al. 2020).

File format: Cat masurca.config

Input files required: Illumina paired end reads supplied as two-character prefix; fragment mean; fragment stdev; forward_reads; reverse_reads; if single-end, do not specify reverse_reads.

For Nanopore, reads must be in a single. fasta or. fastq file with absolute path; it can be Gzipped; if you have both types of reads, supply them both as NANOPORE type.

24.5.5 The De Bruijn Graph Approach

To compensate for the shorter read length and lack of homogeneity in coverage of second-generation sequencing (SGS), de novo assembly initiatives generally generate 100 times as many reads as Sanger sequencing programmes. The de Bruijn graph technique completely eliminates the pairwise overlap computation, which is one of the reasons it has become the dominant method for SGS construction (Iqbal et al. 2012). This method begins by constructing a de Bruijn graph from the read data, as shown later. Every substring of length k (a k-mer) from each read is allocated to a directed edge in a graph linking nodes A and B for a fixed value k. Nodes A and B represent the beginning and last k-1 nucleotides of the original k-mer, respectively.

A draft assembly of the reads is formed by any path through the graph that visits every edge precisely once, properly termed a Eulerian path. In actuality, these graphs are complicated with numerous overlapping cycles and many different Eulerian pathways. Thus, generating the graph is only the first modest step toward creating a successful draft assembly. Complete assembly necessitates the incorporation of mate pair information into the graph, as well as an attempt to untangle the numerous complicated cycles produced by repeating sequences. Because k-mers are shorter than reads, the graph includes less information than reads; thus, the reads must be saved for eventual use in disambiguating routes in the graph.

The fundamental advantage of this strategy is its computational efficiency, which is gained by avoiding computing the enormous number of overlaps. The primary disadvantages are the loss of k-mer adjacency information in the graph and false branching induced by data mistakes.

24.6 DATA ANALYSIS

After sequencing, read quality is assessed and reads are aligned with the reference genome for data analysis. Subsequently, annotation is carried out followed by visualisation.

The data processing may be separated into four distinct processes: (1) raw data quality evaluation, (2) read alignment to a reference genome, (3) annotation, and (4) data display. Long (Nanopore/Pacbio) and short (Illumina) read data are required for hybrid genome assembly. Different tools are better suited to different types of datasets. It is recommended practise testing numerous tools and then select the optimal assembly based on various metrics such as N50, number of contigs, L50, contiguity, and so on. We explain our in-house genome assembly approach, which has been demonstrated to be the most successful for hybrid assembly of microalgal genomes using MaSuRCa and DBG2OLC.

24.6.1 Requirements

1. BBtools, MaSuRCa, Spades, Wengen, Sparseassembler, Blasr, DBG2OLC and Quast installed.
2. High-quality Nanopore and Illumina reads.
3. *Step 1*: Estimate insert size of Illumina reads
4. *Step 2*: Run MaSuRCa (Figures 24.2 and 24.3).

Figures 24.2 and 24.3 describe the assembly and genomic data curation codes.

The algae *Scenedesmus acuminatus* SAG 38.81, *Chlorella variabilis* (Minei et al. 2018), *Ulva compressa*, *Monodopsis* sp., and *Vischeria* sp. were recently sequenced utilising a hybrid sequencing technique with MaSuRCa for assembly.

Step 3: Annotation

The identification and labelling of all significant aspects of the genomic sequence is referred to as genome annotation. It is divided into two parts: Structural annotation (identifying exon locations, start codons, stop codons, etc.) and functional annotation. Whereas structural annotation entails determining the biological function of the gene discovered during structural annotation, functional annotation entails determining the biological function of the gene discovered during structural annotation. First, the open reading frames (ORFs) of potential genes are predicted. ORFs are DNA segments that can be translated into amino acids without the need for stop codons.

The forward (sense) strand of DNA has three reading frames, and the reverse strand has three reading frames (antisense). Only one out of every six reading frames will be free of stop codons throughout hundreds of nucleotides, and such a reading frame is a possibility for a coding gene. To determine the function of ORFs, the gene sequences are compared to multiple databases using a BLAST variant. UniProt, RefSeq at the U.S. National Center for Biotechnology Information (NCBI), and domain databases like Pfam are among the databases. Some ORFs cannot be functionally predicted; hence, they are classified as hypothetical proteins (no match in databases).

```
#set this to 1 if your Illumina jumping library reads are shorter than 100bp
EXTEND_JUMP_READS=0

#this is k-mer size for deBruijn graph values between 25 and 127 are supported, auto will compute the optimal size based on the read data and GC content
GRAPH_KMER_SIZE = auto

#set this to 1 for all Illumina-only assemblies
#set this to 0 if you have more than 15x coverage by long reads (Pacbio or Nanopore) or any other long reads/mate pairs (Illumina MP, Sanger, 454, etc)

USE_LINKING_MATES = 0

#specifies whether to run the assembly on the grid
USE_GRID=0

#specifies grid engine to use SGE or SLURM
GRID_ENGINE=SGE

#specifies queue (for SGE) or partition (for SLURM) to use when running on the grid MANDATORY
GRID_QUEUE=all.q

#batch size in the amount of long read sequence for each batch on the grid
GRID_BATCH_SIZE=500000000

#use at most this much coverage by the longest Pacbio or Nanopore reads, discard the rest of the reads
#can increase this to 30 or 35 if your reads are short (N50<7000bp)
LHE_COVERAGE=25

#set to 0 (default) to do two passes of mega-reads for slower, but higher quality assembly, otherwise set to 1
MEGA_READS_ONE_PASS=0

#this parameter is useful if you have too many Illumina jumping library mates. Typically set it to 60 for bacteria and 300 for the other organisms
LIMIT_JUMP_COVERAGE = 300

#these are the additional parameters to Celera Assembler.  do not worry about performance, number or processors or batch sizes -- these are computed automatically.

#CABOG ASSEMBLY ONLY: set cgwErrorRate=0.25 for bacteria and 0.1<=cgwErrorRate<=0.15 for other organisms.
CA_PARAMETERS =  cgwErrorRate=0.15

#CABOG ASSEMBLY ONLY: whether to attempt to close gaps in scaffolds with Illumina  or long read data
CLOSE_GAPS=1

#number of cpus to use, set this to the number of CPUs/threads per node you will be using
NUM_THREADS = 16
#this is mandatory jellyfish hash size -- a safe value is estimated_genome_size*20 (Aurantiochytrium genome size estimate is ~ 61Mbp))
JF_SIZE = 1236709040

#ILLUMINA ONLY. Set this to 1 to use SOAPdenovo contigging/scaffolding module.

#Assembly will be worse but will run faster. Useful for very large (>=8Gbp) genomes from Illumina-only data
SOAP_ASSEMBLY=0

#If you are doing Hybrid Illumina paired end + Nanopore/PacBio assembly ONLY (no Illumina mate pairs or OTHER frg files).

#Set this to 1 to use Flye assembler for final assembly of corrected mega-reads.

#A lot faster than CABOG, AND QUALITY IS THE SAME OR BETTER.

#Works well even when MEGA_READS_ONE_PASS is set to 1.

#DO NOT use if you have less than 15x coverage by long reads.
FLYE_ASSEMBLY=0
END

# running this will create a shell script assemble.sh
masurca masurca.config

# run that script to start masurca hybrid assembly
./assemble.sh
```

FIGURE 24.2 Bash script containing assembly codes for MaSuRCa assembly of RAW reads received after hybrid sequencing on the Illumina and Nanopore platforms.

In this section, we discuss the genome annotation pipeline using MAKER. MAKER enables the user to mix structural and functional annotations using various techniques. MAKER runs iteratively, doing annotations round by round and improving with each round relative to the preceding one. After four to five rounds, this progress stabilises. MAKER is reliant on control files, which include all information about the MAKER process. Run 'maker-CTL' to generate these files. maker opts.ctl provides input parameters; maker exe.ctl contains absolute paths to executables required by MAKER; maker bopts.ctls contains parameters for executing external tools used by MAKER; and maker evm.ctl contains weights for protein and transcript alignments. In most circumstances, simply modifying the input settings in maker opts.ctl will work for further processes.

```
############################################### WENGEN
wengan.pl -x ontraw \
 -a M \
 -s data/illumina/L_1.fastq.gz,data/illumina/L_2.fastq.gz \
 -l data/nanopore/nanopore.fastq.gz \
 -p asm1 \
 -t 20 \
 -g 70

############################################### SPADES
spades.py --pe1-1 /data/illumina/L_1.fastq.gz \
        --pe1-2 /data/illumina/L_2.fastq.gz \
          -t 30 \
          --careful \
          --nanopore /data/nanopore/nanopore.fasta \
          -o /data/spades/

################################################# Sparseassembler
gunzip L*fastq.gz   # Requires data in uncompressed format

SparseAssembler g 15 \
 k 51 \
 LD 0 \
 GS 88948734 \
 NodeCovTh 2 \
 EdgeCovTh 1 \
 p1 data/illumina/L_1.fastq \
 p2 data/illumina/L_2.fastq


# Run the above several times by changing different values to NodeCovTh and EdgeCovTh
# Use Quast to quickly get the N50 value
# For example, upon optimization, NodeCovTh 5 and EdgeCovTh 4 gave the best N50 = 7409

# Run DBG2OLC by additionally employing nanopore reads (fasta format)
mkdir DBG2OLC; cd DBG2OLC

DBG2OLC k 17 \
 AdaptiveTh 0.0001 \
  KmerCovTh 2 \
   MinOverlap 20 \
    RemoveChimera 1 \
     Contigs ../sparseassembler/Contigs.txt \
      f ../nanopore/nanopore.fasta

# again try different values of AdaptiveTh, KmerCovTh and MinOverlap, get the parameters which give the best N50 and lowest number of
contigs
# For eample in this case AdaptiveTh 0.001, KmerCovTh 2 and MinOverlap 20 gave the best N50 value and the lowest number of contigs

# call consensus
# requires python 2.7, create a separate conda environment for it if it doesn't already exist

mkdir CONSENSUS_DBG2OLC
cd CONSENSUS_DBG2OLC

#Concatenate the contigs and the long reads for consensus
cat ../sparseassembler/Contigs.txt ../nanopore/nanopore.fasta > contigs_and_long_reads.fasta

# we need to open a lot of files to distribute the above file into lots of smaller files
# ulimit unlimited    # not necessary

# run the consensus scripts (present in the utility folder of dbg2olc repository
split_and_run_sparc.sh \
 ../CONSENSUS_DBG2OLC.fasta \
  ../CONSENSUS_DBG2OLCinfo.txt \
   contigs_and_long_reads.fasta ./ 2 >cns_log.txt

################################## QUAST ASSESSMENT
quast.py DBG2OLC_consensus_final_assembly.fasta

# N50 = 139078, L50 = 120, Total length = 55216918; contigs 1144

################################## BUSCO assessment
# against eukaryota
busco --in /data/DBG2OLC_consensus_final_assembly.fasta \
 --cpu 30 \
  --out DBG2OLC_BUSCO_eukaryota \
   --out_path /data/BUSCO_assessment/ \
    --mode genome \
     --lineage_dataset eukaryota_odb10
```

FIGURE 24.3 Bash script with codes for assembly by Wengan, Spades, and Sparseassembler of RAW reads received after hybrid sequencing on the Illumina and Nanopore platforms.

24.6.2 Requirements

1. MAKER and its dependencies, including Perl dependencies; some of them do not come with the MAKER package and need to be installed separately.
2. Mpiexec or openMPI for MAKER compilation with multi-threading support.
3. A soft masked genome.
4. A *de novo* assembled transcriptome of the species (required for good annotation).

5. Transcript/ESTs from closely related organisms (optional in the case of presence of point 4; mandatory in the case of absence of point 4).
6. If proteins from the organism of interest are not available in the database, preference shall be given to closely related organisms.

MAKER ROUND 1

This round includes a masked genome, assembled transcripts, and protein sequences from the organism of interest and other closely related species. Est2 genome and protein2genome options are both enabled, implying that we want MAKER to predict genes directly from the transcript and protein alignment to the genome (Figure 24.4).

MAKER ROUND 2

In this round, the MAKER.gff file from the previous round, the genome, AUGUSTUS trained species, and SNAP hmm model are supplied, but est2genome and protein to genome are disabled (Figure 24.5).

MAKER ROUND 3

Only two things are given here: Gene models from another *ab initio* predictor and the genome (Figure 24.6).

MAKER ROUND 4

AUGUSTUS and SNAP are retrained in this round (Figure 24.7).

MAKER ROUND 5

AUGUSTUS and SNAP are retrained in this round (Figure 24.8).

Functional gene annotation (Figures 24.9a and 24.9b) of microalgae *Nannochloropsis oceanica* CCMP1779, *Scenedesmus obliquus* AS-6-11, and *Chlorella vulgaris* UTEX 395 was done using the MAKER pipeline developed by the respective groups.

Step 4: Data visualization

Data are further refined after annotation by employing visualisation tools. These tools provide several information about reads such as aligned reads, scores of different annotation tools, and annotation information including consequence, mapping quality, etc (Ejigu and Jung 2020).

The Integrative Genomics Viewer (IGV) is a high-performance, user-friendly, interactive application for visualising genetic data. It allows for the flexible integration of all common forms of genomic data and information, whether created by investigators or made publicly available and whether imported from local or cloud sources (Robinson et al. 2011). Circos is a software application that visualises genetic data and metadata. Circos visualises data in a circular arrangement, making it suitable for investigating connections between items or places. Circos is adaptable; it was initially created to visualise genomic data, but it can readily depict relationships or multi-layered annotations of one or more scales. Circos is automatable and managed using plain-text configuration files, making it simple to integrate into data collecting, processing, and reporting pipelines (Krzywinski et al. 2009).

24.6.3 Alignment Visualisation

Assembly analysis and read alignments frequently include examination of the sequencing reads themselves, and all tools show aligned read bases. Read sequences are frequently represented by letter strings stacked vertically and aligned horizontally from left to right. In the case of assemblies, a user can identify the bases contributing to the consensus at a specific place by scanning along the relevant column in the read stack. Base quality (log-transformed error probabilities) is frequently shown with a grey scale, and bases that differ from the consensus are highlighted with colour. By highlighting only discrepancies and hiding all consistent base pairs, some tools, including IGV, Hawkeye6, NCBI Assembly Archive Viewer7, and Text Alignment Viewer in SAMtools8, reduce visual clutter in the read stack (Thorvaldsdóttir et al. 2013).

```
cat maker_opts.ctl

genome=/data/assembly/final_assembly.fasta.masked #genome sequence (fasta file or fasta embeded in GFF3 file)
organism_type=eukaryotic #eukaryotic or prokaryotic. Default is eukaryotic

#-----Re-annotation Using MAKER Derived GFF3
maker_gff= #MAKER derived GFF3 file
est_pass=1 #use ESTs in maker_gff: 1 = yes, 0 = no
altest_pass=1 #use alternate organism ESTs in maker_gff: 1 = yes, 0 = no
protein_pass=1 #use protein alignments in maker_gff: 1 = yes, 0 = no
rm_pass=1 #use repeats in maker_gff: 1 = yes, 0 = no
model_pass=1 #use gene models in maker_gff: 1 = yes, 0 = no
pred_pass=1 #use ab-initio predictions in maker_gff: 1 = yes, 0 = no
other_pass=1 #passthrough anyything else in maker_gff: 1 = yes, 0 = no

#-----EST Evidence (for best results provide a file for at least one)
est=/data/transcriptome_Trinity.fasta #set of ESTs or assembled mRNA-seq in fasta format
altest= #EST/cDNA sequence file in fasta format from an alternate organism
est_gff= #aligned ESTs or mRNA-seq from an external GFF3 file
altest_gff= #aligned ESTs from a closly relate species in GFF3 format

#-----Protein Homology Evidence (for best results provide a file for at least one)
protein=/data/uniprot-myOrganism.fasta,/data/proteomics.fasta #protein sequence file in fasta format (i.e. from mutiple organisms)
protein_gff= #aligned protein homology evidence from an external GFF3 file

#-----Repeat Masking (leave values blank to skip repeat masking)
model_org= #select a model organism for RepBase masking in RepeatMasker
rmlib= #provide an organism specific repeat library in fasta format for RepeatMasker
repeat_protein= #provide a fasta file of transposable element proteins for RepeatRunner
rm_gff= #pre-identified repeat elements from an external GFF3 file
prok_rm=0 #forces MAKER to repeatmask prokaryotes (no reason to change this), 1 = yes, 0 = no
softmask=1 #use soft-masking rather than hard-masking in BLAST (i.e. seg and dust filtering)

#-----Gene Prediction
snaphmm= #SNAP HMM file
gmhmm=#GeneMark HMM file
augustus_species= #Augustus gene prediction species model
fgenesh_par_file= #FGENESH parameter file
pred_gff= #ab-initio predictions from an external GFF3 file
model_gff= #annotated gene models from an external GFF3 file (annotation pass-through)
run_evm=0 #run EvidenceModeler, 1 = yes, 0 = no
est2genome=1 #infer gene predictions directly from ESTs, 1 = yes, 0 = no
protein2genome=1 #infer predictions from protein homology, 1 = yes, 0 = no
trna=0 #find tRNAs with tRNAscan, 1 = yes, 0 = no
snoscan_rrna= #rRNA file to have Snoscan find snoRNAs
snoscan_meth= #-O-methylation site fileto have Snoscan find snoRNAs
unmask=0 #also run ab-initio prediction programs on unmasked sequence, 1 = yes, 0 = no
allow_overlap= #allowed gene overlap fraction (value from 0 to 1, blank for default)

#-----Other Annotation Feature Types (features MAKER doesn't recognize)
other_gff= #extra features to pass-through to final MAKER generated GFF3 file

#-----External Application Behavior Options
alt_peptide=C #amino acid used to replace non-standard amino acids in BLAST databases
cpus=1 #max number of cpus to use in BLAST and RepeatMasker (not for MPI, leave 1 when using MPI)

#-----MAKER Behavior Options
max_dna_len=100000 #length for dividing up contigs into chunks (increases/decreases memory usage)
min_contig=1 #skip genome contigs below this length (under 10kb are often useless)

pred_flank=200 #flank for extending evidence clusters sent to gene predictors
pred_stats=0 #report AED and QI statistics for all predictions as well as models
AED_threshold=1 #Maximum Annotation Edit Distance allowed (bound by 0 and 1)
min_protein=0 #require at least this many amino acids in predicted proteins
alt_splice=1 #Take extra steps to try and find alternative splicing, 1 = yes, 0 = no
always_complete=0 #extra steps to force start and stop codons, 1 = yes, 0 = no
map_forward=0 #map names and attributes forward from old GFF3 genes, 1 = yes, 0 = no
keep_preds=0 #Concordance threshold to add unsupported gene prediction (bound by 0 and 1)

split_hit=10000 #length for the splitting of hits (expected max intron size for evidence alignments)
min_intron=20 #minimum intron length (used for alignment polishing)
single_exon=0 #consider single exon EST evidence when generating annotations, 1 = yes, 0 = no
single_length=250 #min length required for single exon ESTs if 'single_exon is enabled'
correct_est_fusion=0 #limits use of ESTs in annotation to avoid fusion genes

tries=2 #number of times to try a contig if there is a failure for some reason
clean_try=0 #remove all data from previous run before retrying, 1 = yes, 0 = no
clean_up=0 #removes theVoid directory with individual analysis files, 1 = yes, 0 = no
TMP= #specify a directory other than the system default temporary directory for temporary files

# run MAKER this way; this assumes MAKER was compiled with MPI support
mpiexec -n 15 maker -base maker_round_1

# merge the output
cd maker_round_1.maker.output/
gff3_merge -s -d maker_round_1_master_datastore_index.log > round_1.all.maker.gff;
fasta_merge -d maker_round_1_master_datastore_index.log;

# gff without fasta sequences
gff3_merge -n -s -d maker_round_1_master_datastore_index.log > round_1.all.maker.without.sequences.gff;
```

FIGURE 24.4 Annotation of microalgal genome using MAKER; codes for round 1 of MAKER.

```
cat maker_opts.ctl

genome=/data/assembly/final_assembly.fasta.masked #genome sequence (fasta file or fasta embeded in GFF3 file)
organism_type=eukaryotic #eukaryotic or prokaryotic. Default is eukaryotic

#-----Re-annotation Using MAKER Derived GFF3
maker_gff=/data/maker_round_1.maker.output/round_1.all.maker.gff #MAKER derived GFF3 file
est_pass=1 #use ESTs in maker_gff: 1 = yes, 0 = no
altest_pass=1 #use alternate organism ESTs in maker_gff: 1 = yes, 0 = no
protein_pass=1 #use protein alignments in maker_gff: 1 = yes, 0 = no
rm_pass=1 #use repeats in maker_gff: 1 = yes, 0 = no
model_pass=1 #use gene models in maker_gff: 1 = yes, 0 = no
pred_pass=1 #use ab-initio predictions in maker_gff: 1 = yes, 0 = no
other_pass=1 #passthrough anyything else in maker_gff: 1 = yes, 0 = no

#-----EST Evidence (for best results provide a file for at least one)
est= #set of ESTs or assembled mRNA-seq in fasta format
altest= #EST/cDNA sequence file in fasta format from an alternate organism
est_gff= #aligned ESTs or mRNA-seq from an external GFF3 file
altest_gff= #aligned ESTs from a closly relate species in GFF3 format

#-----Protein Homology Evidence (for best results provide a file for at least one)
protein= #protein sequence file in fasta format (i.e. from mutiple organisms)
protein_gff= #aligned protein homology evidence from an external GFF3 file

#-----Repeat Masking (leave values blank to skip repeat masking)
model_org= #select a model organism for RepBase masking in RepeatMasker
rmlib= #provide an organism specific repeat library in fasta format for RepeatMasker
repeat_protein= #provide a fasta file of transposable element proteins for RepeatRunner
rm_gff= #pre-identified repeat elements from an external GFF3 file
prok_rm=0 #forces MAKER to repeatmask prokaryotes (no reason to change this), 1 = yes, 0 = no
softmask=1 #use soft-masking rather than hard-masking in BLAST (i.e. seg and dust filtering)

#-----Gene Prediction
snaphmm=/data/snap/myOrganism_snap.hmm #SNAP HMM file
gmhmm=#GeneMark HMM file
augustus_species=myOrganism #Augustus gene prediction species model
fgenesh_par_file= #FGENESH parameter file
pred_gff= #ab-initio predictions from an external GFF3 file
model_gff= #annotated gene models from an external GFF3 file (annotation pass-through)
run_evm=0 #run EvidenceModeler, 1 = yes, 0 = no
est2genome=0 #infer gene predictions directly from ESTs, 1 = yes, 0 = no
protein2genome=0 #infer predictions from protein homology, 1 = yes, 0 = no
trna=0 #find tRNAs with tRNAscan, 1 = yes, 0 = no
snoscan_rrna= #rRNA file to have Snoscan find snoRNAs
snoscan_meth= #-O-methylation site fileto have Snoscan find snoRNAs
unmask=0 #also run ab-initio prediction programs on unmasked sequence, 1 = yes, 0 = no
allow_overlap= #allowed gene overlap fraction (value from 0 to 1, blank for default)

#-----Other Annotation Feature Types (features MAKER doesn't recognize)
other_gff= #extra features to pass-through to final MAKER generated GFF3 file

#-----External Application Behavior Options
alt_peptide=C #amino acid used to replace non-standard amino acids in BLAST databases
cpus=1 #max number of cpus to use in BLAST and RepeatMasker (not for MPI, leave 1 when using MPI)

#-----MAKER Behavior Options
max_dna_len=100000 #length for dividing up contigs into chunks (increases/decreases memory usage)
min_contig=1 #skip genome contigs below this length (under 10kb are often useless)

pred_flank=200 #flank for extending evidence clusters sent to gene predictors
pred_stats=0 #report AED and QI statistics for all predictions as well as models
AED_threshold=1 #Maximum Annotation Edit Distance allowed (bound by 0 and 1)
min_protein=0 #require at least this many amino acids in predicted proteins
alt_splice=1 #Take extra steps to try and find alternative splicing, 1 = yes, 0 = no
always_complete=0 #extra steps to force start and stop codons, 1 = yes, 0 = no
map_forward=0 #map names and attributes forward from old GFF3 genes, 1 = yes, 0 = no
keep_preds=0 #Concordance threshold to add unsupported gene prediction (bound by 0 and 1)

split_hit=10000 #length for the splitting of hits (expected max intron size for evidence alignments)
min_intron=20 #minimum intron length (used for alignment polishing)
single_exon=0 #consider single exon EST evidence when generating annotations, 1 = yes, 0 = no
single_length=250 #min length required for single exon ESTs if 'single_exon is enabled'
correct_est_fusion=0 #limits use of ESTs in annotation to avoid fusion genes

tries=2 #number of times to try a contig if there is a failure for some reason
clean_try=0 #remove all data from previous run before retrying, 1 = yes, 0 = no
clean_up=0 #removes theVoid directory with individual analysis files, 1 = yes, 0 = no
TMP= #specify a directory other than the system default temporary directory for temporary files

mpiexec -n 15 maker -base maker_round_2
```

FIGURE 24.5 Annotation of microalgal genome using MAKER; codes for round 2 of MAKER.

24.6.4 Genome Browsers

To successfully organise and transmit this massive amount of data, efficient genome browsers accompanied by recent databases are increasing. Genome browsers, in general, offer data and biological annotations from several sources in their genetic context using graphical user interfaces. These tools enable a broad variety of data formats such as cross-species comparisons, genotype variation, and gene expression. Annotations of functionally relevant areas, such as gene locations, transcriptional activity regions, or regulatory elements, are derived from experimental data (for

```
cat maker_opts.ctl

genome=/data/assembly/final_assembly.fasta.masked #genome sequence (fasta file or fasta embeded in GFF3 file)
organism_type=eukaryotic #eukaryotic or prokaryotic. Default is eukaryotic

#-----Re-annotation Using MAKER Derived GFF3
maker_gff=/data/maker_round_2.maker.output/round_2.all.maker.gff #MAKER derived GFF3 file
est_pass=1 #use ESTs in maker_gff: 1 = yes, 0 = no
altest_pass=1 #use alternate organism ESTs in maker_gff: 1 = yes, 0 = no
protein_pass=1 #use protein alignments in maker_gff: 1 = yes, 0 = no
rm_pass=1 #use repeats in maker_gff: 1 = yes, 0 = no
model_pass=1 #use gene models in maker_gff: 1 = yes, 0 = no
pred_pass=1 #use ab-initio predictions in maker_gff: 1 = yes, 0 = no
other_pass=1 #passthrough anyything else in maker_gff: 1 = yes, 0 = no

#-----EST Evidence (for best results provide a file for at least one)
est= #set of ESTs or assembled mRNA-seq in fasta format
altest= #EST/cDNA sequence file in fasta format from an alternate organism
est_gff= #aligned ESTs or mRNA-seq from an external GFF3 file
altest_gff= #aligned ESTs from a closly relate species in GFF3 format

#-----Protein Homology Evidence (for best results provide a file for at least one)
protein= #protein sequence file in fasta format (i.e. from mutiple organisms)
protein_gff= #aligned protein homology evidence from an external GFF3 file

#-----Repeat Masking (leave values blank to skip repeat masking)
model_org= #select a model organism for RepBase masking in RepeatMasker
rmlib= #provide an organism specific repeat library in fasta format for RepeatMasker
repeat_protein= #provide a fasta file of transposable element proteins for RepeatRunner
rm_gff= #pre-identified repeat elements from an external GFF3 file
prok_rm=0 #forces MAKER to repeatmask prokaryotes (no reason to change this), 1 = yes, 0 = no
softmask=1 #use soft-masking rather than hard-masking in BLAST (i.e. seg and dust filtering)

#-----Gene Prediction
snaphmm= #SNAP HMM file
gmhmm=/data/genemark/gmhmm.mod #GeneMark HMM file
augustus_species= #Augustus gene prediction species model
fgenesh_par_file= #FGENESH parameter file
pred_gff= #ab-initio predictions from an external GFF3 file
model_gff= #annotated gene models from an external GFF3 file (annotation pass-through)
run_evm=0 #run EvidenceModeler, 1 = yes, 0 = no
est2genome=0 #infer gene predictions directly from ESTs, 1 = yes, 0 = no
protein2genome=0 #infer predictions from protein homology, 1 = yes, 0 = no
trna=0 #find tRNAs with tRNAscan, 1 = yes, 0 = no
snoscan_rrna= #rRNA file to have Snoscan find snoRNAs
snoscan_meth= #-O-methylation site fileto have Snoscan find snoRNAs
unmask=0 #also run ab-initio prediction programs on unmasked sequence, 1 = yes, 0 = no
allow_overlap= #allowed gene overlap fraction (value from 0 to 1, blank for default)

#-----Other Annotation Feature Types (features MAKER doesn't recognize)
other_gff= #extra features to pass-through to final MAKER generated GFF3 file

#-----External Application Behavior Options
alt_peptide=C #amino acid used to replace non-standard amino acids in BLAST databases
cpus=1 #max number of cpus to use in BLAST and RepeatMasker (not for MPI, leave 1 when using MPI)

#-----MAKER Behavior Options
max_dna_len=100000 #length for dividing up contigs into chunks (increases/decreases memory usage)
min_contig=1 #skip genome contigs below this length (under 10kb are often useless)

pred_flank=200 #flank for extending evidence clusters sent to gene predictors
pred_stats=0 #report AED and QI statistics for all predictions as well as models
AED_threshold=1 #Maximum Annotation Edit Distance allowed (bound by 0 and 1)
min_protein=0 #require at least this many amino acids in predicted proteins
alt_splice=1 #Take extra steps to try and find alternative splicing, 1 = yes, 0 = no
always_complete=0 #extra steps to force start and stop codons, 1 = yes, 0 = no
map_forward=0 #map names and attributes forward from old GFF3 genes, 1 = yes, 0 = no
keep_preds=0 #Concordance threshold to add unsupported gene prediction (bound by 0 and 1)

split_hit=10000 #length for the splitting of hits (expected max intron size for evidence alignments)
min_intron=20 #minimum intron length (used for alignment polishing)
single_exon=0 #consider single exon EST evidence when generating annotations, 1 = yes, 0 = no
single_length=250 #min length required for single exon ESTs if 'single_exon is enabled'
correct_est_fusion=0 #limits use of ESTs in annotation to avoid fusion genes

tries=2 #number of times to try a contig if there is a failure for some reason
clean_try=0 #remove all data from previous run before retrying, 1 = yes, 0 = no
clean_up=0 #removes theVoid directory with individual analysis files, 1 = yes, 0 = no
TMP= #specify a directory other than the system default temporary directory for temporary files

mpiexec -n 15 maker -base maker_round_3

cd maker_round_3.maker.output/

gff3_merge -s -d maker_round_3_master_datastore_index.log > round_3.all.maker.gff;
fasta_merge -d maker_round_3_master_datastore_index.log;

# gff without fasta sequences
gff3_merge -n -s -d maker_round_3_master_datastore_index.log > round_3.all.maker.without.sequences.gff;
```

FIGURE 24.6 Annotation of microalgal genome using MAKER; codes for round 3 of MAKER.

```
cat make_opts.ctl

genome=/data/assembly/final_assembly.fasta.masked #genome sequence (fasta file or fasta embeded in GFF3 file)
organism_type=eukaryotic #eukaryotic or prokaryotic. Default is eukaryotic

#-----Re-annotation Using MAKER Derived GFF3
maker_gff=/data/maker_round_3.maker.output/round_3.all.maker.gff #MAKER derived GFF3 file
est_pass=1 #use ESTs in maker_gff: 1 = yes, 0 = no
altest_pass=1 #use alternate organism ESTs in maker_gff: 1 = yes, 0 = no
protein_pass=1 #use protein alignments in maker_gff: 1 = yes, 0 = no
rm_pass=1 #use repeats in maker_gff: 1 = yes, 0 = no
model_pass=1 #use gene models in maker_gff: 1 = yes, 0 = no
pred_pass=1 #use ab-initio predictions in maker_gff: 1 = yes, 0 = no
other_pass=1 #passthrough anyything else in maker_gff: 1 = yes, 0 = no

#-----EST Evidence (for best results provide a file for at least one)
est= #set of ESTs or assembled mRNA-seq in fasta format
altest= #EST/cDNA sequence file in fasta format from an alternate organism
est_gff= #aligned ESTs or mRNA-seq from an external GFF3 file
altest_gff= #aligned ESTs from a closly relate species in GFF3 format

#-----Protein Homology Evidence (for best results provide a file for at least one)
protein= #protein sequence file in fasta format (i.e. from mutiple organisms)
protein_gff= #aligned protein homology evidence from an external GFF3 file

#-----Repeat Masking (leave values blank to skip repeat masking)
model_org= #select a model organism for RepBase masking in RepeatMasker
rmlib= #provide an organism specific repeat library in fasta format for RepeatMasker
repeat_protein= #provide a fasta file of transposable element proteins for RepeatRunner
rm_gff= #pre-identified repeat elements from an external GFF3 file
prok_rm=0 #forces MAKER to repeatmask prokaryotes (no reason to change this), 1 = yes, 0 = no
softmask=1 #use soft-masking rather than hard-masking in BLAST (i.e. seg and dust filtering)

#-----Gene Prediction
snaphmm=/data/snap/myOrganism_snap.hmm #SNAP HMM file
gmhmm= #GeneMark HMM file
augustus_species=myOrganism #Augustus gene prediction species model
fgenesh_par_file= #FGENESH parameter file
pred_gff= #ab-initio predictions from an external GFF3 file
model_gff= #annotated gene models from an external GFF3 file (annotation pass-through)
run_evm=0 #run EvidenceModeler, 1 = yes, 0 = no
est2genome=0 #infer gene predictions directly from ESTs, 1 = yes, 0 = no
protein2genome=0 #infer predictions from protein homology, 1 = yes, 0 = no
trna=0 #find tRNAs with tRNAscan, 1 = yes, 0 = no
snoscan_rrna= #rRNA file to have Snoscan find snoRNAs
snoscan_meth= #-O-methylation site fileto have Snoscan find snoRNAs
unmask=0 #also run ab-initio prediction programs on unmasked sequence, 1 = yes, 0 = no
allow_overlap= #allowed gene overlap fraction (value from 0 to 1, blank for default)

#-----Other Annotation Feature Types (features MAKER doesn't recognize)
other_gff= #extra features to pass-through to final MAKER generated GFF3 file

#-----External Application Behavior Options
alt_peptide=C #amino acid used to replace non-standard amino acids in BLAST databases
cpus=1 #max number of cpus to use in BLAST and RepeatMasker (not for MPI, leave 1 when using MPI)

#-----MAKER Behavior Options
max_dna_len=100000 #length for dividing up contigs into chunks (increases/decreases memory usage)
min_contig=1 #skip genome contigs below this length (under 10kb are often useless)

pred_flank=200 #flank for extending evidence clusters sent to gene predictors
pred_stats=0 #report AED and QI statistics for all predictions as well as models
AED_threshold=1 #Maximum Annotation Edit Distance allowed (bound by 0 and 1)
min_protein=0 #require at least this many amino acids in predicted proteins
alt_splice=1 #Take extra steps to try and find alternative splicing, 1 = yes, 0 = no
always_complete=0 #extra steps to force start and stop codons, 1 = yes, 0 = no
map_forward=0 #map names and attributes forward from old GFF3 genes, 1 = yes, 0 = no
keep_preds=0 #Concordance threshold to add unsupported gene prediction (bound by 0 and 1)

split_hit=10000 #length for the splitting of hits (expected max intron size for evidence alignments)
min_intron=20 #minimum intron length (used for alignment polishing)
single_exon=0 #consider single exon EST evidence when generating annotations, 1 = yes, 0 = no
single_length=250 #min length required for single exon ESTs if 'single_exon is enabled'
correct_est_fusion=0 #limits use of ESTs in annotation to avoid fusion genes

tries=2 #number of times to try a contig if there is a failure for some reason
clean_try=0 #remove all data from previous run before retrying, 1 = yes, 0 = no
clean_up=0 #removes theVoid directory with individual analysis files, 1 = yes, 0 = no
TMP= #specify a directory other than the system default temporary directory for temporary files

mpiexec -n 15 maker -base maker_round_4

cd maker_round_4.maker.output/

gff3_merge -s -d maker_round_4_master_datastore_index.log > round_4.all.maker.gff;
fasta_merge -d maker_round_4_master_datastore_index.log;

# gff without fasta sequences
gff3_merge -n -s -d maker_round_4_master_datastore_index.log > round_4.all.maker.without.sequences.gff;
```

FIGURE 24.7 Annotation of microalgal genome using MAKER; codes for round 4 of MAKER.

```
cat make_opts.ctl

genome=/data/assembly/final_assembly.fasta.masked #genome sequence (fasta file or fasta embeded in GFF3 file)
organism_type=eukaryotic #eukaryotic or prokaryotic. Default is eukaryotic

#-----Re-annotation Using MAKER Derived GFF3
maker_gff=/data/maker_round_4.maker.output/round_4.all.maker.gff #MAKER derived GFF3 file
est_pass=1 #use ESTs in maker_gff: 1 = yes, 0 = no
altest_pass=1 #use alternate organism ESTs in maker_gff: 1 = yes, 0 = no
protein_pass=1 #use protein alignments in maker_gff: 1 = yes, 0 = no
rm_pass=1 #use repeats in maker_gff: 1 = yes, 0 = no
model_pass=1 #use gene models in maker_gff: 1 = yes, 0 = no
pred_pass=1 #use ab-initio predictions in maker_gff: 1 = yes, 0 = no
other_pass=1 #passthrough anyything else in maker_gff: 1 = yes, 0 = no

#-----EST Evidence (for best results provide a file for at least one)
est= #set of ESTs or assembled mRNA-seq in fasta format
altest= #EST/cDNA sequence file in fasta format from an alternate organism
est_gff= #aligned ESTs or mRNA-seq from an external GFF3 file
altest_gff= #aligned ESTs from a closly relate species in GFF3 format

#-----Protein Homology Evidence (for best results provide a file for at least one)
protein= #protein sequence file in fasta format (i.e. from mutiple organisms)
protein_gff= #aligned protein homology evidence from an external GFF3 file

#-----Repeat Masking (leave values blank to skip repeat masking)
model_org= #select a model organism for RepBase masking in RepeatMasker
rmlib= #provide an organism specific repeat library in fasta format for RepeatMasker
repeat_protein= #provide a fasta file of transposable element proteins for RepeatRunner
rm_gff= #pre-identified repeat elements from an external GFF3 file
prok_rm=0 #forces MAKER to repeatmask prokaryotes (no reason to change this), 1 = yes, 0 = no
softmask=1 #use soft-masking rather than hard-masking in BLAST (i.e. seg and dust filtering)

#-----Gene Prediction
snaphmm=/data/snap/myOrganism_snap.hmm #SNAP HMM file
gmhmm= #GeneMark HMM file
augustus_species=myOrganism #Augustus gene prediction species model
fgenesh_par_file= #FGENESH parameter file
pred_gff= #ab-initio predictions from an external GFF3 file
model_gff= #annotated gene models from an external GFF3 file (annotation pass-through)
run_evm=0 #run EvidenceModeler, 1 = yes, 0 = no
est2genome=0 #infer gene predictions directly from ESTs, 1 = yes, 0 = no
protein2genome=0 #infer predictions from protein homology, 1 = yes, 0 = no
trna=0 #find tRNAs with tRNAscan, 1 = yes, 0 = no
snoscan_rrna= #rRNA file to have Snoscan find snoRNAs
snoscan_meth= #-O-methylation site fileto have Snoscan find snoRNAs
unmask=0 #also run ab-initio prediction programs on unmasked sequence, 1 = yes, 0 = no
allow_overlap= #allowed gene overlap fraction (value from 0 to 1, blank for default)

#-----Other Annotation Feature Types (features MAKER doesn't recognize)
other_gff= #extra features to pass-through to final MAKER generated GFF3 file

#-----External Application Behavior Options
alt_peptide=C #amino acid used to replace non-standard amino acids in BLAST databases
cpus=1 #max number of cpus to use in BLAST and RepeatMasker (not for MPI, leave 1 when using MPI)

#-----MAKER Behavior Options
max_dna_len=100000 #length for dividing up contigs into chunks (increases/decreases memory usage)
min_contig=1 #skip genome contigs below this length (under 10kb are often useless)

pred_flank=200 #flank for extending evidence clusters sent to gene predictors
pred_stats=0 #report AED and QI statistics for all predictions as well as models
AED_threshold=1 #Maximum Annotation Edit Distance allowed (bound by 0 and 1)
min_protein=0 #require at least this many amino acids in predicted proteins
alt_splice=1 #Take extra steps to try and find alternative splicing, 1 = yes, 0 = no
always_complete=0 #extra steps to force start and stop codons, 1 = yes, 0 = no
map_forward=0 #map names and attributes forward from old GFF3 genes, 1 = yes, 0 = no
keep_preds=0 #Concordance threshold to add unsupported gene prediction (bound by 0 and 1)

split_hit=10000 #length for the splitting of hits (expected max intron size for evidence alignments)
min_intron=20 #minimum intron length (used for alignment polishing)
single_exon=0 #consider single exon EST evidence when generating annotations, 1 = yes, 0 = no
single_length=250 #min length required for single exon ESTs if 'single_exon is enabled'
correct_est_fusion=0 #limits use of ESTs in annotation to avoid fusion genes

tries=2 #number of times to try a contig if there is a failure for some reason
clean_try=0 #remove all data from previous run before retrying, 1 = yes, 0 = no
clean_up=0 #removes theVoid directory with individual analysis files, 1 = yes, 0 = no
TMP= #specify a directory other than the system default temporary directory for temporary files

mpiexec -n 15 maker -base maker_round_5;

cd maker_round_5.maker.output/;

gff3_merge -s -d maker_round_5_master_datastore_index.log > round_5.all.maker.gff;
fasta_merge -d maker_round_5_master_datastore_index.log;

# gff without fasta sequences
gff3_merge -n -s -d maker_round_5_master_datastore_index.log > round_5.all.maker.without.sequences.gff;
```

FIGURE 24.8 Annotation of microalgal genome using MAKER; codes for round 5 of MAKER.

(a)

```
# count number of genes in every maker run
grep -c $'\tgene\t' ./maker*/round_?.all.maker.gff

# ./maker_round_1.maker.output/round_1.all.maker.gff:9794
# ./maker_round_2.maker.output/round_2.all.maker.gff:11957
# ./maker_round_3.maker.output/round_3.all.maker.gff:11362
# ./maker_round_4.maker.output/round_4.all.maker.gff:11373
# ./maker_round_5.maker.output/round_5.all.maker.gff:11374

# going ahead with maker round 5 run, it is only going to give false positives from now on

mkdir functional_annotation;

cp maker_round_5.maker.output/maker_round_5.all.maker.proteins.fasta functional_annotation/;
cp maker_round_5.maker.output/maker_round_5.all.maker.transcripts.fasta functional_annotation/;
cp maker_round_5.maker.output/round_5.all.maker.gff functional_annotation/;
cp maker_round_5.maker.output/round_5.all.maker.without.sequences.gff functional_annotation/;
cp maker_round_5.maker.output/round_5.all.maker.gff functional_annotation/;

# change ugly maker gene names to more readable ones
aker_map_ids --prefix MyOrganism --justify 6 round_5.all.maker.gff  > maker5.name.map

# replace names in GFF files
map_gff_ids maker5.name.map round_5.all.maker.gff;
map_gff_ids maker5.name.map round_5.all.maker.without.sequences.gff;

mkdir functional_annotation;

cp maker_round_5.maker.output/maker_round_5.all.maker.proteins.fasta functional_annotation/;
cp maker_round_5.maker.output/maker_round_5.all.maker.transcripts.fasta functional_annotation/;
cp maker_round_5.maker.output/round_5.all.maker.gff functional_annotation/;
cp maker_round_5.maker.output/round_5.all.maker.without.sequences.gff functional_annotation/;
cp maker_round_5.maker.output/round_5.all.maker.gff functional_annotation/

cd functional_annotation

# change ugly maker gene names to more readable ones
aker_map_ids --prefix MyOrganism --justify 6 round_5.all.maker.gff  > maker5.name.map

# replace names in GFF files
map_gff_ids maker5.name.map round_5.all.maker.gff;
map_gff_ids maker5.name.map round_5.all.maker.without.sequences.gff;

# replace names in FASTA headers
map_fasta_ids maker5.name.map maker_round_5.all.maker.proteins.fasta;
map_fasta_ids maker5.name.map maker_round_5.all.maker.transcripts.fasta;

# get the cumulative AED distribution and plot it
AED_cdf_generator.pl -b 0.025 round_5.all.maker.without.sequences.gff > round_5_AED_distribution.tsv

# Make AED curve to check the quality of annotations
R
d <- read.table("round_5_AED_distribution.tsv",
                header = TRUE)

colnames(d) <- c("AED", "cumulative_fraction_of_transcripts")
library(ggplot2)

png(filename = "AED_curve_round_4.png",
    width = 5, height = 5,
    units = "in",
    res = 600)

ggplot(data = d,
       aes(x = AED, y = cumulative_fraction_of_transcripts)) +
  geom_line(color = "steelblue",
            size = 2,
            linetype = 1) +
  xlab("AED score") +
  ylab("Cumulative Fraction of Transcripts") +
  ggtitle("AED curve of genome annotation") +
  scale_y_continuous(breaks = seq(0, 1, by = .05)) +
  theme_minimal()

dev.off()
```

FIGURE 24.9 **(a)** Annotation of microalgal genome using MAKER; codes for functional annotation. **(b)** Annotation of microalgal genome using MAKER; continuation of codes for functional annotation. *(Continued)*

(b)

```
# do blastp against uniprot datbase
wget ftp://ftp.uniprot.org/pub/databases/uniprot/current_release/knowledgebase/complete/uniprot_sprot.fasta.gz;
gunzip uniprot_sprot.fasta.gz;

mkdir uniprot_db;
makeblastdb -in uniprot_sprot.fasta -dbtype prot -out uniprot_db/uniprot_db;

# annotate using BLASTp with uniprot proteins
blastp -query maker_round_5.all.maker.proteins.fasta -db uniprot_db/uniprot_db -evalue 1e-6 -max_hsps 1 -max_target_seqs 1 -outfmt 6 -
out output.blastp -num_threads 20

# intergrate blastp output info in maker output
maker_functional_gff uniprot_sprot.fasta output.blastp round_5.all.maker.gff > myOrganism.renamed.putative_function.gff;

maker_functional_gff uniprot_sprot.fasta output.blastp round_5.all.maker.without.sequences.gff >
myOrganism.renamed.putative_function_without_sequences.gff;

maker_functional_fasta uniprot_sprot.fasta output.blastp maker_round_5.all.maker.transcripts.fasta >
myOrganism.renamed.putative_function_transcripts.fasta;

maker_functional_fasta uniprot_sprot.fasta output.blastp maker_round_5.all.maker.proteins.fasta >
myOrganism.renamed.putative_function_proteins.fasta;

# get GO terms, domains and pathway information from interproscan; if a java error comes up, specify
# java 11 path explictly in the interproscan.sh file

interproscan.sh --cpu 15 -appl pfam -dp -f TSV -goterms -iprlookup -pa -t p -i maker_round_5.all.maker.proteins.fasta -o
output.iprscan

# integrate interproscan output into maker output
ipr_update_gff myOrganism.renamed.putative_function.gff output.iprscan > myOrganism_FINAL_with_sequences.gff;
ipr_update_gff myOrganism.renamed.putative_function_without_sequences.gff output.iprscan > myOrganism_FINAL_without_sequences.gff

mkdir KO_annotation; cd KO_annotation

conda activate

seqkit split ../maker_round_5.all.maker.proteins.fasta -s 4000
# upload these one by one to blastKOALA and later concatenate the results

# also upload protein sequences to eggNOG-Mapper; it will also give KO and EC numbers

# GENOME ANNOTATION COMPLETE
Final files :
1. myOrganism_FINAL_with_sequences.gff
2. myOrganism_FINAL_without_sequences.gff
3. myOrganism.renamed.putative_function_transcripts.fasta
4. myOrganism.renamed.putative_function_proteins.fasta
```

FIGURE 24.9 *(Continued)*

example, sequencing transcripts) or simulation studies (for example, gene model predictions) (Van Dam et al. 2018). In a genome browser, information is frequently organised into 'tracks' that may be downloaded or requested as needed (Wang et al. 2013). Several newly reported genome browsers are intended to serve as a platform for integrating huge genomic datasets, particularly cancer genomic data. These include the National Cancer Institute of the United States' Cancer Molecular Analysis Portal, the IGV, and the UCSC Cancer Genomics Browser (Cerami et al. 2012). The simultaneous presentation of genome data and input data is the key novelty of these new technologies. As heat maps, these browsers offer a whole-genome perspective of genome-wide experimental measures for individual samples and sets of samples. Clinical traits are shown alongside genetic information in a distinctive heat map. Researchers use the browser to group, aggregate, and display data according to clinical traits, biological pathways that have been annotated or user-edited gene collections.

24.7 CONCLUSIONS

Despite advances in sequencing technology and lower prices, the quality of genome assemblies is declining. The quality of genome contiguity, the quality of genome content, the quality of gene annotations, and the accessibility of gene models may all be indicators of this trend. To overcome the aforementioned problems, the field should seek to update and adopt assembly quality standards. Given the fast growth of genomics technology, establishing standards is not easy, yet standards are required to inform analyses and correct application by the scientific community. Furthermore, the community should also look at new standards for evidence-based gene predictions, annotation releases, and ongoing gene prediction and genome evaluation tool development. Similarly, recommendations for mixed sequencing methodologies should be created (exploring combinations of Illumina, PacBio, Nanopore, Hi-C sequencing, and optical mapping), as described in this chapter. Moreover, large gaps in microalgal biodiversity and evolutionary history will undoubtedly be addressed by genome sequencing initiatives, which will reveal emerging principles of microalgal biology and evolution and lead to the discovery of novel proteins, biochemical processes, and natural products.

ACKNOWLEDGMENTS

Support for this research was provided from the Department of Biotechnology (DBT) – BT/PB/Center/03/2011 – Phase II; BT/PR16512/BID/7/647/2016.

REFERENCES

Andersen RA (2004) Biology and systematics of heterokonta and haptophyte algae. *American Journal of Botany* 91: 1508–1522.

Armaleo D, Müller O, Lutzoni F, Andrésson ÓS, Blanc G, Bode HB, Collart FR, Dal Grande F, Dietrich F, Grigoriev IV, Joneson S, Kuo A, Larsen PE, Logsdon JM, Lopez D, Martin F, May SP, McDonald TR, Merchant SS, Miao V, Morin E, Oono R, Pellegrini M, Rubinstein N, Sanchez-Puerta MV, Savelkoul E, Schmitt I, Slot JC, Soanes D, Szövényi P, Talbot NJ, Veneault-Fourrey C and Xavier BB (2019) The lichen symbiosis reviewed through the genomes of *Cladonia grayi* and its algal partner *Asterochloris glomerata*. *BMC Genomics* 20: 1–33.

Armbrust EV, Berges JA, Bowler C, Green BR, Martinez D, Putnam NH, Zhou S, Allen AE, Apt KE, Bechner M, Brzezinski MA, Chaal BK, Chiovitti A, Davis AK, Demarest MS, Detter JC, Glavina T, Goodstein D, Hadi MZ, Hellsten U, Hildebrand M, Jenkins BD, Jurka J, Kapitonov VV, Kröger N, Lau WW, Lane TW, Larimer FW, Lippmeier JC, Lucas S, Medina M, Montsant A, Obornik M, Parker MS, Palenik B, Pazour GJ, Richardson PM, Rynearson TA, Saito MA, Schwartz DC, Thamatrakoln K, Valentin K, Vardi A, Wilkerson FP and Rokhsar DS (2004) The genome of the diatom *Thalassiosira pseudonana*: Ecology, evolution, and metabolism. *Science* 306: 79–86.

Benner S, Chen RJ, Wilson NA, Abu-Shumays R, Hurt N, Lieberman KR, Deamer DW, Dunbar WB and Akeson M (2007) Sequence specific detection of individual DNA polymerase complexes in real time using a nanopore. *Nature Nanotechnology* 2: 718–724.

Bentley DR, Balasubramanian S, Swerdlow HP, Smith GP, Milton J, Brown CG, Hall KP, Evers DJ, Barnes CL, Bignell HR, Boutell JM, Bryant J, Carter RJ, Cheetham R, Cox AJ, Ellis D, Flatbush MR, Gormley NA, Humphray SJ, Irving LJ, Karbelashvili MS, Kirk SM, Li H, Liu X, Maisinger KS, Murray LJ, Obradovic B, Ost T, Parkinson ML, Pratt MR, Rasolonjatovo IM, Reed MT, Rigatti R, Rodighiero C, Ross MT, Sabot A, Sankar SV, Scally A, Schroth GP, Smith ME, Smith VP, Spiridou A, Torrance PE, Tzonev SS, Vermaas EH, Walter K, Wu X, Zhang L, Alam MD, Anastasi C, Aniebo IC, Bailey DM, Bancarz IR, Banerjee S, Barbour SG, Baybayan PA, Benoit VA, Benson KF, Bevis C, Black PJ, Boodhun A, Brennan JS, Bridgham JA, Brown RC, Brown AA, Buermann DH, Bundu AA, Burrows JC, Carter NP, Castillo N, Chiara ECM, Chang S, Neil Cooley R, Crake NR, Dada OO, Diakoumakos KD, Dominguez-Fernandez B, Earnshaw DJ, Egbujor UC, Elmore DW, Etchin SS, Ewan MR, Fedurco M, Fraser LJ, Fuentes Fajardo KV, Scott Furey W, George D, Gietzen KJ, Goddard CP, Golda GS, Granieri PA, Green DE, Gustafson DL, Hansen NF, Harnish K, Haudenschild CD, Heyer NI, Hims MM, Ho JT and Horgan AM (2008) Accurate whole human genome sequencing using reversible terminator chemistry. *Nature* 456: 53–59.

Bhattacharya D, Price DC, Chan CX, Qiu H, Rose N, Ball S, Weber APM, Arias M, Henrissat B, Coutinho PM, Krishnan A, Zäuner S, Morath S, Hilliou F, Egizi A, Perrineau MM and Yoon HS (2013) Genome of the red alga *Porphyridium purpureum*. *Nature Communications* 4: 1941.

Blanc G, Duncan G, Agarkova I, Borodovsky M, Gurnon J, Kuo A, Lindquist E, Lucas S, Pangilinan J, Polle J, Salamov A, Terry A, Yamada T, Dunigan DD, Grigoriev IV, Claverie JM and Van Etten JL (2010) The *Chlorella variabilis* NC64A genome reveals adaptation to photosymbiosis, coevolution with viruses, and cryptic sex. *The Plant Cell* 22: 2943–2955.

Blanc-Mathieu R, Verhelst B, Derelle E, Rombauts S, Bouget FY, Carré I, Château A, Eyre-Walker A, Grimsley N, Moreau H, Piégu B, Rivals E, Schackwitz W, Van de Peer Y and Piganeau G (2014) An improved genome of the model marine alga *Ostreococcus tauri* unfolds by assessing Illumina de novo assemblies. *BMC Genomics* 15: 1–12.

Bogen C, Al-Dilaimi A, Albersmeier A, Wichmann J, Grundmann M, Rupp O, Lauersen KJ, Blifernez-Klassen O, Kalinowski J, Goesmann A, Mussgnug JH and Kruse O (2013) Reconstruction of the lipid metabolism for the microalga *Monoraphidium neglectum* from its genome sequence reveals characteristics suitable for biofuel production. *BMC Genomics* 14: 1–18.

Bowler C, Allen AE, Badger JH, Grimwood J, Jabbari K, Kuo A, Maheswari U, Martens C, Maumus F, Otillar RP, Rayko E, Salamov A, Vandepoele K, Beszteri B, Gruber A, Heijde M, Katinka M, Mock T, Valentin K, Verret F, Berges JA, Brownlee C, Cadoret JP, Chiovitti A, Choi CJ, Coesel S, De Martino A, Detter JC, Durkin C, Falciatore A, Fournet J, Haruta M, Huysman MJ, Jenkins BD, Jiroutova K, Jorgensen RE, Joubert Y, Kaplan A, Kröger N, Kroth PG, La Roche J, Lindquist E, Lommer M, Martin-Jézéquel V, Lopez PJ, Lucas S, Mangogna M, McGinnis K, Medlin LK, Montsant A, Secq MO, Napoli C, Obornik M, Parker MS, Petit JL, Porcel BM, Poulsen N, Robison M, Rychlewski L, Rynearson TA, Schmutz J, Shapiro H, Siaut M, Stanley M, Sussman MR, Taylor AR, Vardi A, von Dassow P, Vyverman W, Willis A, Wyrwicz LS, Rokhsar DS, Weissenbach J, Armbrust EV, Green BR, Van de Peer Y and Grigoriev IV (2008) The *Phaeodactylum tricornutm* genome reveals the evolutionary history of diatom genomes. *Nature* 456: 239–244.

Brodie J and Lewis J (2007) Unravelling the Algae: The Past, Present, and Future of Algal Systematic, First Ed. CRC Press: Boca Raton.

Cantarel BL, Korf I, Robb SM, Parra G, Ross E, Moore B, Holt C, Sánchez Alvarado A and Yandell M (2008) MAKER: An easy-to-use Annotation pipeline designed for emerging model organism genomes. *Genome Research* 18: 188–196.

Cavalier-Smith T (1999) Principles of protein and lipid targeting in secondary symbiogenesis: Euglenoid, dinoflagellate, and sporozoan plastid origins and the eukaryote family tree. *Journal of Eukaryotic Microbiology* 46: 347–366.

Cerami E, Gao J, Dogrusoz U, Gross BE, Sumer SO, Aksoy BA, Jacobsen A, Byrne CJ, Heuer ML, Larsson E, Antipin Y, Reva B, Goldberg AP, Sander C and Schultz N (2012) The cBio cancer genomics portal: An open platform for exploring multidimensional cancer genomics data. *Cancer Discovery* 2: 401–404.

Chen CY (2014) DNA polymerases drive DNA sequencing-by-synthesis technologies: Both past and present. *Frontiers in Microbiology* 5: 305.

Cock JM, Sterck L, Rouzé P, Scornet D, Allen AE, Amoutzias G and Wincker P et al. (2010) The *Ectocarpus* genome and the independent evolution of multicellularity in brown algae. *Nature* 465: 617–621.

Collén J, Porcel B, Carré W, Ball SG, Chaparro C, Tonon T and Boyen C et al. (2013) Genome structure and metabolic features in the red seaweed *Chondrus crispus* shed light on evolution of the Archaeplastida. *Proceedings of the National Academy of Sciences USA* 110: 5247–5252.

Compeau PE, Pevzner PA and Tesler G (2011) How to apply de Bruijn graphs to genome assembly. *Nature Biotechnology* 29: 987–991.

Corteggiani EC, Telatin A, Vitulo N, Forcato C, D'Angelo M, Schiavon R, Vezzib A, Giacomettib GM, Morosinotto T and Valle G (2014) Chromosome scale genome assembly and transcriptome profiling of *Nannochloropsis gaditana* in nitrogen depletion. *Molecular Plant* 7: 323–335.

Curtis BA, Tanifuji G, Burki F, Gruber A, Irimia M, Maruyama S and Archibald JM et al. (2012) Algal genomes reveal evolutionary mosaicism and the fate of nucleomorphs. *Nature* 492: 59–65.

Delwiche CF (1999) Tracing the thread of plastid diversity through the tapestry of life. *The American Naturalist* 154: 164–177.

Ejigu GF and Jung J (2020) Review on the computational genome annotation of sequences obtained by next-generation sequencing. *Biology* 9: 110–117.

Gao C, Wang Y, Shen Y, Yan D, He X, Dai J and Wu Q (2014) Oil accumulation mechanisms of the oleaginous microalga *Chlorella protothecoides* revealed through its genome, transcriptomes, and proteomes. *BMC Genomics* 15: 1–14.

Glenn TC (2011) Field guide to next-generation DNA sequencers. *Molecular Ecology Resources* 11: 759–769.

Gobler CJ, Berry DL, Dyhrman ST, Wilhelm SW, Salamov A, Lobanov AV and Grigoriev IV et al. (2011) Niche of harmful alga *Aureococcus anophagefferens* revealed through ecogenomics. *Proceedings of the National Academy of Sciences USA* 108: 4352–4357.

Gonzalez-Esquer CR, Twary SN, Hovde BT and Starkenburg SR (2018) Nuclear, chloroplast, and mitochondrial genome sequences of the prospective microalgal biofuel strain *Picochlorum soloecismus. Genome Announcements* 6: e01498–e01417.

Goodwin S, McPherson JD and McCombie WR (2016) Coming of age: Ten years of next-generation sequencing technologies. *Nature Reviews Genetics* 17: 333–351.

Gray MW and Doolittle WF (1982) Has the endosymbiont hypothesis been proven? *Microbiology Reviews* 46: 1–42.

Grigoriev IV, Hayes RD, Calhoun S, Kamel B, Wang A, Ahrendt S, Dusheyko S, Nikitin R, Mondo SJ, Salamov A, Shabalov I and Kuo A (2021) PhycoCosm, a comparative algal genomics resource. *Nucleic Acids Research* 49: 1004–1011.

Grumbt B, Eck SH, Hinrichsen T and Hirv K (2013) Diagnostic applications of next generation sequencing in immunogenetics and molecular oncology. *Transfusion Medicine and Hemotherapy* 40: 196–206.

Hamaji T, Kawai-Toyooka H, Uchimura H, Suzuki M, Noguchi H, Minakuchi Y, Toyoda A, Fujiyama A, Miyagishima S, Umen JG and Nozaki H (2018) Anisogamy evolved with a reduced sex-determining region in volvocine green algae. *Communications Biology* 1: 17.

Head SR, Komori HK, LaMere SA, Whisenant T, Van Nieuwer Burgh F, Salomon DR and Ordoukhanian P (2014) Library construction for next-generation sequencing: Overviews and challenges. *BioTechniques* 56: 61–77.

Heather JM and Chain B (2016) The sequence of sequencers: The history of sequencing DNA. *Genomics* 107: 1–8.

Hodges E, Rooks M, Xuan Z, Bhattacharjee A, Benjamin Gordon D, Brizuela L, Richard McCombie W and Hannon GJ (2009) Hybrid selection of discrete genomic intervals on custom-designed microarrays for massively parallel sequencing. *Nature Protocols* 4: 960–974.

Holtgrewe M, Messerschmidt C, Nieminen M and Beule D (2020) DigestiFlow: From BCL to FASTQ with ease. *Bioinformatics* 36: 1983–1985.

Iqbal Z, Caccamo M, Turner I, Flicek P and McVean G (2012) De novo assembly and genotyping of variants using colored de Bruijn graphs. *Nature Genetics* 44: 226–232.

Jain K, Krause K, Grewe F, Nelson GF, Weber APM, Christensen AC and Mower JP (2015) Extreme features of the *Galdieria sulphuraria* organellar genomes: A consequence of polyextremophily? *Genome Biology and Evolution* 7: 367–380.

Jain M, Olsen HE, Paten B and Akeson M (2016) The Oxford nanopore MinION: Delivery of nanopore sequencing to the genomics community. *Genome Biology* 17: 1–11.

Jurkowski TP (2020) Technologies and applications for the assessment of 5-hydroxymethylcytosine. In Epigenetics Methods (Eds. T. Tollefsbol). Academic Press, pp. 261–278. ISBN: 978-0-12-819414-0

Keeling PJ (2004) Reduction and compaction in the genome of the apicomplexan parasite *Cryptosporidium parvum. Developmental Cell* 6: 614–616.

Korostin D, Kulemin N, Naumov V, Belova V, Kwon D and Gorbachev A (2020) Comparative analysis of novel MGISEQ-2000 sequencing platform vs. Illumina HiSeq 2500 for whole-genome sequencing. *PLOS ONE* 15: e0230301.

Krzywinski M, Schein J, Birol I, Connors J, Gascoyne R, Horsman D, Jones SJ and Marra MA (2009) Circos: An information aesthetic for comparative genomics. *Genome Research* 19: 1639–1645.

Lawson D, Arensburger P, Atkinson P, Besansky NJ, Bruggner RV, Butler R and Dialynas E et al. (2007) VectorBase: A home for invertebrate vectors of human pathogens. *Nucleic Acids Research* 35: 503–505.

Lee J, Yang EC, Graf L, Yang JH, Qiu H, Zelzion U, Chan CX, Stephens TG, Weber APM, Boo GH, Boo SM, Kim KM, Shin Y, Jung M, Lee SJ, Yim HS, Lee JH, Bhattacharya D and Yoon HS (2018) Analysis of the draft genome of the red seaweed *Gracilariopsis chorda* provides insights into genome size evolution in rhodophyta. *Molecular Biology and Evolution* 35: 1869–1886.

Li Z, Chen Y, Mu D, Yuan J, Shi Y, Zhang H and Fan W (2012) Comparison of the two major classes of assembly algorithms: Overlap-layout-consensus and de-Bruijn-graph. *Briefings in Functional Genomics* 11: 25–37.

Lieberman KR, Cherf GM, Doody MJ, Olasagasti F, Kolodji Y and Akeson M (2010) Processive replication of single DNA molecules in a nanopore catalyzed by phi29 DNA polymerase. *Journal of the American Chemical Society* 132: 17961–17972.

Lommer M, Roy AS, Schilhabel M, Schreiber S, Rosenstiel P and LaRoche J (2010) Recent transfer of an iron-regulated gene from the plastid to the nuclear genome in an oceanic diatom adapted to chronic iron limitation. *BMC Genomics* 11: 1–13.

Margulies M, Egholm M, Altman WE, Attiya S, Bader JS, Bemben LA and Rothberg JM et al. (2005) Genome sequencing in microfabricated high-density picolitre reactors. *Nature* 437: 376–380.

Matsuzaki M, Misumi O, Shin-i T, Maruyama S, Takahara M, Miyagishima S and Kuroiwa T et al. (2004) Genome sequence of the ultrasmall unicellular red alga *Cyanidioschyzon merolae* 10D. *Nature* 428: 653–657.

McFadden GI (2001) Primary and secondary endosymbiosis and the origin of plastids. *Journal of Phycology* 37: 951–959.

McKernan KJ, Peckham HE, Costa GL, McLaughlin SF, Fu Y, Tsung EF and Blanchard AP et al. (2009) Sequence and structural variation in a human genome uncovered by short-read, massively parallel ligation sequencing using two-base encoding. *Genome Research* 19: 1527–1541.

Merchant SS, Prochnik SE, Vallon O, Harris EH, Karpowicz SJ, Witman GB and Grossman AR et al. (2007) The *Chlamydomonas* genome reveals the evolution of key animal and plant functions. *Science* 318: 245–250.

Metzker ML (2010) Sequencing technologies the next generation. *Nature Reviews Genetics* 11: 31–46.

Miller JR, Zhou P, Mudge J, Gurtowski J, Lee H, Ramaraj T and Denny R et al. (2017) Hybrid assembly with long and short reads improves discovery of gene family expansions. *BMC Genomics* 18: 1–12.

Minei R, Hoshina R and Ogura A (2018) De novo assembly of middle-sized genome using MinION and Illumina sequencers. *BMC Genomics* 19: 1–12.

Minoche AE, Dohm JC and Himmelbauer H (2011) Evaluation of genomic high-throughput sequencing data generated on Illumina HiSeq and genome analyzer systems. *Genome Biology* 12: 1–15.

Mock T, Otillar RP, Strauss J, McMullan M, Paajanen P, Schmutz J and Grigoriev IV et al. (2017) Evolutionary genomics of the cold adapted diatom *Fragilariopsis cylindrus. Nature* 541: 536–540.

Moreau H, Verhelst B, Couloux A, Derelle E, Rombauts S, Grimsley N, Bel MV, Poulain J, Katinka M, Hohmann-Marriott MF, Piganeau G, Rouzé P, Silva CD, Wincker P, De Peer YV and Vandepoele K (2012) Gene Functionalities and genome structure in *Bathycoccus prasinos* reflect cellular specializations at the base of the green lineage. *Genome Biology* 13: 1–16.

Navarro-Díaz M, Aparicio-Trejo V, Valdez-Vazquez I, Carrillo-Reyes J, Avitia M and Escalante AE (2022) Variations in microbial diversity affect the stability and function of dark fermentation bioreactors. *bioRxiv* 4: 497814.

Ohan JA, Hovde BT, Zhang XL, Davenport KW, Chertkov O, Han C, Twary SN and Starkenburg SR (2019) Nuclear genome assembly of the microalga *nannochloropsis salina* CCMP1776. *Microbiology Resource Announcements* 8: e00750-19.

Palenik B, Grimwood J, Aerts A, Rouzé P, Salamov A, Putnam N and Grigoriev IV et al. (2007) The tiny eukaryote *Ostreococcus* provides genomic insights into the paradox of plankton speciation. *Proceedings of the National Academy of Sciences USA* 104: 7705–7710.

Palmer JD (2003) The symbiotic birth and spread of plastids: How many times and whodunit? *Journal of Phycology* 39: 4–12.

Pierella Karlusich JJ, Ibarbalz FM and Bowler C (2020) Phytoplankton in the Tara ocean. *Annual Review of Marine Science* 12: 233–265.

Polle JEW, Barry K, Cushman J, Schmutz J, Tran D, Hathwaik LT and McKie-Krisberg Z et al. (2017) Draft nuclear genome sequence of the halophilic and β-carotene accumulating green alga *Dunaliella salina* strain CCAP19/18. *Genome Announcements* 5: 43.

Prochnik SE, Umen J, Nedelcu AM, Hallmann A, Miller SM, Nishii I and Rokhsar DS et al. (2010) Genomic analysis of organismal complexity in the multicellular green alga *volvox carteri*. *Science* 329: 223–226.

Read BA, Kegel J, Klute MJ, Kuo A, Lefebvre SC and Maumus F et al. (2013) Pan genome of the phytoplankton *Emiliania* underpins its global distribution. *Nature* 499: 209–213.

Rhoads A and Au KF (2015) PacBio sequencing and its applications. *Genomics Proteomics Bioinformatics* 13: 278–289.

Rice ES and Green RE (2019) New approaches for genome assembly and scaffolding. *Annual Review of Animal Biosciences* 7: 17–40.

Robinson JT, Thorvaldsdóttir H, Winckler W, Guttman M, Lander ES, Getz G and Mesirov JP (2011) Integrative genomics viewer. *Nature Biotechnology* 29: 24–26.

Rothberg JM, Hinz W, Rearick TM, Schultz J, Mileski W, Davey M and Bustillo J et al. (2011) An integrated semiconductor device enabling non-optical genome sequencing. *Nature* 475: 348–352.

Sanger F, Nicklen S and Coulson AR (1977) DNA sequencing with chain-terminating inhibitors. *Proceedings of the National Academy of Sciences USA* 74: 5463–5467.

Schoch CL, Ciufo S, Domrachev M, Hotton CL, Kannan S, Khovanskaya R and Karsch-Mizrachi I et al. (2020) NCBI taxonomy: A comprehensive update on curation, resources and tools. *Database (Oxford)* 2020: baaa062.

Shaikh KM, Kumar P, Nesamma AA, Abdin MZ and Jutur PP (2020) Hybrid genome assembly and functional annotation reveals insights on lipid biosynthesis of oleaginous native isolate *Parachlorella kessleri*, a potential industrial strain for production of biofuel precursors. *Algal Research* 52: 102118.

Shendure J, Balasubramanian S, Church GM, Gilbert W, Rogers J, Schloss JA and Waterston RH (2017) DNA sequencing at 40: Past, present and future. *Nature* 550: 345–353.

Simpson JT and Pop M (2015) The theory and practice of genome sequence assembly. *Annual Review of Genomics and Human Genetics* 16: 153–172.

Slatko BE, Gardner AF and Ausubel FM (2018) Overview of next-generation sequencing technologies. *Current Protocols in Molecular Biology* 122: 59.

Starkenburg SR, Polle JEW, Hovde B, Daligault HE, Davenport KW, Huang A, Neofotis P and McKieKrisberg Z (2017) Draft nuclear genome, complete chloroplast genome, and complete mitochondrial genome for the biofuel/bioproducts feedstock species *Scenedesmus obliquus* strain DOE0152z. *Genome Announcements* 5: e00617–17.

Thorvaldsdóttir H, Robinson JT and Mesirov JP (2013) Integrative genomics viewer (IGV): High-performance genomics data visualization and exploration. *Briefings in Bioinformatics* 14: 178–192.

Van Dam S, Võsa U, van der Graaf A, Franke L and de Magalhães JP (2018) Gene Co-expression analysis for functional classification and gene-disease predictions. *Briefings in Bioinformatics* 19: 575–592.

Voelkerding KV, Dames S and Durtschi JD (2010) Next generation sequencing for clinical diagnostics-principles and application to targeted resequencing for hypertrophic cardiomyopathy: A paper from the 2009 William Beaumont Hospital symposium on molecular pathology. *The Journal of Molecular Diagnostics* 12: 539–551.

Wang J, Kong L, Gao G and Luo J (2013) A brief introduction to web-based genome browsers. *Briefings in Bioinformatics* 14: 131–143.

Worden AZ, Lee JH, Mock T, Rouzé P, Simmons MP, Aerts AL and Grigoriev IV et al. (2009) Green evolution and dynamic adaptations revealed by genomes of the marine picoeukaryotes *Micromonas*. *Science* 324: 268–272.

Ye N, Zhang X, Miao M, Fan X, Zheng Y, Xu D, Wang J, Zhou L, Wang D, Gao Y, Wang Y, Shi W, Ji P, Li D, Guan Z, Shao C, Zhyang Z, Gao Z, Qi J and Zhao F (2015) *Saccharina* genomes provide novel insight into kelp biology. *Nature Communications* 6: 6986.

Zhang Z, Qu C, Zhang K, He Y, Zhao X, Yang L, Zheng Z, Ma X, Wang X, Wang W, Wang K, Li D, Zhang L, Zhang X, Su D, Chang X, Zhou M, Gao D, Jiang W, Leliaert F, Bhattacharya D, Clerck OD, Zhong B and Miao J (2020) Adaptation to extreme Antarctic environments revealed by the genome of a sea ice green alga. *Current Biology* 30: 3330–3341.e3337.

Index

T

U

V

W

Z